Hydrodynamic Control of Wave Energy Devices

Umesh A. Korde is a professor of Mechanical Engineering at the South Dakota School of Mines and Technology. He has had a long career working on wave energy conversion. He has published widely in the areas of design, dynamics, and control of wave energy devices, floating bodies, and deformable structures. He is an associate editor for the journal *Ocean Engineering and Marine Energy*, serves on the editorial boards of other journals, and is a fellow of the American Society of Mechanical Engineers.

John V. Ringwood is a professor in the Department of Electronic Engineering and Director of the Centre for Ocean Energy Research at Maynooth University, Ireland. His research background is in dynamics and control, covering a wide variety of application areas, including robotics, human physiology, bioreactors, semiconductor manufacture, boiler systems, vehicle dynamics, steel rolling, food processing and electricity demand, as well as ocean energy systems. John is a senior member of the IEEE, a fellow of Engineers Ireland, and a chartered engineer.

Hydrodynamic Control of Wave Energy Devices

UMESH A. KORDE
South Dakota School of Mines and Technology

JOHN V. RINGWOOD
Maynooth University

CAMBRIDGE
UNIVERSITY PRESS

CAMBRIDGE
UNIVERSITY PRESS

University Printing House, Cambridge CB2 8BS, United Kingdom

Cambridge University Press is part of the University of Cambridge.

It furthers the University's mission by disseminating knowledge in the pursuit of education, learning and research at the highest international levels of excellence.

www.cambridge.org
Information on this title: www.cambridge.org/9781107079700

First published 2016

Printed in the United States of America by Sheridan Books, Inc.

A catalogue record for this publication is available from the British Library

Library of Congress Cataloging-in-Publication data
Names: Korde, Umesh A. | Ringwood, John V.
Title: Hydrodynamic control of wave energy devices / Umesh A. Korde, South Dakota School of Mines and Technology, John V. Ringwood, Maynooth University.
Description: Cambridge : Cambridge University Press, [2016] | Includes bibliographical references and index.
Identifiers: LCCN 2016011206 | ISBN 9781107079700 (Hardback)
Subjects: LCSH: Ocean wave power–Research. | Hydraulic engineering–Research. | Energy conversion–Research. | Tidal power-plants–Research. | Wave resistance (Hydrodynamics)–Research. | Water-power–Research. | Renewable energy sources–Research.
Classification: LCC TC147 .K67 2016 | DDC 621.31/2134–dc23
LC record available at http://lccn.loc.gov/2016011206

ISBN 978-1-107-07970-0 Hardback

Contents

Preface

Waves seem to speak to something deep in our souls. For some of us, they are best enjoyed from a distance, as if they somehow connected us with dreams and places far away. Some of us draw strength and recreation from an actual "discourse" with them, surrounded by them, riding over them, or sailing through them. They are among nature's most beautiful gifts, capable of both delighting and destroying us. Trying to utilize waves to provide energy for earthly uses might almost seem like attempting to teach some stately wild beast to perform silly tricks. Yet that is precisely what this book is about. That is precisely what some of the most brilliant and imaginative minds in engineering and mathematics have been working on the last four decades and, in some cases, longer. This book is about the work of giants who were building the engineering and mathematical foundations of wave energy converters (WECs) long before the authors could even have begun to understand their work. Some of this work has a beauty all its own, hard to find elsewhere.

This is an ambitious book. It attempts to merge the exciting but refined worlds of hydrodynamics, rigid-body dynamics, vibrations, and control systems with the complex and ambiguous world of design. Volumes enough to fill entire libraries have been written about these disciplines. So what makes this book worth picking up and opening? If you are interested in using waves as an energy source but find the literature daunting in its quality and quantity, this book will give you the tools to work your way through more complex material. If you want to be efficient with the technology you develop and if you are, like the rest of us, operating on a limited budget, this book will first introduce you to techniques that may lead to a better return on investment, and then illustrate some of these techniques through more detailed case-study applications. We speak, of course, of the natural marriage between wave energy and control engineering. To the wave energy engineer, controls provide a way to perform efficiently in ever-changing wave fields, and to do so with smaller structures. To the controls engineer, wave energy conversion is a uniquely challenging problem that demands a rare combination of established and new techniques. Thus, if you are a controls person looking for tough nuts to crack, this book will give you the hydrodynamics and related fundamentals to move forward with confidence. It will also acquaint you with recent research, as seen from a controls perspective. If you are an undergraduate student, parts of this book will give you a gentle introduction based on your vibrations and controls work so far, and hopefully whet your appetite to learn more from your wave energy–related projects than you would otherwise learn. Perhaps you will even want to invent something new and/or design and complete interesting new projects of your own. If you are a graduate student with a penchant for technical depth, this book will provide a bridge to the current literature by helping you understand some of the thinking underlying the mathematical techniques and arguments you will work through in your research. If you are a sustainable energy person, this book will give you a sense for where the challenges of wave energy lie, how one gives order to apparent disorder, and not only how "performance" is "designed into" a converter but also how it can be altered simply by making clever changes at the "software level." You may also see why there is such room for a diversity of ideas in wave energy conversion (as witnessed in the different devices being tested today), and why all may work equally well depending on where they are supposed to operate and how they are controlled.[1] If you are someone simply interested in the notion of wave energy conversion, you could get a good sense of the excitement of the field simply by reading the verbal discussions and skipping over the mathematical arguments.

We have tried to maintain a reasonable level of mathematical precision in this book, although at times we have sacrificed rigor in favor of clarity of explanation and accessibility. We have tried to explain the mathematics at least as clearly as we understand it, particularly in the early, foundational, chapters. The reader interested in a more rigorous treatment of the hydrodynamics of waves and WECs will enjoy the book by Professor Falnes,[2] which continues to be a "beacon of reason" for us in our work.

Chapter 1 provides a largely nontechnical and completely nonmathematical view of waves and wave energy conversion. It also describes some "iconic" devices and some recently sea-tested devices. An informal introduction to control engineering is given in Chapter 2, with only a small amount of mathematics to give meaning to the words. In that chapter we begin a discussion of the special challenges of controlling wave energy devices. Chapters 3 and 4 explain these challenges in the context of oscillating systems, with a particular emphasis on the difference between oscillations in air and oscillations on the water surface. There is a significant amount of mathematics in those chapters, mainly because it paves the way for later material that deals more directly with device hydrodynamics and energy conversion. Chapter 5 discusses waves. Our purpose there is to introduce the reader to the mathematical descriptions of waves on the water surface (which do not readily fall into the more general mathematical treatments of waves, such as the "wave equation"), beginning with the most elemental regular wave and ending with a treatment of irregular waves, spectral density functions, group wave statistics, and evaluation of device-response statistics within the spectral density formalism. Chapter 6 takes the reader "to the trenches" of wave energy work, with a review of the important aspects of device hydrodynamics. Control is important and potentially profitable, but it is the essential hydrodynamic response of devices that is more critical to their success.

[1] Such a diversity is conspicuously absent in wind and solar energy. Whether this reflects a certain level of technological maturation that wave energy is yet to reach, or whether there are other, more interesting, factors at play, is not a question we address in this book. Our emphasis here is on the unifying principles and techniques that help us understand and work with classes of devices.

[2] J. Falnes, *Ocean Waves and Oscillating Systems: Linear Interactions Including Wave Energy Extraction*, Cambridge University Press, UK, 2002.

Chapters 7 and 8 and Chapters 9 and 10 describe two approaches to controlling the hydrodynamic response. These are essentially two complementary ways to design control systems for efficient wave energy conversion (this point is discussed further in Chapter 2). One is ground up, where the control forces are synthesized entirely via step-by-step hydrodynamic modeling, from wave propagation to device response without control and the desired response. For best conversion efficiency, the device needs to move at the "hydrodynamic velocity optimum" (defined later in the text). Chapter 7 discusses the more ambitious reactive control, especially its application in real time in irregular waves. In that chapter, we illustrate the hydrodynamics-based approaches that can be applied directly to the inner workings of a device, modeling to reasonable detail what we know in the time domain about the propagation of waves approaching the device and the dynamics of the device. Chapter 8 imposes the realistic restriction, where only current and past knowledge are required, in order to synthesize a causal control algorithm, which has its roots in the development of Chapter 7. The approach in Chapters 9 and 10 is top down and uses established techniques from control theory and engineering to design control "inputs" for a "system" comprised of a wave energy device, such that a performance index (e.g., the absorbed energy over a time interval) is maximized. In the first approach, the user builds a controller around the hydrodynamics, so to speak, while in the second approach, the hydrodynamic model for the device forms an element within a larger analysis and model-based design structure drawn from control theory/engineering. Thus, in Chapter 9, we discuss one of the most commonly used control approaches, namely, "latching," and in Chapter 10 we show a more general method to efficiently numerically optimize the velocity profile of a wave device.

In rounding out Part V, Chapter 11 extends the algorithm of Chapter 10 to an array of devices and demonstrates, via some numerical examples, some idea of the benefit of basing the control on a complete WEC array hydrodynamic model, which can articulate the interactions resulting from control actions, particularly radiation. Finally, closing Part V, some algorithms for short-term wave forecasting are presented in Chapter 12, addressing the needs of noncausal control from Chapters 7, 9, and 10.

Part VI examines a number of WEC and WEC array optimization issues that directly impact the control problem. In Chapter 13, the issue of constraints is addressed. In particular, the perspective is broadened to consider the constrained space in which the hydrodynamic control problem is solved, and the relationship of this space to the excitation force. The framework can be used as a design tool, where the effect of a change in displacement or force constraints can be understood in terms of the impact on power capture. Finally, Chapter 14 examines other aspects of WEC system optimisation that have a strong interaction with the control problem. In particular, it is demonstrated that both the optimal WEC geometry and the optimal WEC array layout are strongly influenced by the control algorithm utilized, with the often sequential approach of geometry/layout design, followed by the control design, likely to result in an overall suboptimal system solution.

This book concentrates exclusively on hydrodynamic control of WECs, underpinned by linear hydrodynamic models. The focus on linear models is, of course, a simplification, but one that may be (partly, at least) defended by the desire to focus on the fundamentals and take advantage of the generality that a linear representation offers. At the time this book was written, the predominant approach to modeling WEC was via linear or, at best, mildly nonlinear models based on the popular linear Cummins' equation. However, there is an increasing focus on nonlinear hydrodynamic models that, though beyond the scope of this current volume, might be the subject of future books. One issue that is becoming clear is that, while linear hydrodynamic models may well describe uncontrolled floating bodies, the increased relative fluid/body motion under controlled conditions may heighten the need to consider nonlinear hydrodynamic effects. Ironically, the presence of control expands the motion/force space, which, in turn, calls into question the validity of the linear hydrodynamic models on which the control algorithm itself is based. In any case, the nature of the nonlinear hydrodynamic effects is likely to vary considerably between WECs utilizing different principles and geometries, making the possibility to have a general nonlinear treatment highly unlikely.

A number of areas crucial to the success of wave energy devices are not covered in this book. We recognize that the ocean is an extremely complex environment and that our treatment here is at a level of abstraction that some ocean scientists and seafarers may find amusing. We do not discuss here how devices respond to extremely large waves, what survival strategies we would use, and what role control could play in this area. However, one could imagine that such supervisory control would sit on top of the hydrodynamic control system and would monitor the sea state, watching for the point at which system displacement/velocity/force constraints cannot be simultaneously satisfied, and switch from a power production mode to a survival mode. We do not delve into questions related to the potential environmental footprint of wave energy devices, though we recognize that the intimate interaction between control and device/array designs could significantly influence the impact large-scale wave energy conversion will have on marine life and coastal processes. We do not consider the details of actuator design (a critical element to ensuring that the control discussed in this book is practical or even worthwhile) and include only some peripheral material on the "lower loop" control, namely, the implementation of particular velocity or force setpoints derived from hydrodynamic control calculations. We do not address structural design or mooring systems design, nor do we enter into the materials, reliability, manufacturing, and deployment aspects of wave energy conversion. All these factors clearly influence the economics of WECs. For the purpose of this book, we assume that these aspects have been, or can be, sufficiently well addressed. With this understanding, the question we address in this book is: how best can we combine our hydrodynamic design choices with our control design choices to seek the most efficient solution?

Acknowledgments

I would like to begin by expressing my gratitude to Larry and Linda Pearson, the South Dakota School of Mines and Technology Foundation, and Professor Mike Langerman for supporting my work through the Pearson endowment. I want to thank South Dakota School of Mines and Technology student Cole Conrad for his highly professional work on several iterations of the numerous computer drawings he produced for me.

The idea for this book was first suggested, almost 18 years ago, by Professor Mike McCormick. Mike also helped me with the development of our ideas for the first two chapters. I am especially grateful to Mike for his invaluable mentorship over the last few decades.

I was permanently drawn to wave energy in 1983 on first meeting Professor Stephen Salter and listening to his description of the Salter duck. For what I have learned from Stephen and from my long-time good friend Jamie Taylor over the last few decades, I feel special gratitude. One other creative genius I want to thank here is my long-time good friend Dr. Axel Ruoff, for the inspiration he has given me over the last few decades, through his thoughts, music, and paintings. To my good friend Dr. Wolfgang Faller, who has inspired me with his engineering brilliance and no-nonsense approach since the mid-1980s, I owe equal thanks. I am pleased to thank Professor Adrian Long, Professor Trevor Whittaker, and Dr. Frank McPeake for giving me the benefit of their knowledge in the early 1980s. To Professors V.S. Raju and M. Ravindran, I owe many thanks for giving me my first opportunity to work on wave energy (in 1982).

I am particularly grateful to Professor Hisaaki Maeda for his supervision of my early work and for his mentorship since the mid-1980s. I want to thank Mr. Fumihiro Suzuki and Mr. Koichiro Eguchi, whose help proved crucial to my work in the late 1980s. I am indebted to Dr. Takeaki Miyazaki, Dr. Hitoshi Hotta, Dr. Yukihisa Washio, and Dr. Hiroyuki Osawa for sharing their knowledge and experience and for letting me get involved in their research on the Mighty Whale device. I want to give particular thanks to Dr. David Pizer, Peter Karlsson-Ahlm, and Elisabet Flinkberg, with whom I have shared many happy moments working on wave energy. I want to thank Professors Marco Schoen, John Congdon, and Steve Carr for the work we did together. A word of appreciation also goes to Professor Takeshi Kinoshita, for a chance to work with him. Much gratitude is due to Ben Wilcox, my friend and collaborator of nearly 10 years, for the opportunity to work with him on a number of projects.

There are others who have been my important teachers over my decades in wave energy, to whom I am indebted: Professor Johannes Falnes, for what I have been able to learn from him (mostly through his papers, though also through personal interactions); Professor David Evans, for what I have learned from the lucid explanations in his papers; Professor C.C. Mei, for what I have learned from his books, and Professor Antonio Falcao, for the many discussions I have been able to have with him.

I am grateful to Professor R. Cengiz Ertekin, for his friendship and for the opportunity to collaborate over several years. I am pleased to acknowledge the productive collaboration and good friendship of Professor Rush Robinett. To my other wave energy friends, Dr. Giorgio Bacelli, Dr. David Wilson, and Professor Ossama Abdelkhalik, I am similarly grateful. There is one particular friend and collaborator to whom I owe much gratitude: Professor John Ringwood. It was only after John agreed to join this adventure that the book really got off the ground. It has been a privilege for me to learn from John's insights. Whenever the road did rise to meet us during our journey on this book; it was largely thanks to John's ability and energy.

I want to thank some government organizations that have supported my journey so far: the British Council; the Ministry of Education, Japan; the Science and Technology Agency, Japan; the Japan Society for Promotion of Science; the U.S. National Science Foundation; the U.S. Office of Naval Research; and the U.S. Defense Advanced Research Projects Agency. I am pleased to thank my many students through the years, for the motivation they have provided me so far.

I want to thank my wife, Michelle, and express my gratefulness to her for her unconditional love, support, and encouragement, without which my work on this book would not have been possible. Finally, to my parents and family, to Hobo, and to others, who have walked with me, taught me, supported me, and cared for me at various times during my work on wave energy, I will always owe an immeasurable debt of gratitude.

Umesh A. Korde

A book is a significant undertaking requiring an effort in writing but also representing a significant effort in the production of the results reported in the book. Many people and organizations have accompanied us on this journey and contributed a not insignificant amount of inspiration, resources, and expertise to our own industry. When Umesh invited me to join in this book adventure, I had little hesitation in accepting, and it has been quite a journey of self-reflection and an eye opener into many other aspects of wave energy science for me. My sincere gratitude to Umesh for his patience, expertise, and good humor, which has served to keep us both sane throughout the writing process!

I would particularly like to acknowledge the many research students and postdoctoral students who have inspired and motivated me throughout my research career and particularly those who have directly contributed to the work reported in this book, including Giorgio Bacelli, Francesco Fusco, Francesco Paparella, and Paula Garcia-Rosa. These and other researchers at the Centre for Ocean Energy Research at Maynooth University provide a constant source of ideas, motivation, and interaction and a dynamic environment within which to indulge ourselves in theoretical, applied, and commercializable ocean energy research.

I am grateful for the motivation and perseverance of Micheál Ó Catháin, who introduced me to the exciting, new (to me, at least!) topic of wave energy, became my first research student in this area, and was instrumental in securing initial research funding. I was fortunate to have the wise counsel of Professors Mike McCormick and Johannes Falnes in my early "wave energy" years and am also grateful to Jean-Christophe Gilloteaux, for giving me some much needed fundamentals in hydrodynamics.

I owe considerable gratitude to a number of funding agencies who provided the resources to carry out much of the research reported in this book, including Enterprise Ireland and Science Foundation Ireland, even at times when funding wave energy looked like a long shot. I would particularly like to mention Jim Lawlor, who, during his time at Enterprise Ireland, was a great champion for wave energy. I have also been fortunate to have a number of key industrial collaborators who have provided much needed real-world input and have done their best to keep my feet on the ground. They include Joe Murtagh and the staff at Ocean Power Technology, Aquamarine Power, and the Wave Energy Corporation of America. Though Wavebob. is no longer in existence, the close bilateral collaboration with the former staff there, including Jochem Weber, Ronan Costello, Andrew Parish, and William Dick, provided an invaluable education at the cutting edge of commercial wave energy development.

Last, and by no means least, I owe a debt of gratitude to my wife, Laurina, and sons Rory, and Seán, for their unconditional love and support.

John V. Ringwood

We are grateful to Mr. Peter Gordon of Cambridge University Press for his early interest in this book and for his continued encouragement thereafter. We also thank Mr. Steven Elliot, for his interest in this work since Peter's retirement. Much gratitude goes to Ms. Stephanie Sakson, Ms. Clare Dennison, Mr. Jonathan Ratcliffe, Ms. Sara Werden, Mr. Anand Shanmugam, and others at Cambridge University Press for their competent and professional support throughout.

We know there are people not mentioned above to whom we are grateful but whom we have been remiss not to acknowledge by name; and for this, we ask your forgiveness. It is a rare book on a scale such as our present effort that is completely free of errors. We hope that the reader will communicate to us any errors he or she discovers in this book. We hope also that readers will feel free to share any other comments or questions on this book with us and to suggest improvements.

Umesh A. Korde and John V. Ringwood

Part I

Introduction

This part introduces waves, wave energy, and some select wave energy conversion techniques so that we may learn something about the thought processes and insights that drove not only some of the exciting early developments, but also some of the interesting new developments that build on the early work.

1 Wave Energy Conversion

Who can contemplate the surface of the ocean and not notice its incessant, seemingly random motion? This motion represents the net effect of long swells bearing memories of distant storms, shorter waves driven up by winds blowing nearby and over the surrounding area, and very small ripples caused by surface tension, with the pattern constantly changing and its constituents continuously traveling in different directions. Closer to shore, the ocean surface seems dominated by waves of steeper slopes with aligned crests mostly traveling towards the shore. As the crests approach the shore they get steeper, slightly disintegrating near their tops but still getting taller until they turn into plunging, roaring breakers, finally giving up their energy as they dissipate (with some reluctance) over sandy beaches or rocky shores. Whether maneuvering (or simply observing) a small craft on its way through the waves, or watching the waves crash loudly onto the beach, one can hardly fail to be impressed by their unrelenting display of energy.

1.1 Waves as Energy Carriers

Waves carry energy from one point to another. One observes an excellent illustration of this phenomenon when the relatively calm surface of a pond or a small lake is disturbed by a pebble or a small stone thrown into the water. A pattern of waves emanates from where the stone fell, traveling radially out and creating ever-expanding circles with alternating crests and troughs that get smaller, the farther the disturbance travels. In this case, waves carry away a portion of the kinetic energy of the stone over the water surface. A roughly similar phenomenon occurs when wind blows over a region of the ocean surface and generates waves that are higher and longer, the faster and longer the wind blows. In deep waters, the dissipation rate of the energy contained in waves is very small, enabling them to last longer and travel farther than the wind that created them. This happens for wind blowing over different regions, leading to an averaging, storing, and concentration of energy, both spatially and temporally. Because the density of sea water is about three orders of magnitude greater than that of air, energy delivered by waves is also more concentrated than wind energy. Fig. 1.1 shows some waves grown by distant and local wind action, getting too large to sustain their forms as they drive themselves to the coast over the diverse shapes and features of the sea floor.

Figure 1.1 Waves as carriers of energy. Large waves will lose energy to friction with the sea floor and ultimately break on the way to and on the shore. Their continual pounding action on the shore can break rocks and lead to beach formation and sometimes erosion of the coast line. Images provided by the National Oceanic and Atmospheric Administration (NOAA).

1.2 Nature of Wave Motion

With some exceptions outside the scope of this book, an interesting aspect of wave motion is that, although the energy propagates along the sea surface (the "free surface"), the surface itself only undulates, with an up-down motion that is a function of location and time. Distilled down to the most basic description, the most straightforward wave to think about is a sinusoidal train of peaks ("crests") and valleys ("troughs") that travel in a single direction. For such a wave, while the water surface oscillates up and down, the water particles under the surface perform a circular motion about a central mean position that remains unaltered in time. The top of the particle's circular trajectory coincides with the wave crest, while the bottom defines the wave trough. The frequency of a wave equals the angular velocity of particle motion, and the amplitude equals the radius of particle motion. The phases of various particle motions in the direction of wave propagation are such that the distance between successive peaks (troughs), or wavelength, is related to the frequency through the so-called dispersion relation, which in deep water also contains gravity acceleration. After all, it is really the restoring action of gravity to counter any disturbing action of wind (or any other surface forcing) that sustains the waves and water particle motion that will be addressed in this book. In shallower water, water depth also becomes increasingly important and plays a prominent role in the dispersion relation. Particle motion becomes increasingly elliptical, with the horizontal excursion progressively exceeding the vertical.

A compact mathematical description of the relation between particle motion and water surface deflection is possible if one ignores the relatively small contribution of viscous effects affecting wave motion (as we shall see in Chapter 5). Water is for most purposes incompressible and the particles travel without deforming in shear, so methods of ideal fluid hydrodynamics are by and large sufficient. Specifically, the particle velocity vector components under the wave can be defined as spatial derivatives (along the three coordinate directions) of a scalar "velocity potential." The water surface "elevation" or "wave profile" (more accurately referred to as "free-surface elevation") changes at a rate such that no individual particles leap out of or dive into the surface. More precisely, the rate of change of surface elevation equals the vertical velocity of individual particles over the free surface. The time rate of change of surface elevation thus equals the derivative of the velocity potential (at any given point) perpendicular to the free surface. Clearly, such a representation cannot be used to model sprays produced when a wave strikes a structure or the actual dynamics near the top of a breaking wave.

Wave forces and moments on the device can then be found using the velocity potential, which can be determined analytically or, more commonly, numerically, such that it satisfies the boundary conditions at the device. When the device is able to oscillate in response to the waves, its oscillation velocity determines the boundary conditions on it. When it is held stationary, the fact that fluid cannot flow across the device walls provides the necessary boundary conditions. Solving for the velocity potential in the presence of the device enables one to solve for the device oscillations and forces/moments, and then to use these solutions to determine device's energy conversion abilities. These effects are reviewed in greater detail in Chapter 6.

Excellent discussions of water waves can be found in the timeless works of Lamb [1], Stoker [2], Kinsman [3], and Whitham[4]. For a deeper study of the motion of floating bodies in waves, the reader may wish to consult Wehausen [5], Newman [6], Lewis [7], Mei [8], and others.

1.3 Regular versus Irregular Waves

The description above, though simplistic, provides an excellent basic mathematical/engineering model for realistic wave motion in many situations, as discussed below. It is simplistic in that it applies only to sinusoidal (regular, or harmonic) waves on the face of it, while ocean waves are almost never purely harmonic. Recall that harmonic waves have a single frequency (or its reciprocal, period), a single amplitude or height, a single wavelength (or its reciprocal wave number), and a single phase angle relative to some chosen location and time. Recall also that frequency is often "angular frequency" (i.e., the reciprocal of period times 2π), and wave number is generally 2π times the reciprocal of wavelength.

large amplitudes and short wavelengths tend to break and lose energy (and amplitude). Thus, over time, the wave energy content in a spectrum tends to be dominated by longer waves that can support greater amplitudes. A wave spectrum dominated by longer waves has a longer "energy period" and provides more energy to a device. As wavelength increases with the square of the wave period in deep water, and as wave period is easier to measure than wavelength (and wave height easier to measure than wave amplitude), wave power is frequently expressed just in terms of the square of the wave height and wavelength. Wave power spectra are plotted with period (or frequency) on the abscissa and power spectral density (proportional to wave height squared) on the ordinate.

Wave energy conversion technology is designed to utilize a large portion of the energy in waves for useful purposes. The wave energy availability off a given coast or for a country or the world as a whole is often quantified in terms of power (kW) per unit crest length. Per unit crest length, wave power is proportional to wave period and to wave height squared, meaning that waves longer and taller carry more energy. To get a sense for the available energy levels [10], about 37 kW/m may be seen off the U.S. Northern Pacific coast, compared with about 33 kW/m off the U.S. Northern Atlantic coast. Over 70 kW/m can be found off the west coast of Scotland, about 50 kW/m off Norway, about 40 kW/m off Portugal, and about 12–15 kW/m off the Pacific coast of Japan, while further south, amounts approaching or exceeding 70 kW/m may be seen off southern Australia and southern Chile. It is estimated that between 2000 and 4000 TWh/year of energy can be extracted worldwide from waves, with about 230 TWh/year from the East Coast and 590 TWh/year from the West Coast of the United States alone. In general, as indicated by these figures, more wave power can be found in the high northern or southern latitudes, especially in the winter months. As the energy demand is greater in the winter in these latitudes due to higher heating loads, locally generated power from waves may be a natural fit in these regions. As we see in this section, there is a compelling case to be made for wave energy conversion and for significantly increasing wave energy generation capacity through the present century. Careful research and development is thus called for. There is potential to make more deep water conversion techniques commercially attractive and, further, to make wave energy conversion more cost effective even in areas or seasons in which the available power is on the order of 15 kW/m. Boxes 1.1 and 1.2 consider waves as a potential energy source for the United States. Similar arguments are possible for other countries and regions with coastlines.

Finally, it should be noted that the practical techniques to capture the energy transported by waves require that the waves be large enough and long enough to carry sufficient energy for it to seem worthwhile to develop, construct, and deploy a converter system. Equally, it is impractical to attempt to utilize waves that are so large and so energetic that they will produce damaging forces and unsustainably large motions of device components. Most technologies therefore are designed to stop generation and go into survival mode (e.g., temporary submergence, relaxing of constraints and joints) in severe storms, and for similar reasons most devices are not deployed in regions where mostly only breaking waves are present. The type of sea conditions of interest to wave energy conversion are not so much fierce breakers or wall-like storm seas but often

Box 1.1 Energy and the Future: A Case Study for the United States, I

We encourage the reader to verify the figures and estimates below. Some readers may even wish to perform a similar analysis within their own geographic areas.

- U.S. energy production during 2013 was about 110 quad, (1 quad = 10^{15} Btu (British thermal unit); 1 EJ (exajoule) = 10^{18} J; 1 TWy (terawatt year) = 8.76×10^{12} kWh; etc.) which translates to 116 EJ or 3.7 TWy, [11]. Of this amount, about 9.0%, or 0.33 TWy was supplied by renewable energy. Currently, the United States produces about 20% of the world's energy [12]. At the current rate, the United States will produce about 258 quad (8.65 TWy) in 2050. Many believe such an increase could far surpass energy consumption if conservation measures are in place. Nevertheless, if it can be attained without jeopardizing the earth system and current lifestyles, it could represent a useful vision to guide research and development that seeks to match current economic growth and prosperity expectations worldwide.
- Many energy models show that keeping within some stringent climate mitigation and pollution objectives could require [13] that 28%–74% of the world's energy in 2050 be generated from renewable sources, while the rest could be provided by hydrocarbons or nuclear reactors. An "average" figure of 50% for the United States would mean generation of about 4.3 TWy from renewables, which would require an addition of over 3.3 quad (100 GWy) of renewable energy every year starting in 2014 (selected here just as an example).
- For the present U.S. renewable energy portfolio and at roughly the present respective contributions, a 3.3 quad (100 GWy) per year increase will imply on average an annual addition of 1.16 quad (38 GWy) from new hydroelectric, 0.08 quad (2.75 GWy) from new geothermal, 0.06 quad (1.9 GWy) from new solar/photovoltaic (PV), 0.43 quad (14 GWy) from new wind, and 1.6 quad (53 GWy) from new biomass sources.
- This translates into new reservoirs with about 90 billion tons of water falling from 1000 m, about 1580 km^2 of new PV panel area, and about 180 km^2 of new turbine swept area, *all added annually*. Clearly, such rates of increase are impossible or impractical, based on currently estimated resource availability. *Significant further growth must therefore come from offshore resources, if the expectations outlined above are to be met.*

long swells of relatively small amplitudes that carry energy in sizable amounts without generally producing excessive impact forces and motion amplitudes. Fig. 1.3 shows examples of such waves.

A highly readable, nontechnical account of the early developments of wave energy conversion can be found in the book by Ross [15]. Important early developments of wave energy conversion, along with recent additions, are described in the book by Professor Michael McCormick [16]. Although serious efforts to develop wave energy conversion technologies had to await the oil crisis of the 1970s and the early work of Professor Stephen Salter [17], interest in wave energy utilization dates back at least to

Box 1.2 Energy and the Future: A Case Study for the United States, II

- Recent estimates [14] show that, about
  - 237 TWh (0.8 quad/27 GWy)/year of wave energy is available along the outer continental shelf off the East Coast:
    - 16 TWh (1.71 GWy)/year off New York,
    - 14 TWh (1.6 GWy)/year off New Jersey,
    - 45 TWh (5.1 GWy)/year off Massachusetts,
    - 57 TWh (6.1 GWy)/year off North Carolina,
    - 41 TWh (4.7 GWy)/year off the Florida east coast,
    - 23 TWh (2.6 GWy)/year off the Florida Gulf coast, etc.
  - 590 TWh (2 quad/67 GWy)/year off the West Coast, including about 293 TWh (34 GWy)/year off California
  - 130 TWh (0.56 quad/18.8 GWy)/year off Hawaii
  - 1550 TWh (4.96 quad/166 GWy)/year off Alaska
- In terms of available potential for the United States, wave energy ranks between offshore wind and tidal resources. However, the additional promise of wave energy lies in the opportunity to design converters for deeper waters with yielding joints and/or ability to submerge temporarily for a short time before the approach of a storm, so energy production may be resumed quicker (compared with onshore power plants or offshore wind farms). Furthermore, the converters can be designed to utilize retired ships and submarines and shipbuilding infrastructure for overall reduction of cost and time to deployment. Finally, since wave power is nearly an order of magnitude denser than wind power, a well-designed wave energy device may enable more efficient space utilization.

1799 when a patent was issued to Girard and son in France [15]. Other early implementations include a U.S. patent to Wright in 1898 [16], as well as early operating systems in California and France in the 1900s (see Figs. 1.4 and 1.5). Commander Yoshio Masuda developed a wave-powered navigational buoy in Japan in the 1940s, and similar developments by the Japanese Navy continued through the 1960s [18]. However, the pace of development altered dramatically in the 1970s when a number of device concepts began receiving attention and resources worldwide. Much of the rapid development of the subsequent decades was aided by mathematical results of general applicability that were discovered in parallel, notably, through the work of Professor Masatoshi Bessho in Japan [19], Professor David Evans in the United Kingdom [20], Professor J.N. Newman [21], Professor C.C. Mei in the United States [22], and others.

WECs in which approaching wave crests are directed to spill over into a stationary or floating barrier and then to drive a water turbine (as in a tidal head energy converter), are often thought of as "passive devices." A wider majority include an oscillating body

Figure 1.3 Wave fields of interest to wave energy conversion efforts. Courtesy of the National Oceanic and Atmospheric Administration (NOAA), La Jolla Shores, CA.

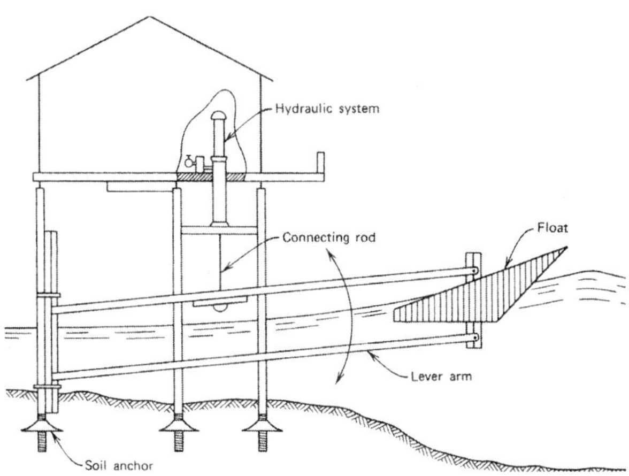

Figure 1.4 Early development in wave energy, by Wright, patented in the United States in 1898 [16]. Courtesy Dover Publications Inc., reproduced by permission.

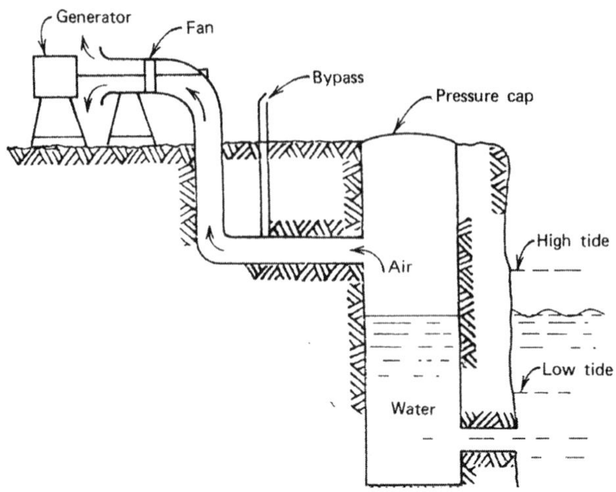

Figure 1.5 Another device from the early twentieth century, designed by Palme in 1920. Note its unmistakable resemblance to the many of the more recent shore-based oscillating water column devices [16]. Courtesy Dover Publications Inc., reproduced here by permission.

(sometimes water level in a chamber), whose oscillations in response to waves drive the subsequent energy conversion stages. Such devices are thought of as "active devices." A number of different approaches saw rapid development in the 1970s. Devices intended to be placed in sea with their lengths parallel to the approaching wave crests and breadths parallel to the wave propagation direction were found effective in locations where long-crested waves predominated. A single line of such "beam-sea devices" placed end to end could theoretically absorb most of the energy brought forth by the approaching wave fronts, and hence came to be known as "terminator" devices. Other devices preferred "head-sea operation" in a ship-like configuration, with their lengths along the wave propagation direction and breadths along wave crests. When placed side to side in one line, such devices were referred to as "attenuators," with the waves getting progressively absorbed as they traveled down the length of each unit. The performance of both configurations depends on the wave direction, however, and far enough from the coast single units designed for beam-sea configuration will sometimes find it necessary to function in the head-sea configuration. Deep sea devices in fact could find themselves working in waves approaching from any direction between head seas and beam seas. "Omnidirectional" devices typically have identical absorbers arranged around a vertical axis, so that they are less sensitive to wave direction. When such devices have very small diameters compared with the approaching predominant wavelengths at a location and predominantly oscillate vertically, they are termed "point absorbers." Fig. 1.6 schematically shows these basic configurations.

The energy conversion task from wave (more specifically, orbital velocity of water particles in a wave) to oscillating body for an active device is generally understood as "primary energy conversion." This oscillation may drive a hydraulic pump or cylinder or more directly a linear/rotary generator. This next stage in energy conversion is the so-called secondary energy conversion. The component responsible for secondary energy conversion is also commonly referred to as the power take-off (PTO).

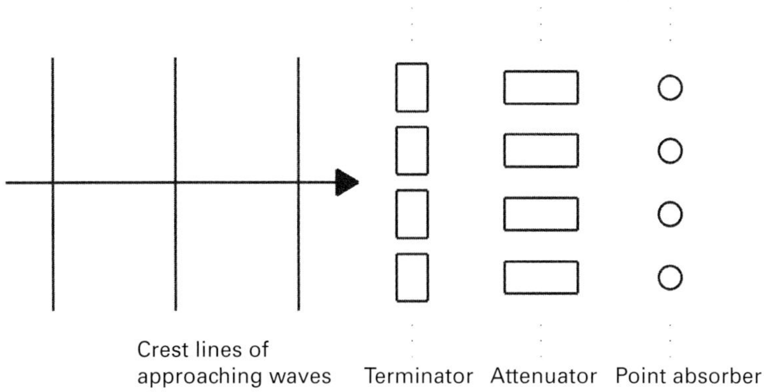

Figure 1.6 The three configurations and device arrangements that have been studied since the early 1970's. Each brings its unique advantages and challenges.

1.5 Primary Energy Conversion

The primary energy converter may be comprised of two oscillating bodies, with the relative oscillation between the two being used by the PTO for secondary energy conversion. Often, one of the oscillating bodies may be water in a chamber, and often, one of bodies may simply serve to provide a structure or support. It is therefore convenient to think of active devices in terms of an oscillator and a reference. The reference is expected to react against the oscillator to enable power absorption, and its presence is critical for further conversion of the oscillator motion into usable power. The reference may be the sea floor, a sea wall, or another floating body. The oscillator and reference together represent the primary energy converter of active devices. In our discussion in the subsequent chapters, "device oscillations" or "device response" typically will refer to oscillator response relative to the reference, unless the two are identified specifically.

This step is naturally different for passive devices whose aim is to use just the rising surface elevation in successive approaching wave crests in the manner of a tidal-head energy converter. Thus, the approaching wave crests are allowed to spill over or "overtop" a barrier, enabling a progressive build-up of a water head in a reservoir. The water head is used to drive a water turbine coupled to a generator for electricity generation. Because only the rise in surface elevation is used, and because efficiency of water turbines is sometimes limited, passive devices have so far not been pursued as vigorously as active devices. On the other hand, certain bathymetries lead to a natural refractive focusing of approaching waves in shallower waters (where crests tend to be larger than troughs). In such locations, the overtopping passive devices may be advantageous for both conversion performance and implementation. Utilization of refractive focusing in shallow water was proposed in the 1970s. [23]. In some sea locations where a natural narrowing of channels is present, such narrowing could be used to overtop approaching wave crests into a reservoir to build an increased water head, and then to allow water to flow back into the sea, driving a turbine in the process. Such a system was tested in Norway in the 1980s (Fig. 1.7). The recent implementation Wave Dragon achieves a

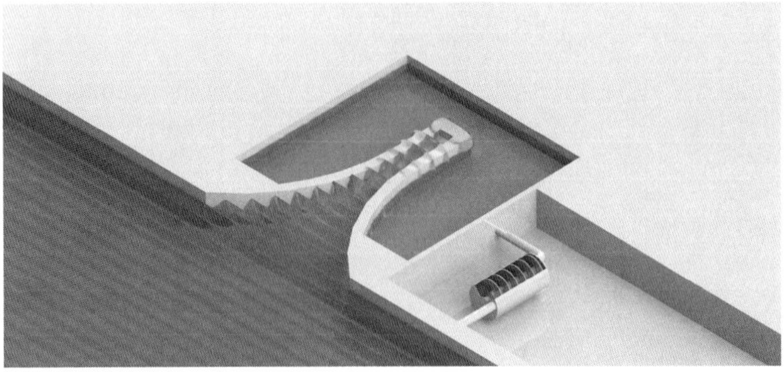

Figure 1.7 An illustration of the concept behind the TAPCHAN (tapered channel) installation in Norway, utilizing the narrowing of a channel to cause crests to spill over a wall into a reservoir, and then to use the increased head to drive a turbine.

similar focusing effect in deeper waters, through channeling of approaching waves with a series of baffles forming a narrowing mouth leading up to the overtopping barrier/wall of the reservoir. Primary energy conversion in such overtopping passive devices is not discussed in detail in this book, which focuses mainly on active devices.

1.6 Secondary Energy Conversion

The challenge of secondary energy conversion in active devices is to convert the oscillatory, irregular mechanical movement of the oscillator relative to the reference to another form of energy that can drive an electric generator. Three types of secondary conversion have been studied: (1) direct electric, (2) hydraulic, and (3) pneumatic. The most straightforward secondary converter is one where oscillations directly drive a linear or rotary generator. In typical wave climates, the period of mechanical oscillations is on the order of 5–15 s. In general, direct electric conversion at such low frequencies incurs excessive power loss. For this reason, the secondary converter (PTO) frequently attempts to "step up" the frequency to a point where electric generation losses are not too great. However, recently developed magnetic materials such as neodinium boron iron (NdBFe) have made direct low-frequency generation more practical, and hence a few recent designs have relied on direct conversion using linear or rotary generators based on NdBFe permanent magnets.

Because the wave input is variable and irregular, however, an intermediate storage system such as a battery is generally required, so that the electric energy supplied to the point of use is smooth. An additional inverter is used where an alternating current supply is needed.

Hydraulic secondary energy converters use high-pressure oil hydraulic pumps or rams to convert the wave-induced reciprocating mechanical oscillation to an oil hydraulic pressure head, which can drive hydraulic motors at the constant, high speeds ideal for electricity generation. The electric generator driven by the hydraulic motor may be a high-speed induction or synchronous generator. The advantages of hydraulic secondary conversion are: (1) the high efficiency of subsequent electric generation, (2) smooth electric power delivery with the help of hydraulic accumulators for short-term energy storage, and (3) the speed and precision with which the hydraulic pumps can be used to control the response of the device to waves. Disadvantages include: (1) lower efficiencies of the hydraulic stages, (2) the need for additional hardware, and (3) the need to design a system capable of clean and reliable long-term operation.

Pneumatic secondary energy conversion is common to devices where waves cause a reciprocating airflow under pressure (e.g., see Fig. 1.8 for an early design by Masuda). In this case, energy conversion is through the oscillation of the water level within a semi-enclosed chamber that is open to wave action. In such devices the oscillator is really an "oscillating pressure" or an "oscillating pressure distribution." The reference body in Fig. 1.8 is the structural chamber containing the oscillating pressure distribution. Such devices are commonly referred to as "oscillating water column" devices. Here the reciprocating air flow drives an air turbine, which may be self-rectifying with symmetric blade profiles as in the case of the Wells turbine, developed by Professor Alan Wells of Queen's University, Belfast. Some devices have used more efficient "impulse turbines" with asymmetric blade profiles and guide vanes along with pneumatic valve/duct arrangements to "rectify" the air flow direction. Because the air turbines spin at high speeds, they can drive electric generators efficiently. Some battery storage or pneumatic

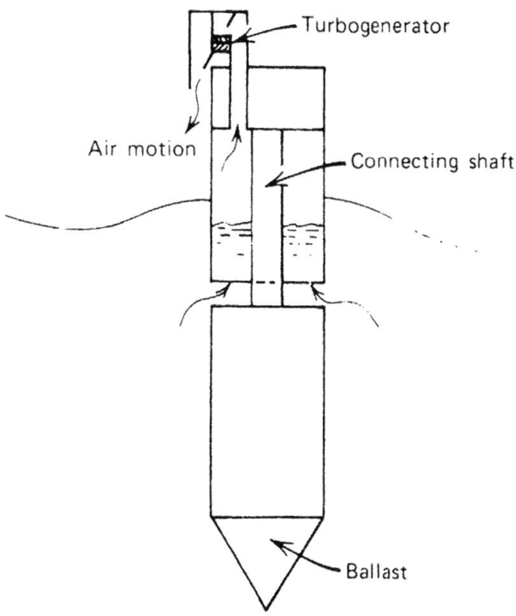

Figure 1.8 One of Masuda's early devices for using wave energy to power navigational and oceanographic buoys; patented in the United States in 1965 [16]. Courtesy Dover Publications Inc, reproduced here by permission.

storage using high-pressure air reservoirs may be used with these devices to enable smooth electric output.

The energy conversion stage between a prime mover such as a hydraulic motor or an air turbine and electric generator is sometimes considered "tertiary energy conversion." For most design and control purposes it is helpful to consider this stage as integral to secondary energy conversion. The part of the secondary energy converter that actively interacts with the motion of the oscillator is often thought of as the PTO device/mechanism/system. In mechanical or dynamical systems terms, the PTO can be considered an "actuator" that applies a "mechanical load" on the oscillator. This is done relative to the reference, and hence any motion of the reference plays a direct role in the overall energy conversion. The simplest type of load required of the PTO for energy absorption is resistive, which the oscillator sees as damping. However, some PTO mechanisms may be designed to provide reactive (spring or inertia) loads in addition. This point will be revisited in the subsequent chapters.

The overall efficiency of energy conversion from the wave input to the electric output is the product of all energy conversion efficiencies involved in the overall system. Thus, wave energy devices with hydraulic/pneumatic secondary conversion could be more efficient overall than those with direct energy conversion. The ultimate choices must therefore depend on the design of all component systems, primary to tertiary, and the overriding design goals.

Below we briefly examine some specific devices that well illuminate the problems and solutions of wave energy conversion. Some of these devices also embody fundamental principles and techniques that have found wider appeal and have been adopted into other devices, including those currently being pursued for commercial use worldwide.

1.7 Tail-Tube or Pneumatic Buoy

This was one of the early wave energy inventions, and the principle was used by Commander Yoshio Masuda to power navigation/instrument buoys in Japan [16] from the late 1940s through the 1970s. Masuda's device provided an early illustration of the use of relative oscillations for energy conversion. Professor Michael McCormick presented analyses of this principle in 1974–1976 [24]. The buoy consisted of a cylindrical "collar" surrounding a deeply immersed central tube that is open at the bottom (Fig. 1.9). The collar provides buoyancy, and its axisymmetric geometry permits a predominant heave oscillation in response to waves from any direction. If the tube is deeply submerged, the water level inside remains roughly stationary and serves as a piston to provide a pumping action as the buoy heaves in response to waves. This forces pressurized air in and out of the chamber and over an air turbine, which can drive an electric generator. A version of this device was used with the self-rectifying Wells turbine by Whittaker et al. [25] in a full-scale test in 1976. The axisymmetric buoy geometry has provided the basis for many wave energy devices from the mid-1970s to the present. In addition, Masuda's tail tube buoy was also one of the earliest oscillating

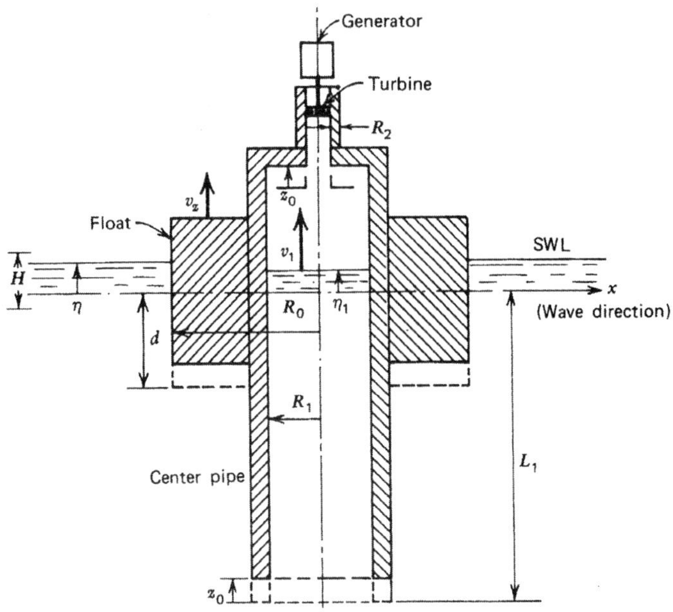

Figure 1.9 One of the early wave energy devices widely studied around the world [16]. Courtesy Dover Publications Inc., reproduced here by permission.

water columns, and the underlying principle has continued to be used worldwide to the present day. For the tail-tube buoy, the oscillator is the buoy with the internal water level as the reference near the heave resonant frequency of the buoy. If the frequency at which the internal water column resonates is distant from the buoy resonant frequency, the buoy may oscillate at a negligible amplitude and provide a reference for the water column oscillation. In either case, the relative oscillation between the oscillator and the reference drives the air turbine for energy conversion. This interesting dynamic behavior was later used on other devices, notably, the floating oscillating water columns Kaimei and Mighty Whale tested in Japan.

1.8 Edinburgh Duck

This device was developed in the early 1970s by Professor Stephen Salter at Edinburgh University in Scotland and remains exemplary because of the depth of technology and engineering perfectionism that have been integral to its development. Working alongside Professor Salter through many of the developments at Edinburgh University were David Jeffery and Jamie Taylor. The duck was one of the earliest "terminator" type devices in that it was designed to be "crest-spanning," that is, to receive waves in a beam-sea configuration and to absorb most energy in waves approaching from a predominant direction (Fig. 1.10), although subsequent laboratory development made it more versatile to wave directions. Multiple sections were laid end to end along the

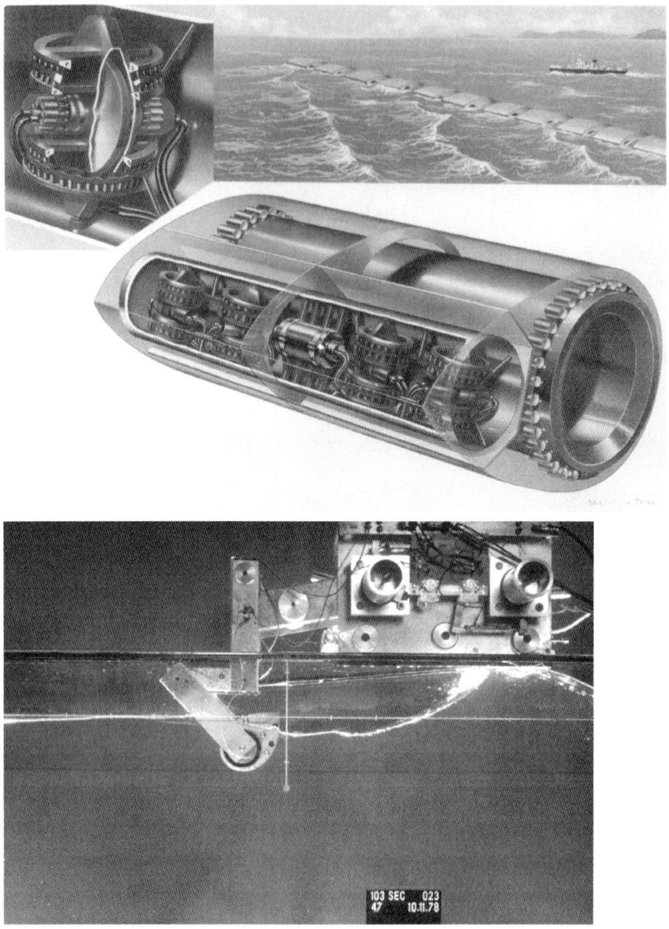

Figure 1.10 An overall design from 1979 that used high-pressure oil hydraulics for power conversion relative to an on-board reference. Duck design figure provided by the University of Edinburgh; duck narrow tank test photograph was taken and provided by Jamie Taylor. Early laboratory work on the Salter duck may have played a major role in shaping the development of wave energy devices. Reproduced by permission of Jamie Taylor, University of Edinburgh.

predominant crest lines in deep waters. Each section consisted of the duck that performed pitch oscillations about a central cylindrical "spine" unit. Neighboring units were jointed together to allow relative oscillation. The jointed central sections formed a long spine, and the phase lag between the different wave crests (and troughs) reaching different parts of the spine caused it to remain approximately in place to reduce mooring loads. The development of the duck demonstrated some new insights of lasting significance.

First, the best energy conversion of the duck approached 100%. The duck geometry leading to efficient conversion was arrived at experimentally using the reasoning that the back of the duck must be shaped like a circular cylinder, so that the pitching oscillations of the duck did not produce any waves behind the duck. On the other hand, the front was shaped so that an incident wave produced the maximum pitch amplitude. This was also found to be the shape that created the largest waves traveling away to the front of the duck. In short, it was discovered that for waves approaching from the front, best conversion was achieved when the duck was designed to radiate most energy to the front, and none behind. This finding was later proved mathematically by other workers in the field (see e.g., [26]).

Second, the duck was one of the first devices to use active control. The load force opposing the duck pitching motion was controlled such that its amplitude and phase could be adjusted independently. Although energy absorption requires just a damping (dissipative or resistive) force, independent control of amplitude and phase meant that the applied load force also included a non-dissipative or reactive component having the same effect as a spring or inertia force. Thus, the natural frequency of the pitching duck could be adjusted to maximize performance in wave spectra with different predominant frequencies.

For full-scale operation, high-pressure hydraulic systems were developed for power absorption/loading. The relative heave and surge oscillations of the duck and the spine segments (i.e., duck relative to spine and spine elements relative to each other) were used for energy conversion. The need was felt therefore for an internal reference for the duck pitch motion. This was provided by internal high-speed rotors that maintained their angular orientation through gyroscopic action. Hydraulic ring cam pumps were used to convert energy from duck pitch relative to the gyroscopically stabilized platform. The principle and geometry of the spine along with the hydraulic PTO between segments led to the development of the Pelamis WEC versions installed and tested since 2005 off the coastline of Scotland. In addition, the idea of a stabilized internal reference was also adopted for use in other devices in subsequent years.

1.9 Contouring Rafts

This was one of the earliest wave energy devices and was proposed by Sir Christopher Cockerell, inventor of the Hovercraft (Fig. 1.11). This device also illustrates an interesting design, consisting entirely of lightweight bodies meant to oscillate in response to waves; power absorption was through relative oscillation between adjacent rafts with no single body or aggregate of bodies serving as a general reference for other oscillations. The device was comprised of a row of rafts receiving an approaching long wave in "head-sea" configuration. Different rafts performed pitch oscillations at different amplitudes and phases in "contouring" the entire wave profile as it passed under. In some versions, this device operation illustrates the "attenuator" mode where the long axis of the device is parallel to the wave direction. Recall that the action of an attenuator mode in broad terms can be visualized as a progressive decrease in the energy of a wave as it travels over the length of the device. This mode is much more like a ship in head-sea condition, and the structural forces applied by extreme waves are smaller than in the beam-sea condition of the terminator mode. For this reason, the raft-type devices may lead to better structural efficiency.

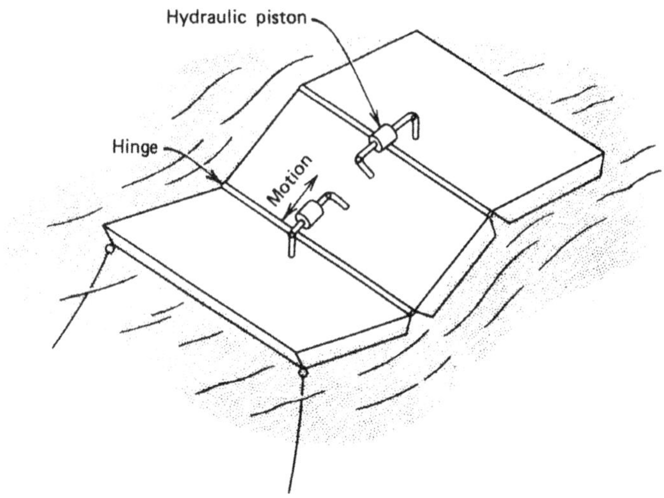

Figure 1.11 The design of the Cockerell raft-type converter. Energy was absorbed as waves passed along the raft length [16]. Courtesy Dover Publications, reproduced here by permission.

Different elements of the contouring raft were hinged together so that the design of the hinges proved a major challenge for this device. Relative oscillation between the elements drove hydraulic pumps for energy conversion. Use of joint technology as in the Pelamis device could conceivably make the raft-type devices attractive.

1.10 Submerged Cylinder

This device was proposed by Professor David Evans at Bristol University in the 1970s. It embodied important hydrodynamic insights that led to nearly 100% energy absorption. Evans analytically discovered an alternative way to produce efficiencies similar to those achieved experimentally by Salter. Recall that the Edinburgh duck was shaped to radiate waves only to the front and not behind and demonstrated efficiencies close to 100%. Evans' approach was inspired by the work of Dean [27] and Ursell [28] on submerged cylinders held fixed in approaching waves and on Ogilvie's work [29] showing that the same effect could be achieved by a cylinder oscillating in heave and surge, essentially so its long axis traced a circular orbit around its original position. No reflected wave was created, and all of the incident wave was simply transmitted with only a phase change. This occurred at all frequencies. This meant that if the cylinder could be controlled to radiate waves behind it so they cancelled the transmitted wave, then all of the incident wave energy could be converted. As discussed in [30] and summarized later in Chapter 6, this in turn meant that (1) if the submerged cylinder responding to waves was constrained to surge and heave oscillations with a phase difference of $\pi/2$ radians and (2) if the rate of energy absorption and the mooring forces could be correctly controlled, then all of the incident wave energy could be converted. For the cylinder, the constraint was provided by a seabed-supported mooring system (Fig. 1.12).

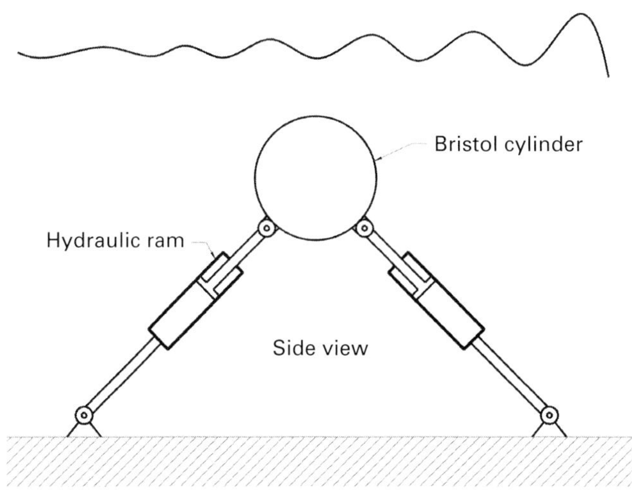

Figure 1.12 The submerged cylinder device with heave and surge oscillations driven at the correct phase difference. This device is capable of highly efficient conversion when the orbital motion is correctly tuned.

The mooring system was designed to incorporate hydraulic pumps for secondary energy conversion and heave-surge phase constraint application. The overall design was comprised of a row of submerged cylinders with their longitudinal axes parallel to the approaching wave crest lines in shallower waters, as the secondary conversion and motion control both required the seabed as the reference. This device was different from many other designs of the early 1970s in that it was fully submerged, which lessened the slamming and wave-impact loads in extreme waves.

Scaled models of this device were tested in the Edinburgh University narrow tank with the surge-heave-pitch rig simulating the controlled moorings and PTO system [30]. Although the Bristol cylinder in its original form was not pursued as vigorously in the following years as some other devices, it remains important in that it illustrates fundamental insights behind efficient wave energy conversion with terminator devices. Additional work on the duck spine and the Sloped IPS buoy at Edinburgh University further demonstrated that the predominant radiation direction could be dictated by combining a symmetric motion with an antisymmetric motion at the correct phases. For a terminator device, heave is a symmetric motion because waves radiated to the front and to the back both have the same phase. On the other hand, surge is antisymmetric because waves radiated in front and behind are opposite in phase.

1.11 Flexible Bag–Type Devices

Professor Michael French of Lancaster University proposed a device in the 1970s designed to perform in the attenuator mode, and each unit used a central spine-like framework along the long axis supporting a number of “flexible bags” made of marine

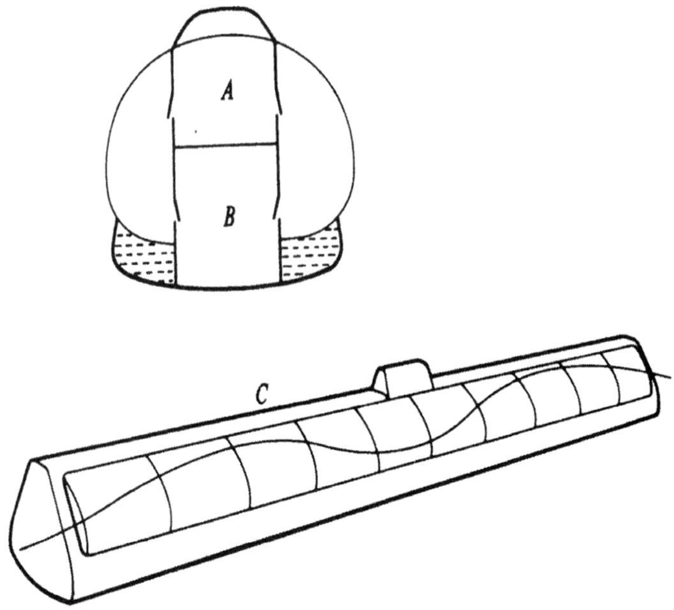

Figure 1.13 The French flexible air bag device working in the attenuator mode [31]. Bag compression causes an airflow between chambers *B* and *A*, driving an air-turbine. *C* represents the "spine" supporting the flexible bags. (Reprinted from French MJ. *Invention and Evolution: Design in Nature and Engineering*. Cambridge University Press, Cambridge, UK; 1988.)

fabric. As waves passed along the unit, they alternately compressed and released the bags, which pumped air into a high-pressure reservoir and back from a low-pressure reservoir (Fig. 1.13). A unidirectional air flow drove an air turbine and in turn an electric generator.

Another flexible bag-type device pursued in the 1970s was the Lanchester sea clam. This device was designed to operate mainly in the terminator mode. The oscillating element was essentially a hinged flap pumping air into a high-pressure reservoir and drawing air from a low-pressure reservoir. The space between the flaps and the central support structure was sealed by means of a flexible fabric. The air flow between the high- and low-pressure reservoirs drove an air turbine coupled to a generator.

1.12 Omnidirectional Buoys

Although WECs based on axisymmetric buoys such as the tail-tube buoys had been tested prior to the seventies, the early and mid seventies saw systematic developments in Norway that focused on heaving buoys with cylindrical and hemispherical shapes, some with internal hydraulic pumps that used the seabed as a reference (Fig. 1.14). The late Professor Kjell Budal (1933–1989) and Professor Johannes Falnes of the Norwegian University of Technology in Trondheim led the work on these devices, and through analytical and experimental work showed that an axisymmetric heaving buoy can absorb

Figure 1.14 The N2 version of the heaving point absorber buoy undergoing tests at 1:10 scale in 1983. These tests examined the latching control approach. The original development of point absorber devices (including the N2 version shown) and latching control is due to Professor Kjell Budal [34]. The photograph was taken by Professor Johannes Falnes in 1983, and is reproduced with his permission.

more energy from an approaching wave than is directly incident on its diameter (at a maximum, all energy incident over a width $\lambda/(2\pi)$ times diameter, λ being the wave length) [32]. This, coupled their small sizes led to the term "point absorber" for such devices (strictly speaking, a point absorber is small enough in diameter that it can be approximated as a "point" in relation to the length of approaching waves).

Budal and Falnes also showed that mutual interactions among adjacent point absorbers in a row facing a long wave crest led to an improvement in the absorbed power by a factor of π at a maximum [33]. Excellent structural economy and smaller mooring costs are expected with these devices. Point absorbers have continued to be of interest, and have formed the basis for many sea-tested WECs and small-scale wave energy devices for niche applications such as mid-sea battery charging and powering of navigational devices. The smaller sizes imply shorter resonance periods, however, while it is the longer wave periods that bring in more energy for the same wave amplitudes. For this reason, Budal proposed and developed the phase-control technique of "latching," which involved locking the buoy heave oscillation for part of the wave period and coordinating the release so that the heave velocity was in phase with the wave force for each approaching wave[34]. This effectively meant a "switching control" to delay buoy oscillation such that resonance-like conditions were produced at least for the part of the period over which motion was released. We will study this control technique in detail in Chapter 9.

1.13 Attenuator and Terminator Oscillating Water Column Devices

Many oscillating water column devices are in a sense descendants of Masuda's tail-tube buoy device. Oscillating water columns have enjoyed worldwide popularity through the last four decades partly because shore-based oscillating water columns can function without moving parts in water, and partly because the overall concept is straightforward to understand and implement. All oscillating water column devices essentially consist of an air chamber open at the bottom or on the side, which allows incident waves to cause the water level within the chamber to rise and fall in response. The resulting reciprocating air flow inside the chamber causes significant air velocities in a smaller duct fitted to an opening near the top of the chamber. A system of two to four valves may be used to "rectify" the reciprocating air flow before it reaches an air turbine with asymmetric blade profiles optimized for a single air flow direction. Alternatively, a self-rectifying air turbine with a symmetric blade profile (e.g., the Wells turbine) may be used. This turbine spins in the same direction for either direction of air flow, enabling the device to dispense with the rectification valve system. In the mid-1970s, Masuda led (for the Japan Marine Science and Technology Center [JAMSTEC]) the development and sea testing of the Kaimei ship, which consisted of 22 oscillating water column chambers open at the bottom [35] (Fig. 1.15). The oscillation of the water level inside the individual chambers relative to the floating hull provided the air flow to the turbines. Being a ship-like structure, the device operated in the attenuator mode. Tests were conducted in two phases in the late 1970s and mid-1980s in which secondary conversion was by means of unidirectional turbines with valve systems and, to a lesser extent, by self-rectifying Wells turbines. The Kaimei was connected to the grid and was able to supply power for household use to the coastal community of Yura in Yamagata prefecture in Japan.

A number of studies were carried out in the mid- to late 1970s on the oscillating water column principle [16]. Both floating and fixed versions were investigated. In the

Figure 1.15 An early large-scale implementation of the oscillating water column device, the Kaimei used a retired ship and was designed to work primarily in head-seas, thus illustrating the attenuator configuration. Copyright JAMSTEC, reproduced by permission.

former, the reference to the oscillating water column/pressure distributions in the chamber was a floating structure, while in the latter, a structure anchored to the sea wall or seabed provided the reference. The latter designs were easier to integrate into existing breakwater technology and received greater attention in the early years of wave energy conversion. Using the mathematically established insight that best energy absorption from an approaching wave was possible with a device that radiated most energy into the direction of the approaching wave, it was found that an oscillating water column with a side opening facing the predominant approaching wave direction had the best conversion efficiency. Such oscillating water columns could thus easily be incorporated into breakwaters.

Shore-fixed oscillating water columns may be a natural choice in certain shoreline formations where a natural gully may be present, such as the Sanze site in Yamagata prefecture in Japan, where a shore/gully fixed oscillating water column was used to drive a Wells turbine connected to a generator [36]. A similar site was used off Scotland for the Isle of Islay oscillating water column device LIMPET, tested during the 1990s and 2000s. This device also used the Wells turbine, and its development was steered by Queen's University, Belfast [37]. Other shore-fixed oscillating water column devices tested in the 1980s and 1990s included three devices in Japan [36], one in India [38], one in the Azores off Portugal [39], and one in Norway [40]. The Azores device was based on the work of Professors Antonio Falcao and Antonio Sarmento, and employed a Wells turbine, with the University of Edinburgh providing a system for controlling the blade pitch and a valve for potentially applying hydrodynamic control. Jamie Taylor at Edinburgh University (as mentioned, a key figure in many early and later developments on the Edinburgh duck) spearheaded the development of this system. Although the controllable pitch turbine was not tested, the Azores device remains operational and continues to provide performance data. The Norwegian device was built by Kvaerner Brug A/S and used projecting side walls referred to as "harbor" to increase the bandwidth of efficient energy absorption. (This effect is naturally available in shoreline gully formations.) The Wells turbine was the more commonly used secondary energy converter in the oscillating water column sea tests of the 1980s and 1990s, though one of the devices tested in Japan (Takenaka Komuten) used a high-pressure manifold with a radial air turbine, while another used an impulse turbine (Tohoku Power and Mitsui Engineering) [41]. The device developed by the Ocean Engineering Department at the Indian Institute of Technology, Madras, was a caisson-type seabed-supported device that tested both a Wells turbine and an impulse turbine [38].

While the devices above were tested as single units, their operation was akin to the terminator mode in the way the side openings faced the approaching waves. A floating version of the oscillating water column with side openings was developed by JAMSTEC as a follow-up to the Kaimei tests and was named the "Mighty Whale" since its floating hull resembled a whale (Fig. 1.16). The device contained three oscillating water columns at the forward end with openings facing the approaching wave direction. The device was shaped to maximize the relative oscillation between the water columns and the hull, which served as the reference. All three oscillating water columns drove tandem Wells turbines coupled to induction generators and an air compressor.

Figure 1.16 The Mighty Whale device developed by the Japan Marine Science and Technology Center, which illustrated a floating terminator-like operation and transmitted performance data and power to shore through its sea tests. It was tested farther from the shore and in deeper water than most devices of its time. Copyright JAMSTEC, reproduced by permission.

The Mighty Whale was tested in the late 1990s in Gokasho Bay off Mie Prefecture in Japan [42].

One of the major contributions of the oscillating water column sea tests has been the demonstration of reliable operation of the primary energy conversion process and the Wells turbine–based secondary energy conversion. The natural ability of this principle to exploit favorable shoreline configurations such as gullies is also likely to prove useful in future applications.

1.14 Other Recent Sea-Tested Devices

A number of other interesting device ideas have evolved since the 1990s which draw on the insights and lessons derived since the 1970s. The Pelamis device, for instance, is an offshoot of the large body of work performed at Edinburgh University on the spine elements designed for the duck (Fig. 1.17). A number of implementations have been tested at various locations off the Scottish coastline since 2005 [43]. This device is moored in the open sea and being slack-moored may operate in either the terminator or the attenuator mode. Wave action on the different cylindrical elements produces relative oscillation between the elements. This is used to drive hydraulic pumps and ultimately electric generators. The several sea tests of Pelamis have demonstrated reliable long-term operation of the biaxial joints and the hydraulic pump/motor circuits. In addition, the tests have also demonstrated the ability of a fully floating device with moving parts for primary energy conversion and of the mooring system to support device operation and survival in changing directional wave climates and extreme waves.

The Wave Dragon tests off the coast of Denmark demonstrated the passive converter approach where incident waves are channeled to the mouth so as to overtop a wall,

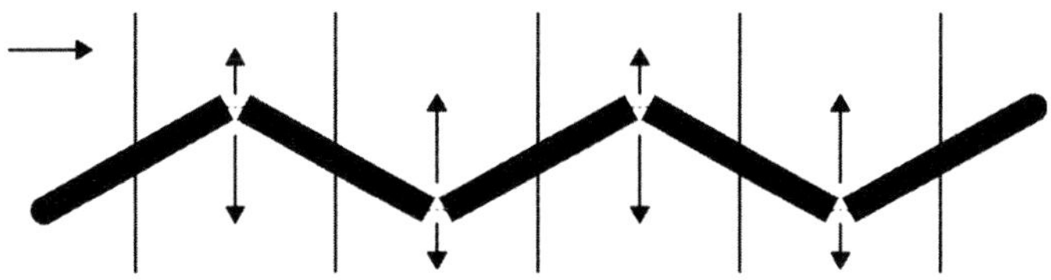

Figure 1.17 A schematic view of the Pelamis device during tests off the coast of Scotland. This device has also seen sea tests in a number of other locations.

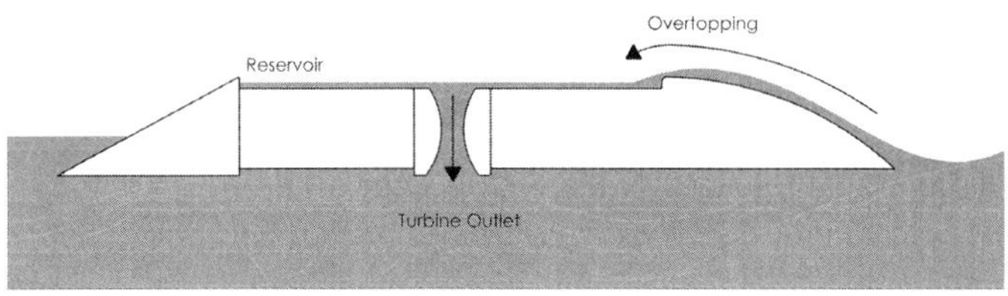

Figure 1.18 A schematic view of the Wave Dragon device recently tested off the coast of Denmark. This device uses baffles to focus waves toward a chamber. The water head then drives a turbine.

causing build-up of a water head in a reservoir, and the water head in turn driving a low-head water turbine (Fig. 1.18). In this respect, this device is an adaptation of tidal-head energy converter systems. However, the Wave Dragon is remarkable in that it operates in the open Wave and uses a floating structure, together with the natural power smoothing and energy storage enabled by its driving principle [44].

Also being tested currently off the coast of Denmark is the Wave Star device [45], which uses heaving and surging hemispherical buoys as primary energy converters (Fig. 1.19). Buoy oscillation forces reciprocating oscillation of a pitching arm that pumps hydraulic fluid into a common high-pressure manifold. A hydraulic motor then drives an electric generator. An offshore "pier" anchored in position by four (or more at full implementation) legs on gravity-based foundations provides the stationary reference. Full implementation is capable of using favorable interactions within a row of axisymmetric buoys. The sea tests have demonstrated a two- to three-fold improvement in the overall conversion efficiency via smooth load control, where a reactive load component is controlled to tune the resonance frequency of buoy oscillation through changing wave spectra. The selective tuning control is non-real-time and is based on periodic measurements of the approaching wave spectral parameters (e.g., energy period and spectral density). The pitching arms can be used to draw up and stow buoy units in extreme wave conditions. A number of other concepts have been tested at sea, including small-scale wave-powered sensors that drift through the ocean collecting data [46].

1.15 Need for Control

Over the last four decades the field of wave energy conversion has made great advances in fundamental understanding of the primary energy conversion principles. Secondary

Figure 1.19 The Wave Star floats undergoing sea tests at one-quarter scale. The device uses both surge and heave modes, though these are constrained by the PTO hardware and absorption mode. Courtesy: Wave Star A/S, reproduced here by permission.

energy conversion has received considerable development as well. Structural and mooring system design methods have developed to a point where several prototypes have operated long term through seasonal variations and survived extreme storms. Improvements continue to be made in support technologies to optimize device transportation to site and installation. While wave energy devices are smaller than solar or wind energy converters, with the offshore devices occupying no space on land, large-scale wave-generated electric power is still more expensive than wind or solar power. Hydrodynamic response control can enable a three- to five-fold improvement in the overall conversion efficiency in changing wave conditions, thereby maximizing power production all year. This can bring about a three- to five-fold improvement in cost effectiveness as measured in currency units/kWh of electric power to the consumer. Such gains could make wave energy commercially attractive even in weaker wave climates and in parts of the world where more options exist. We recall that hydrodynamic response control was first implemented in the 1970s, in early laboratory tests of the duck, the point absorber heaving buoys, and, later, on devices such as the oscillating water column (e.g., [47]). Developments on the Pico/Azores oscillating water column and more recent implementations on the Wave Star device in Denmark demonstrate a strong interest in control on the part of the wave energy developers and commercial entities, including some utilities. Control brings some unique challenges, however, some of which are fundamental (in a manner resembling the second law of thermodynamics). The rest of this book aims to provide a solid background on control systems applications to wave energy devices and highlights precisely how control can and cannot be implemented.

1.16 Conclusion

This chapter provided a "semi-technical" review of waves and wave energy conversion technology. We tried to summarize the major historical developments and, from among a multitude of conversion devices, chose ones that in some way represented the basic principles driving a number of other devices not directly discussed here. The diversity of designs simply illustrates multiple approaches to addressing an immense technological challenge, with no one design being the perfect solution; however, each design idea brings its own rewards. Control is almost always essential, and the principles underlying most control techniques being pursued today provide some degree of horizontal integration over classes of device designs. We conclude this chapter by listing some of the commonly used terms in the wave energy literature.

1.17 Commonly Used Wave Energy Terminology

Device: Wave energy converter.

WEC: Wave energy converter.

Active devices: Devices containing a movable element designed to absorb energy by oscillating in response to waves.

Passive devices: Devices in which waves build a pressure head that drives a water turbine.

Reference or reference frame: Part of an active device that provides the reaction forces for the oscillating element ("oscillator") and relative to which the oscillator moves to enable energy generation through the relative motion.

Fixed devices: Devices for which the sea bottom ("bottom fixed" or "bottom-standing") or the shore ("shore fixed") provides the reference frame.

Floating devices: Devices in which both the oscillator and the reference are floating, and their relative motion is used for energy conversion.

Rigid devices: Fixed devices.

Compliant devices: Floating devices.

Primary energy conversion: Conversion of wave energy to mechanical energy, typically of the oscillator element.

Secondary energy conversion: Intermediate-stage energy conversion prior to electricity generation, for example, mechanical to hydraulic or air turbine.

Tertiary energy conversion: Hydraulic to electric or air turbine to electric generation; sometimes secondary and tertiary conversion may be combined into a single unit.

Oscillator/oscillating elements: Heaving plates, buoys, pitching flaps, ducks ("backless flaps"), rafts, oscillating water columns.

Beam sea configuration: When the long dimension of the device is parallel to the approaching wave crests and the breadth is parallel to the predominant wave propagation direction; a line of devices in beam-sea configuration constitutes a "terminator."

Head-sea configuration: When the long dimension of the device is parallel to the approaching wave direction and the breadth dimension is parallel to the wave crest lengths. A line of such devices in head-sea configuration comprises an "attenuator."

Omni-directional devices: When absorbing units are arranged around a vertical axis to reduce performance dependence on wave direction. When the predominant approaching wavelengths are more than an order of magnitude greater than device diameter, and the predominant oscillation mode is heave, such devices form "point absorbers."

Terminator configuration: When the long dimension of the oscillator (and reference) is perpendicular to the dominant wave direction.

Attenuator configuration: When the long dimension of the oscillator (and reference) is parallel to the dominant wave direction.

Point absorber: Characteristic size of the device is some orders of magnitude smaller than the dominant wavelength; response is in the heave mode and independent of wave direction.

Power take-off (PTO): The secondary or secondary plus tertiary power converter, that applies the energy absorbing load on the primary energy converter.

Wave climate: The group of wave spectra expected during a particular season.

Hydrodynamic efficiency: Efficiency with which incident wave power (over device width) is converted into mechanical power (i.e., efficiency of primary energy conversion).

Secondary conversion efficiency: Efficiency of the secondary energy converter.

Overall efficiency: Efficiency of the overall system, that is, the efficiency with which electric power is produced from the wave power incident on the device.

Capture length/width: Effective wave crest length from which energy is absorbed by the primary converter of a point absorber.

Maximum capture width: The highest capture width value, typically found at resonance, for the primary converter of a point absorber.

Capture width ratio: Ratio of capture width to device width (point absorber diameter in most cases).

Bandwidth: The frequency range over which "high" efficiency is achieved.

Part II

The Basics

In this part we get acquainted with the very basics of control engineering (Chapter 2) and the dynamics of mechanical oscillations, in air and in water, in response to some oscillatory external force (Chapters 3 and 4). In Chapter 2, our interest is in first developing an intuitive understanding of control engineering, and then using this to begin to explore control theory, as it pertains to our goal in this book. Since all "active" wave energy devices rely on wave-induced oscillations for power capture from waves, we study oscillating systems next, in Chapters 3 and 4. Our particular focus is on answering the question: how do we capture most energy from the external oscillatory force by controlling the oscillations of our mechanical system? We realize in the process that control of mechanical oscillations on or just below the water surface is much more complicated than when such oscillations take place in air.

2 Introduction to Control Engineering

Having considered the challenges and solutions of wave energy conversion systems and the important role of hydrodynamic response control, a brief review may now be in order of the fundamental principles and analysis techniques associated with control systems design. Here we will see the techniques as used in a systems-level analysis of multiple connected mechanisms, energy converters, actuators, and sensors, where a number of forces and responses shape the overall behavior of the combined system. We will also see how one may use measurements of certain response variables in feedback to influence the overall behavior in a known and favorable manner. Altering behavior based on measurements of that behavior is the essential notion of feedback control. This is almost never as straightforward as it sounds, however, and control systems design is generally a nontrivial task, especially for a wave energy converter. Analysis is integral to the design process in control systems, and we will review design techniques that draw on analysis of system behavior under several design scenarios with and without feedback control. Mathematics forms an inseparable part of the analysis and design processes, as all too frequently, systems that combine multiple mechanical, hydraulic, and electric elements behave in ways that challenge intuition. Moreover, control systems design purely via experimental trial and error is often too time consuming and too expensive to be practical, unless guided by strong fundamental understanding. For complex systems, systematic mathematical techniques are an effective way to gain such understanding. Notwithstanding this, the value of qualitative reasoning and design adjustments through experimental testing cannot be overemphasized and has been proven repeatedly through the history of wave energy conversion.

In this chapter, we seek to understand the nature of control systems at a more conceptual level using purely verbal and intuitive arguments. We begin by summarizing the commonly used techniques and terminology in the area of controls. Textbooks such as [48–50] provide excellent explanations and illustrations for most of the basic controls material discussed in this chapter.

2.1 Techniques and Terminology

The entire objective of control is to impose a desired behavior or response on a system in the presence of its normal sources of excitation and in spite of any internal or external disturbances. For a mechanical system such as a primary energy converter in a wave energy device, the normal excitation of the oscillator is the forcing due to approaching waves. In this case, a possible control objective could be to force the oscillating body such that its resonant frequency matches the predominant wave frequency in the incident wave spectrum anticipated at a given time. The purpose of such control would be to enhance energy conversion in that spectrum. A hydraulic ram/cylinder may be placed between the reference and the body for energy absorption from the relative oscillation between them. The same cylinder, or a separate dedicated cylinder between the reference and the body, could be used to apply a "control force" on the body such that it resonates at the desired frequency. It will be shown in a future chapter that for a wave energy oscillator, the net energy spent in forcing this desired behavior is theoretically zero, but practically just equal to the energy lost in the cylinder and any other device involved in the application of the control force. The hydraulic cylinder applying the control force is referred to as the "actuator" in controls terminology. The body and reference combination forms the "plant." In order to determine the correct control force to apply, some measure of the plant behavior needs to be used. This may be provided by a real-time measurement of the relative velocity (i.e., body velocity relative to reference) or displacement. This measurement serves as the "feedback," and is used to determine and/or alter the control force applied by the actuator.

A system operating without feedback allows no determination or alteration of the control force based on its present behavior, and is referred to as an "open-loop system." In such a system, the control force may simply be based on a mathematical model of the plant or some other form of prior knowledge. This may work if the model and/or prior knowledge are and remain perfect at all times. Feedback allows the system to adjust the "control force" depending on its actual present behavior, and so it "closes the loop," leading to a "closed loop system." A closed-loop system uses feedback to adjust its behavior such that the error between the "desired" and "actual" output values is minimized. The system just described can be represented in a block diagram where each functional element of the system is shown as a block, and each exchange/transfer of information or force is shown as an arrow. For clarity of explanation, most of this chapter considers only bodies oscillating in air. A skeletal block diagram for such systems is shown in Fig. 2.1, where the goal is to control the plant responding to an external force and to do so at a desired velocity. Here the controller block determines the correct control force to apply on the oscillator based on the difference between the desired and the actual velocity. The "feedback loop" is seen as going into a summing junction along with the external, and input force (over which we have no control). The summing junction is essentially a comparator and captures the fact that the oscillator feels the resultant of the two forces. This diagram is a systems-level representation of the overall process, and, of course, it masks the possibility that each block in itself could be a set of connected blocks each with its own feedback loop. A more operationally detailed block diagram for this system is shown in Fig. 2.2. Note that this diagram is more "practical" in that velocity is directly related to the voltage typically driving electric actuators (motors) or servo valves controlling hydraulic actuators, while forces are linked to current, which is often less straightforward to control than voltage. More practical also because the verbal descriptions in the successive blocks in this diagram can now be replaced by "transfer

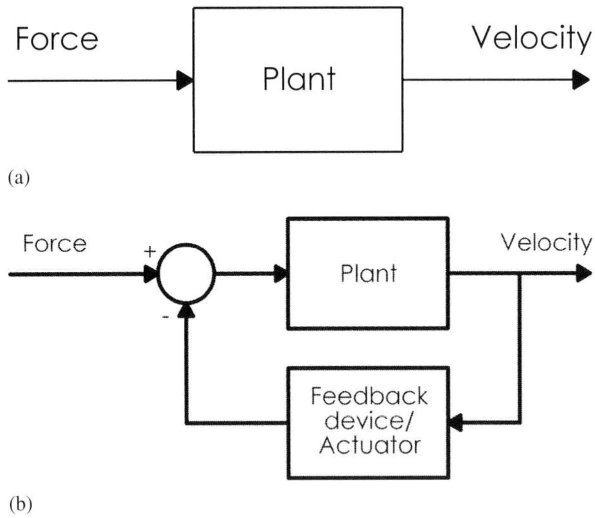

Figure 2.1 Block diagrams for a basic open-loop system (a) and a skeletal closed-loop system (b).

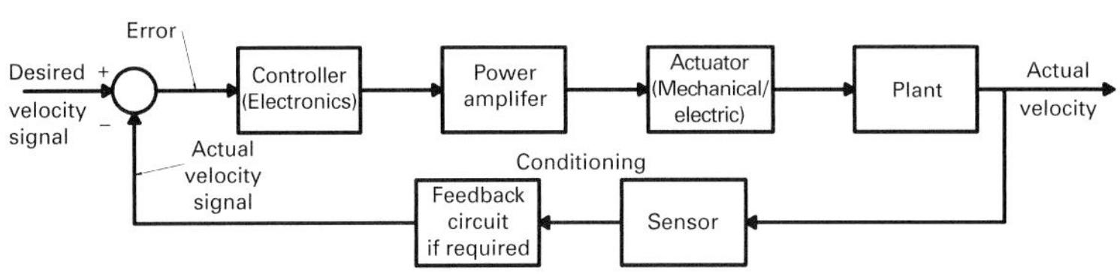

Figure 2.2 Another example of a functional block diagram.

functions" which are algebraic functions that multiply the input quantity brought in by the incoming arrow to produce the output of the block at the outgoing arrow.

When the entity within a block is a "static system," its action can be quantified fully by a single constant value multiplying the input variable to the block, and the output of the block is simply proportional to (and in phase with) the input at all times. When the block represents a dynamic system (such as most systems one meets with in practice), the picture is more complicated. The action of the block can no longer be quantified by a constant. The output is no longer necessarily in phase with the input, and the action of the block may be different for faster-varying inputs than for slowly varying inputs. The action of the block thus must be represented by a complex quantity (with an amplitude and a phase) that is a function of frequency. This is the transfer function that multiplies incoming inputs to produce the output for each block, though the input and output are now represented as functions of frequency rather than time. This type of block diagram modeling is possible and especially helpful for dynamic systems that satisfy the conditions for linearity.[1] For instance, a simple spring–mass–damper

[1] Broadly speaking, a system is linear if its response to the sum of two or more inputs equals the sum of its individual responses to inputs if applied separately. Also for a linear system, the response to a multiple of the input equals that multiple times the response to original input. Sinusoidal inputs produce sinusoidal

system may be considered a linear system if (1) the spring rate, damping rate, and mass are all constant and independent of the deflection, velocity, and acceleration and (2) in addition the external force amplitude is unaffected by the deflection, velocity, and acceleration. We would quantify the behavior of such a system (linear or not) using Newton's laws (especially the second, which relates external forcing, including spring and damping forces in addition to the applied force, to the acceleration produced by the mass times the mass) and using a second-order differential equation with force as the "input" and "displacement" as the "output." For a linear system this would be a second-order differential equation with coefficients that are independent of the displacement, velocity, and acceleration. The block diagram technique allows one to capture the action of such a system within a single block, wherein resides the transfer function for this system.

The technique of Laplace transformation enables a simple representation as long as the system is linear and has constant parameters (i.e., also independent of time) (see Box 2.1). For many controls applications, it is sufficient to consider the situation when the initial conditions on position and velocity are set to zero. The input and output variables of a block are the Laplace transforms of the actual variables, and the quantity within the block then is the algebraic function of the "Laplace-frequency variable" (or simply "s" variable) that "maps" the time domain action of the differential equation into an algebraic relation in the Laplace frequency domain (or s-domain).

Blocks can be combined following some basic rules of block diagram algebra, and the action of an intricate system comprised of multiple interconnected dynamic systems can be expressed using a single algebraic function acting on the input to the first block (perhaps a voltage) to produce the output of the last (typically, this would be a transducer voltage proportional to displacement or velocity). The behavior of the actual output is then inferred by inverting the Laplace transform operation (or inverse-Laplace transforming the block output). While the technique of Laplace transformation works well for systems with a well-defined initial point in time at which all action starts, the analogous method of Fourier transformation is geared for handling systems with inputs and outputs that began a long time ago and will continue a long time into the future. When signals and systems are represented using their Laplace transforms, they are supposed to be in the Laplace frequency domain. When represented using their Fourier transforms, they are said to be in the Fourier frequency domain or, simply, frequency domain (the equivalent of a transfer function in the Fourier frequency domain is the "frequency response function," or often just "frequency response").

When all variables are expressed as functions of time and all system actions as differential equations, matters are being handled in the "time domain." Time domain treatments are the only straightforward recourse for nonlinear systems or for systems that cannot be modeled using differential equations alone. All wave energy devices strictly speaking fall into the latter category. Even for those whose behavior is sometimes studied using linear differential equations, important behavior

outputs of the same frequency for linear systems. Linear differential equations with constant coefficients multiplying the derivatives are examples of linear systems.

Box 2.1 Laplace Transforms and Transfer Functions

The Laplace transform "maps" a function of time $x(t)$ on to the complex-frequency domain, often referred to as the "s-domain." In the s-domain derivatives and integrals with respect to time look like multiplications or divisions by the complex frequency s. Our willingness to go up one level of abstraction thus allows us to express differential equation models as algebraic functions. The Laplace transform $\mathcal{L}[x(t)]$ or $X(s)$ of $x(t)$ is defined as

$$X(s) = \int_0^\infty x(t)\mathrm{e}^{-st}\,dt \tag{2.1}$$

The Laplace transform of the derivative of $x(t)$, written here as $dx/dt = \dot{x}(t)$, is simply $sX(s) - x(0^-)$, where $x(0^-)$ is the initial condition for x and denotes the value of x at $t = 0$ just before any inputs were applied. The Laplace transform of $\ddot{x}(t)$ is $s^2X(s) - sx(0^-) - \dot{x}(0^-)$. In control analysis the initial conditions are all frequently taken to be zero. Successive differentiations thus generally imply successive multiplications by s. The Laplace transform of $\int x(\tau)d\tau$ is $X(s)/s$, with successive integrations implying successive divisions by s. As long as the system is linear and contains parameters that do not change with time – that is, has constant coefficients – the Laplace transform can be used to obtain algebraic equations in the s-domain. Thus a system can be modeled as

$$a_n\frac{d^nx}{dt^n}+a_{n-1}\frac{d^{n-1}x}{dt^{n-1}}+\cdots+a_1\frac{dx}{dt}+a_0x=b_m\frac{d^mf}{dt^m}+b_{m-1}\frac{d^{m-1}f}{dt^{m-1}}+\cdots+b_0f \tag{2.2}$$

For zero initial conditions 0 that is, when both x and f and their derivatives up to order $n - 1$ are zero an instant before the process starts – we can apply the Laplace transform to both sides to obtain an algebraic equation

$$\left(a_ns^n + a_{n-1}s^{n-1} + \cdots + a_0\right)X(s) = \left(b_ms^m + b_{m-1}s^{m-1} + \cdots + b_0\right)F(s), \tag{2.3}$$

allowing the definition

$$G(s) = \frac{X(s)}{F(s)} = \frac{b_ms^m + b_{m-1}s^{m-1} + \cdots + b_1s + b_0}{a_ns^n + a_{n-1}s^{n-1} + \cdots + a_1s + a_0}. \tag{2.4}$$

The function $G(s)$ is the transfer function for the system described by Equation (2.2).

cannot correctly be modeled by constant parameters. Typically, an integro-differential equation is required to model the response of wave-activated bodies, as will be seen in the following chapters. An analysis based on a straightforward transfer function or frequency response type treatment is possible only in a small restricted set of cases surrounding wave energy devices. However, the overall approach and the thought processes of this chapter are important and form the basis for the actual methods used in the control of wave energy devices that we examine later.

2.2 Benefits and Pitfalls of Feedback

Closed-loop systems have the ability to make corrections so that their output maintains a particular value or follows a particular variation. Feedback is a necessary element in closed-loop systems. We quantify its effect on system behavior in the following chapters. Here we just discuss this in qualitative terms. First, feedback can be positive or negative, but most useful applications prefer negative feedback. As pointed out earlier, most closed-loop systems are at the very basic level designed around the goal of making the output equal as closely as possible a specified reference input representing the desired output variation. Feedback enables the system to compare the output with the reference input, and then to take necessary corrective action so that the output will equal the reference input. This essentially translates into subtracting the output from the reference input to derive an "error" signal and using the error to drive system action. In very basic terms, when the output is smaller than the reference input, the error is positive, and the system receives a driving signal that increases the output. This process will continue until the error becomes negligible. Similarly, when the output exceeds the reference input, a negative error is produced, which drives the system so as to decrease the output until a negligible error remains. Thus, in its most basic form, feedback attempts to reduce the error between the reference input and the output.

Often, however, in addition to the reference input and the anticipated forcing, the system is also subject to unaccounted, unknown environmental disturbances. We can verify quantitatively that, all quantities remaining within practical limits, feedback reduces the system's sensitivity to such disturbances, enabling the output to follow the reference closely despite unknown disturbances. Thus, one important advantage of feedback is that it alleviates the need for precise knowledge of the operating environment and all of the possible disturbances it might present.

It can also be verified that feedback reduces sensitivity to errors in the plant and actuator models used to design the control system. This is significant, as only an approximate knowledge of the dynamic behavior of the system to be controlled is needed for successful operation. This is why linear models often work well in the presence of feedback even when the system makes excursions into nonlinearity during its operation.

With these benefits comes the practical pitfall that a control designer must avoid. This is the issue of stability. By definition, a dynamic system is described by a differential equation. In input–output terms, therefore, not only the amplitude but also the phase of the output is generally different from the phase of the input. As the reference input or the fed-back output or the error signals course through multiple connected dynamic systems, considerable changes in amplitude and phases occur.

We recall that the comparison between "reference" or "desired" and "actual" implied in a feedback implementation is usually performed by subtracting the actual output from the reference input, and letting the error signal drive the system/components that follow the summing junction. Feedback is thus generally negative, and the error signal is generally smaller than either the input or the output. With correct design, the error can be driven to zero or made very small. Consider the opposite situation, however, where instead of subtracting, we add the output to the input at the summing junction. A larger signal now drives the components that follow, with the system output getting larger, which further increases the signal out of the summing junction, which will still further increase the system output. Clearly, instead of getting smaller, the error is getting larger. Quickly, all of the signals running through the system will become too large to sustain, with the system failing or incurring serious damage. When the error between the reference and the output keeps increasing indefinitely instead of decreasing or approaching a constant or zero value, the system is thought to be unstable. When the error approaches asymptotically a zero value, the system is "asymptotically stable." On the other hand, the output might approach a constant value or keep oscillating at a constant amplitude, in which case the system is thought to be "marginally stable."

For a dynamic system with phase changes at every block, it is possible that at some combinations of signal frequency, system parameter values, and amplifier gains, feedback may increase the error through system operation, potentially leading to destruction of the apparatus. Care must be taken therefore at the design stage to ensure that for the foreseeable operating conditions, the system does not become unstable. Design for stability is considerably easier for linear systems than for nonlinear systems, because in the latter case, frequency domain analysis and design techniques based on straightforward Laplace or Fourier transformation are unavailable (however, see, e.g., [51]).

One way to understand stability for linear dynamic systems with multiple elements and feedback loops is to begin with the realization that the entire and ultimate action of that system is described by a single differential equation, however complicated. Such a differential equation may be Nth order (i.e., contain the Nth derivative of the output, or the ultimate dependent variable), and therefore must have N linearly independent solutions that must be added together to be able to describe the output response to an input variation over time. Stability requires that all of these solutions must remain finite throughout (see Boxes 2.2 and 2.3). Often, they will decay to zero or to a constant value, or sometimes to an oscillation at a constant amplitude. If even one of them grows indefinitely with time, it renders the entire system unstable, as the overall response will eventually approach infinity.

One of the classical techniques for obtaining a definitive verdict on the stability of an overall control system, when linear, is to apply the Routh–Hurwitz criterion to the combined transfer function for the overall system. In a few steps, this criterion provides the number of solutions that will grow to infinity over time. For an asymptotically stable system, this number will be zero. For linear systems with feedback, another classical technique based on transfer functions is also applicable. This is the root locus method, which uses the input–output transfer function with the feedback loop open (i.e., the open-loop transfer function) to predict the behavior of the system with the loop closed for all possible values of a "gain" parameter that determines how strong an influence feedback will have on the behavior of the closed-loop system. This method is thus able to predict the range of gain values for which the closed-loop system will be unstable, so these can be ruled out at the design stage.

Box 2.2 Poles and Stability

A transfer function such as that represented by Equation (2.4) can alternatively be expressed using factorized forms of the numerator and denominator as

$$G(s) = \frac{(s - z_1)(s - z_2)\cdots(s - z_m)}{(s - p_1)(s - p_2)\cdots(s - p_n)} \tag{2.5}$$

where $z_i, i = 1, \ldots, m$ may be complex-valued, and are the so-called zeros of $G(s)$, as $G(s)|_{s=z_i} = 0$. In a similar vein, $p_i, i = 1, \ldots, n$ may also be complex-valued, and they are called the poles of $G(s)$, as $G(s)|_{s \to p_i} \to \infty$.

The poles are indicative of the "natural" behavior of the system and are important to the way the system represented by $G(s)$ will respond. Each pole represents a first-order differential equation solution with a time-dependence of the form $\sim e^{p_i t}$. In general p_i are complex, or $p_i = \sigma_i + i\omega$. Thus, $e^{p_i t} = e^{\sigma_i t}e^{i\omega_i t} = e^{\sigma_i t}(\cos \omega_i t + i \sin \omega_i t)$. If $\sigma_i = 0$, the response will oscillate forever at frequency ω_i. If $\sigma_i < 0$, the response will decay to zero exponentially at a rate determined by the magnitude of σ_i. However, if $\sigma_i > 0$, the response will increase exponentially with time, until, in a practical situation, actuators or sensors saturate, fuses trip, and/or systems are damaged or destroyed. This is unstable behavior. Such instability is guaranteed for a system if any of its n poles has a $\sigma_i > 0$ (i.e., a positive real part). Conversely, all of the poles for all stable systems must have negative real parts for full "asymptotic" stability. In other words, all stable systems must have all of their poles in the left half of the s-plane, or "in the left half plane." $\sigma_i = 0$ means the ith pole lies on the imaginary axis and will lead to an oscillatory response with a constant amplitude. This is the case of marginal stability.

Closing the loop through feedback changes the overall transfer function of the system with respect to the same input and output. In other words, feedback shifts the poles and zeros of the system from their open-loop values and places them at their closed-loop values. For this reason, an open-loop unstable system can be turned into a closed-loop stable system, and equally, an open-loop stable system can be changed into a closed-loop unstable system. This is when feedback "goes wrong." An important part of design is to ensure that this does not happen in any of the realistic operating conditions that the system is likely to face.

2.3 Control Design

The root locus method is a classical technique of some inner beauty, as with one diagram, it captures the entire range of possible behaviors of a given closed-loop dynamic system. An important element of a closed-loop system is the "controller" block in the block diagram of Fig. 2.3. The input to this block is the error between the reference and the output. The function within this block determines precisely how the error will try to correct the plant behavior. This block can contain just a constant, so that a corrective signal proportional to the error drives the actuator. This is in fact the so-called proportional

Box 2.3 Closed-Loop Systems and Controllers, I

Most closed-loop systems include the so-called controller block placed between the summing junction ("comparator") and the actuator driving the plant. Feedback of the output could then directly be supplied to the comparator (ignoring amplifier gain, etc.) with "unity gain." Fig. 2.3 shows such an implementation, where the controller action is modeled by the transfer function $G_c(s)$. The closed loop transfer function $T(s)$ for this system is

$$T(s) = \frac{G(s)G_c(s)}{1 + G(s)G_c(s)}. \tag{2.6}$$

In the absence of feedback, that is, with feedback supplied with zero gain, the system transfer function is $G(s)G_c(s)$. The poles of this system with the loop open are s values where $G(s)G_c(s) \rightarrow 0$. Note that the denominator of the closed-loop transfer function is considerably different, and the poles of $T(s)$ are the s values where $1 + G(s)G_c(s) = 0$. Both the stability and regular performance of the system can thus alter drastically with the inclusion of feedback. Given the importance of the poles in determining system behavior (as seen in its "transient response"), the controller $G_c(s)$ can be used to great effect in taking the system poles from their original locations and placing them where the system has a more desirable response for a given application (pole placement).

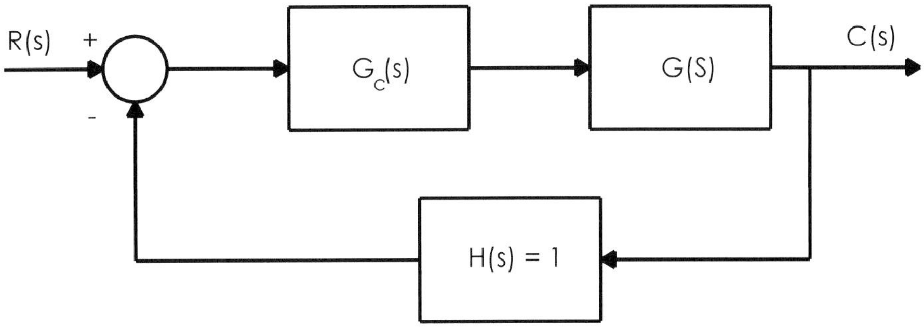

Figure 2.3 A block diagram realization showing the various transfer functions including the controller $G_C(s)$.

(P) controller. When, in addition, the controller also records and sums the error variation over time and produces an additional signal proportional to the integral of the error up to the current instant, a proportional integral (PI) controller results, which enables both the error at the current instant and its integral to correct plant behavior. Another possibility in frequent use is the "proportional derivative" (PD) controller, which incorporates a signal proportional to the derivative of the error at the current instant, so that not only the error but the rate of change of error will now try to influence plant behavior. The constants of proportionality associated with each case represent the controller gains that determine the relative contribution of each in the corrective signal. Finally, if all three

Box 2.4 Closed-Loop Systems and Controllers, II

Finding a controller transfer function $G_c(s)$ that will place the poles where needed is much of the crux of controller design. In practice, perfect pole placement is not always possible due to actuator and other hardware performance limits, and controller design is a matter of compromise involving desired performance, hardware limitations, and controller "cost" or "effort" in terms of the energy required for control.

Given their ability to alter specific transient response parameters, the following controller types are described in the main section. A P-controller corresponds to a $G_c(s)$ given by

$$G_c(s) = K_P. \tag{2.7}$$

recall that $G_c(s)$ multiplies $E(s) = X_r(s) - X(s)$, which is the error between the reference and actual outputs. A PD controller is

$$G_c(s) = K_P + sK_D \tag{2.8}$$

with the PI controller given as

$$G_c(s) = K_P + \frac{K_I}{s}. \tag{2.9}$$

The PID controller combines the three, so that

$$G_c(s) = K_P + \frac{K_I}{s} + sK_D. \tag{2.10}$$

terms, proportional, integral, and derivative, are added together in different proportions (as determined by the respective gain values), a "proportional-integral-derivative" (PID) controller results. These are the commonly used controller types, and the root locus method enables design of all three.

We consider below how these controllers can be designed, and how the prescribed response requirements for the closed-loop system determine the appropriate controller for a given situation and the associated gain parameters for that controller. Because the system in its lifetime is likely to face many different input variations and disturbances that are hard to predict and plan for in advance, certain "test inputs" are used in controller design. Design based on these inputs enables the system to meet precise measures for how fast the system responds to changes, how quickly the output reaches its desired value, whether and how much and how long it oscillates on its way to a steady value, how close the steady value is to the desired value, and so on. In case the output settles to a steady value that does not match the desired value, there is said to be a "steady-state error" in the system. One commonly used test input is the unit step function, which at ideal realization represents an instantaneous "step change" from an old desired value. The "Heaviside step function" (instantaneously changing from a zero value at clock start to unity in the appropriate physical units) typically is used to describe this function mathematically. The system's response to this input gives an excellent quantitative measure of its attributes and is referred to as its "transient response." Designing the controller so the closed-loop system will meet its prescribed transient response characteristics is frequently an important step in the overall control system design. The time it takes the output to reach and thereafter remain within 98% of its steady, "final" value is called "settling time." Also used in design is the "rise time," the time it takes for the response to increase from 10% of its final value to 90% of its final value. The difference between unity (representing the desired value) and the final value of the output is the steady-state error. Often, the output overshoots to a value exceeding its final value and then oscillates about the final value. The difference, at its largest, expressed as a percentage of the final value, is the "percent overshoot." This first tallest peak the output reaches also defines its "peak value," and the time it takes to reach this is the "peak time" (see Fig. 2.4). The root locus method enables a controller design (i.e., suggests values for the gain parameters involved) so that the desired transient response is achieved. Typically this procedure concerns the most dominant terms in the system response, and hence frequently an iterative approach coupled with simulation is followed. P-controller design is the most straightforward and is based on a direct, explicit determination of the P-gain on the root locus. However, a P-controller enables control only of one transient response attribute (percentage overshoot or peak time or settling

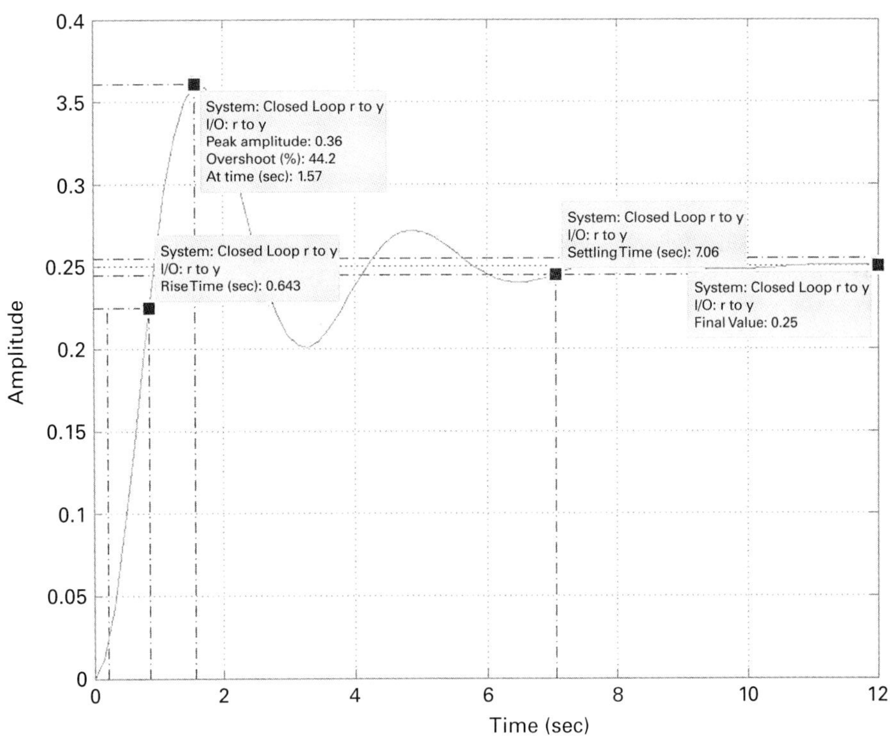

Figure 2.4 The response of a second-order system with two complex-conjugate poles to a unit-step input. The parameters referred to in the text are indicated with black squares. Note that stable linear systems with negative real poles do not exhibit an oscillatory response (such as shown here) to a step input.

time, etc.). A PD controller allows one independently to dictate two transient response attributes, while a PI controller is used when a steady-state error of zero is required. The PID controller is thus a good general-purpose controller and is used frequently in actuator control. Use of the root locus for any of the three controllers eventually reduces to finding a single gain parameter, with the derivative and integral elements often being treatable with informed additions and changes to the original system root locus. PI, PD, and PID controller design requires some practice and iteration, but the overall effort is still significantly less time consuming and more satisfying than an arbitrary trial and error, which can quickly turn into a frustrating exercise the more complex a system is.

A direct counterpart to the root locus (equivalent in theory but quite different in implementation) is based more on the system's amplitude and phase response to sinusoidal inputs, and is known as the "frequency response design" technique using Bode plots and Nyquist diagrams. This is not discussed in the present chapter in the interest of clarity and focus.

2.4 State Space Modeling

This is the so-called modern approach to modeling and design (as opposed to the classical approach above) that was preferred more in the Russian literature before gaining widespread use in the Western literature in the 1960s [52]. This approach does not rely critically on the algebraic transfer functions in the Laplace or Fourier frequency domains and for the most part is confined to the time domain. Further, it recognizes that any complex Nth order differential equation describing the behavior of a system (or an element of a system) can be represented as a set of N first-order differential equations. The intermediate variables that must be defined in the process are known as "state variables" or "states" of the system. Many of the state variables are hard to observe directly, and to distinguish them from measurable system outputs, an additional set of variables is defined that consists of just the directly measurable quantities (which often can be used in feedback). The full mathematical model for a system under this formulation is thus comprised of (1) a "state equation," which is a first-order differential equation relating the rate of change of the state vector to the internal dynamics or the inner workings of the system and the control or reference inputs applied to the system, and (2) an observation equation, which relates the measurable output variables (arranged in a vector) to the state vector. In straightforward cases, this is an algebraic equation in which a rectangular matrix multiplies the longer state vector to give the shorter output vector. Any environmental disturbances can be added to either or both of these equations. This approach can be used for nonlinear systems where both vector equations are functional relationships rather than simple products of vectors with matrices. These reduce to matrix equations with constant coefficient matrices for linear systems whose parameters are independent of time (i.e., linear time-invariant systems).

The stability of the overall system can be assessed by computing the eigenvalues of the "system matrix" representing the inner operation of the system. Any eigenvalue with a real part greater than zero represents a solution that will grow indefinitely with time

Box 2.5 State Space Models, I

A system with n poles, that is, an nth order system, can be expressed in the state space form by introducing n state variables so that the nth order differential equation in the time domain becomes a system of n first-order differential equations in the time domain. A state space model for linear systems with constant parameters (a "linear time invariant system") has the form,

$$\dot{\mathbf{x}} = \mathbf{A}\mathbf{x} \tag{2.11}$$

where $\mathbf{x}$ is the state vector for the system, and $\mathbf{A}$ is the system matrix. $\mathbf{X}$ may be an $n \times 1$ vector and $\mathbf{A}$ an $n \times n$ matrix. When the system is driven by an external reference input $\mathbf{X}_r$ through a drive/actuator system the states evolve according to

$$\dot{\mathbf{x}} = \mathbf{A}\mathbf{x} + \mathbf{E}\mathbf{X}_r \tag{2.12}$$

where $\mathbf{E}$ is an $n \times r$ matrix. In addition, the system may also supplied m control inputs in the vector $\mathbf{u}$. If the interaction of the control input vector with the system is a linear relation expressed by an $n \times m$ matrix $\mathbf{B}$, then

$$\dot{\mathbf{x}} = \mathbf{A}\mathbf{x} + \mathbf{B}\mathbf{u} + \mathbf{E}\mathbf{X}_r. \tag{2.13}$$

Not all of the states may always be directly available for measurement, and the actual measurable outputs of the system may be related to its states as

$$\mathbf{y} = \mathbf{C}\mathbf{x} \tag{2.14}$$

where $\mathbf{y}$ may be a $p \times 1$ vector, with $\mathbf{C}$ being a $p \times n$ matrix.

and therefore lead to overall instability of the system. Feedback control methods often are in essence designed to "place" the eigenvalues at locations that lead to the desired state and output variations in time domain. The approach extends to nonlinear systems in a restricted way when applied "locally" (i.e., over small mathematical neighborhoods) in state space (which essentially is a multidimensional mathematical space with as many axes, or dimensions, as the number of states, with each point representing a combination of different state values). Many nonlinear systems have multiple "equilibrium points" where the system or its state values may "settle." Linearizing the nonlinear relations using a Taylor series expansion at each equilibrium point, the eigenvalues of the resulting state matrix can be computed to evaluate the stability of that equilibrium point with respect to small changes. Controller design for such linearized systems is specific to particular equilibrium points.

2.5 Challenges for Wave Energy Conversion

As we shall see in the subsequent chapters, the control system for a wave energy converter presents unique challenges. The classical or modern design techniques above

Box 2.6 State Space Models, II

If all or some of the states can be measured and used in feedback so as to determine the control input $\mathbf{u}$,

$$\mathbf{u} = -\mathbf{F}\mathbf{x}. \tag{2.15}$$

Substitution of Equation (2.15) into Equation (2.13) gives

$$\dot{\mathbf{x}} = [\mathbf{A} - \mathbf{BF}]\mathbf{x} + \mathbf{E}\mathbf{X}_r. \tag{2.16}$$

The matrix $\mathbf{A} - \mathbf{BF}$ is the closed-loop system matrix. Asymptotic stability of the open-loop system is guaranteed if all eigenvalues of the matrix $\mathbf{A}$ lie in the negative half-plane, that is, have negative real parts. Similarly, if all eigenvalues of the closed-loop system matrix $\mathbf{A} - \mathbf{BF}$ have negative real parts, then the closed-loop system is also asymptotically stable. Controller design here in part amounts to choosing the gain matrix $\mathbf{F}$ so as to "place" the system eigenvalues where their combined effect is to give the system the desired response characteristics. The parallels between this approach and the classical approach are easy to see here.

A nonlinear system in state space form can be represented as

$$\begin{aligned} \dot{\mathbf{x}} &= \mathbf{f}(\mathbf{x}; \mathbf{u}; \mathbf{X_r}) \\ \mathbf{y} &= \mathbf{g}(\mathbf{x}) \end{aligned} \tag{2.17}$$

where $\mathbf{x}$ may be $n \times 1$. An equilibrium point in n-dimensional space is defined as a point $\mathbf{x}_e$ where

$$\mathbf{f}(\mathbf{x}_e; \mathbf{u}) = \mathbf{0}, \tag{2.18}$$

provide an excellent insight into the thought processes prevalent in the control systems area. As far as wave energy conversion goes, the techniques of the previous sections need to be extended significantly. For the primary energy converter (i.e., the oscillating body and the reference) whose control is our main goal in this text, we summarize the following challenges. The next four chapters should help us understand these challenges better. Subsequent chapters describe different ways to overcome the challenges so that the full potential of wave energy devices can be utilized.

1. The dynamics of a body/reference interacting with the free surface (where the waves are) cannot be described by a differential equation in general. An integro-differential equation combining derivatives and a convolution integral is required for analysis in irregular waves. It is only when the analysis is forced to consider purely regular sinusoidal wave of a single frequency (i.e., harmonic waves) that the dynamics can be described by a differential equation that is valid at that particular frequency only. This approach works in a limited way for spectral analysis of body motions (ships, offshore platforms, even wave energy converters), but fails when simultaneous control of both amplitude and phase of that motion is desired, as we see in the subsequent chapters.

2. For this reason, the transfer function/Laplace transform–based or Bode plot–based control design approaches should be applied with caution for bodies directly interacting with irregular waves.
3. Determining the reference/desired velocity for the primary converter is a challenge in itself and requires knowledge of the approaching wave field a distance up-wave of the converter.
4. Designing a controller for the actual velocity of the primary converter to follow the above reference velocity (to close the loop) is a challenge, requiring prediction of the actual velocity or wave measurement up-wave of the converter.
5. We see then that there is a "feedforward" element to the design of control systems for wave energy devices and there is a feedback element. Feedforward will enable us to determine the correct control forces to apply to the wave energy converter so as to produce a desired velocity. This is at best based on a model of the entire system that we consider most expedient. There may thus be errors in the model, which would cause the actual velocity to be different from the desired velocity. There may be disturbances from the device environment, measurement errors, unaccounted for wave effects, and small uncertainties in our knowledge of, and slow changes in, device and actuator parameters. Actual implementations will therefore require feedback. Small deviations in the velocity variation from the desired variation could be corrected by means of "inner" feedback loops, which themselves could range in complexity (P, PD, or PID controllers being some of the simpler examples).
6. Challenges above are of a fundamental nature rooted in the physics of wave energy absorption, though valid practical approximations can be used to design control systems for near-maximum primary energy conversion. The history of wave energy device control reflects evolving attempts to seek good performance in the face of these challenges.
7. Finally, the discussion of this chapter, as well as the arguments and methods in the material in the following chapters, derive from an understanding based on linear theory. Actual behavior of devices and control systems may sometimes depart from the region of applicability of assumptions implicit in linear theory (small-amplitude waves, small-amplitude deflections, small forces, all so that descriptions based on linear differential equations are valid). The tasks of understanding nonlinear behavior and control system designs based on nonlinear dynamic modeling are not tackled in this book. Our problems are challenging enough already.

Figs. 2.5 and 2.6 summarize possible control approaches applicable to control of wave energy devices. Implicit in the "controller + actuator" block of Fig. 2.5 and the force resulting from it could be the feedforward portion in Fig. 2.6. Fig. 2.7 explains the system differently. In response to the external forcing, one calculates the desired velocity variation with respect to time ("optimal set point") which drives the controller/actuator, which will apply a control force on the device. The response of the device, in terms of its actual velocity variation, is compared with the desired velocity variation, and any

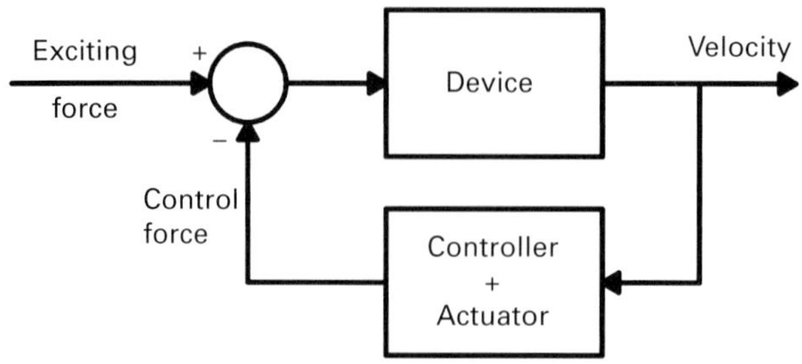

Figure 2.5 A basic block diagram summarizing the role of control in a wave energy device

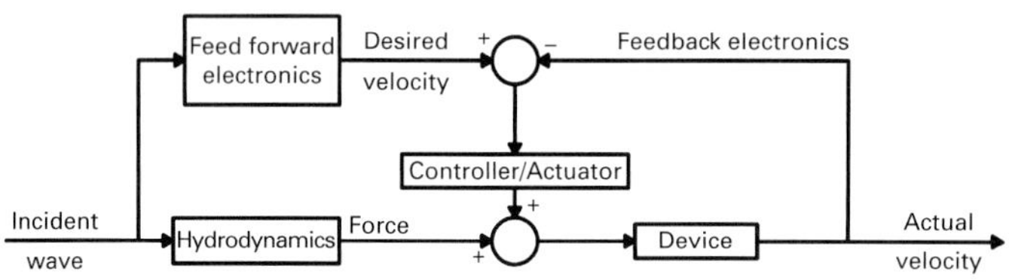

Figure 2.6 A block diagram realization showing more of the elements involved in control application on a wave energy device.

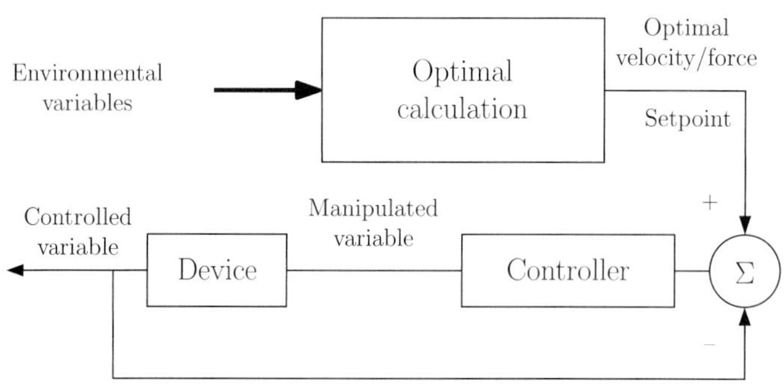

Figure 2.7 A block diagram realization illustrating the feedforward and feedback elements in a control system for a wave energy device.

deviations are used to generate an additional control force that will attempt to reduce these deviations. Chapters 3–6 address the fundamentals that work together to determine the parts of these diagrams that are unique to wave energy devices. Chapters 7 and 8 discuss two approaches to deriving the desired velocity variation. Chapters 9 and 10 are concerned with finding the velocity variation that maximizes converted power when the control force consists entirely of braking and releasing the body oscillation (on-off, switching, or bang-bang control). This velocity variation may be used as the "reference trajectory," which could provide a way to adjust the on-off force sequence to be applied. Chapter 7 builds on the results of Chapter 6 to derive the approximate hydrodynamically optimum velocity variation and uses this as the reference velocity trajectory for the device to follow.

2.6 Control of Wave Energy Devices

Three points should be recognized when trying to understand the role for control in wave energy conversion. Note that our discussion below assumes linearity at all levels.

- It is not easy to derive the velocity variation that leads to maximum energy conversion, as this requires knowledge or prediction of the future incident waves at the device.
- Such a variation may not be practical, if it calls for physically insupportable velocities/oscillations, or if the forces required are too large to be practical with the machinery available for control.
- It may not be possible for the device to track the desired variation even if it falls within practical bounds, unless additional adjustments are made on a continuing basis.

Two complementary approaches seem possible.

- Use hydrodynamics fundamentals to derive the velocity variation that leads to maximum power capture; that is, obtain the hydrodynamic velocity optimum over the time span of interest, utilizing real-time wave measurements and predictions as necessary. This provides the "reference trajectory" for the actual device velocity to track. Next use hydrodynamic and dynamic models for device response to derive the control forces that will be intended to produce the desired velocity variation in the device response. Correct any deviations ("tracking errors") using simple feedback techniques. Incorporate knowledge of oscillation constraints into the hydrodynamic response formulation. We may refer to this approach as the hydrodynamicist's way, for want of a better description.
- Recognize that the optimum control design is one that will seek the best trajectory and the best tracking performance under practical mechanical limitations, not forgetting that wave predictions are not easy to obtain, nor that real-time control requires fast computations, nor that precise hydrodynamic modeling is difficult. Once one recognizes that the goal is to maximize energy conversion within constraints (and that the hydrodynamic velocity optimum within constraints would be a means to achieving this), the focus of the effort can shift to optimizing a certain index of high performance (such as absorbed energy over a period) within practical constraints such as oscillation limits, random disturbances, measurement errors, unmodeled dynamics and so on. Almost the entire machinery of control theory and its general results now become available. We refer to this approach as the control engineer's way.

Both approaches are evident in this book, and we feel the reader will agree that both are essential. A hydrodynamics-based approach cannot succeed in practical situations without incorporating control engineering. Similarly, approaches based purely on controls formalisms risk sliding into the "black-box mode" where the device hydrodynamics are highly simplified and important effects are ignored. Such strategies may lead to inaccurate designs and thwarted expectations. The two approaches are thus complementary. Some of the interesting general techniques from control theory are summarized in the following discussion, which should provide a gateway to the methods used in this book. Chapter 6 provides the theory that allows one to compute the hydrodynamically optimum velocity that will enable power conversion at the best efficiency as predicted by linear theory. One approach to deriving the constrained hydrodynamic velocity optimum is also discussed in Chapter 6 [53]. This velocity constraint is not an "inequality constraint." Chapter 7 discusses how the hydrodynamic velocity optimum may be approached in the time domain (i.e., in real time) for a device operating in irregular waves. As argued in detail in Chapters 4 and 7, this step requires knowledge or prediction of the future excitation (essentially, wave profiles). Thus, almost the entire focus of Chapter 7 is on deriving the hydrodynamic velocity optimum in the time domain and the wave prediction aspect. The control force variations are derived and specified in an open-loop manner, and the design of feedback elements (to ensure correct tracking in the presence of disturbances and measurement errors) is not considered. The expectation is that if the necessary control forces (as we shall see, there are "reactive" components required in addition to the energy absorbing components) can be provided, power absorption will be maximized within the constraints embedded in the determination of the velocity optimum. The control engineer's approach begins with an expression for absorbed power (as defined in terms of an as yet unknown velocity and the energy-absorbing force), and following mathematical arguments integral to the "optimization procedure," arrives at the control forces that maximize the absorbed energy over a desired time duration. Constraints are "adjoined" to the "performance index" along the way, where a typical performance index for a wave energy device is the absorbed energy over a specified time duration. If the performance index is set up so that it is a more direct measure of the error in tracking the hydrodynamic velocity optimum, a feedback loop (the most straightforward being P, PI, etc.) can be designed to optimize tracking performance over a desired time duration (see Box 2.8). The hydrodynamic velocity optimum in such a situation may be evaluated using the methods of Chapters 7 or 8. On the other hand, by using a control design procedure that begins with a performance index more directly related to the energy absorption by a device, one could also derive a velocity variation that is different from the hydrodynamic optimum,

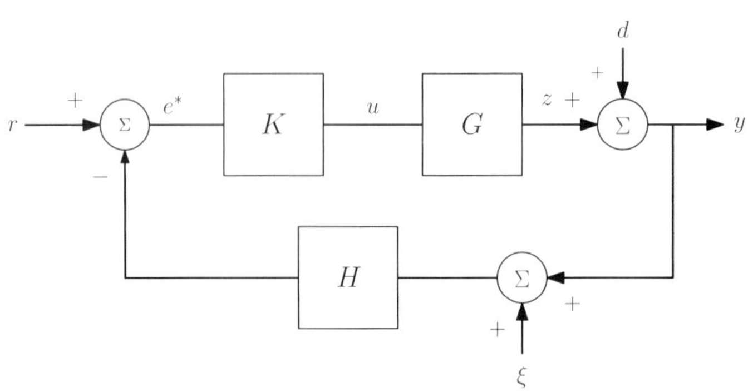

Figure 2.8 A block diagram realization for a feedback system used to track a reference trajectory r.

Box 2.7 Feedback for Trajectory Tracking

A block diagram implementation of a closed-loop system is shown in Fig. 2.8, where K, G, and H are all transfer functions, with G representing the plant to be controlled, H a feedback transfer function, often just equal to unity (as seen in the discussion above). r is the "reference trajectory" to be determined using one of the two complementary approaches above (for instance, either the hydrodynamically optimum velocity variation or a trajectory that satisfies a different optimization criterion). e^* is the instantaneous trajectory tracking error. e^* is not equal to the actual error $e = r - y$, when H is not a constant unity function. u is the control input. d denotes external/environmental disturbances and model errors, and so on, while ξ is just the measurement error.

A suitable control design (i.e., the transfer functions H and K) can be obtained for particular frequencies, depending on any limits on steady-state error, peak time, overshoot, and so on. The desired tracking performance can be specified using one or more of the following criteria:

1. Desired transient response (peak time, percent overshoot, settling time, etc.)
2. Desired steady-state response (steady-state error, phase lead/lag, etc.)
3. Desired closed-loop pole locations (which determine the transient response)
4. Trade-off between control effort/energy u^2 and tracking error e^2
5. Sensitivity of the closed-loop system to variations in system description, that is, model error
6. Sensitivity of the closed-loop system to external/environmental disturbances

but is desirable from other points of view. Such a procedure is evident in Chapters 9 and 10, where the control force can be only a braking or clutching force, which is either fully switched on or fully switched off. The body velocity variation resulting from the "switching" sequence is not the hydrodynamic velocity optimum. It is still an optimum variation, however, relative to the performance index chosen and the type of control forces allowed.

Box 2.7 describes the control engineer's approach to designing a feedback system that tracks a specified reference trajectory (such as the hydrodynamic velocity optimum in irregular waves). Box 2.8 summarizes ways to build robustness into the controller design, so that good performance can be obtained despite uncertainties in our understanding of the system or the environment in which it operates. Boxes 2.9 and 2.10 introduce the general techniques of optimal control design.

2.7 Conclusion

In the first few sections of this chapter, our goal was to provide an informal introduction of dynamic system analysis and control engineering. Just the basics of classical control

Box 2.8 Robust Control

A systematic procedure evolved during and following the 1970s, which explicitly attempted to build controller designs that were robust against errors or variations in system description (i.e., the plant, actuator models used) due to time variance, linearization, and so on, as well as external disturbances [54, 55]. Robustness implies insensitivity of the closed-loop performance to the uncertainties in our knowledge of the system to be controlled and the disturbances acting on it. The system model used in design and a measure of the uncertainty are required for robust control design. Depending on the way the uncertainty is described ("structured," e.g., known to be a diagonal matrix, or "unstructured," could be nondiagonal), either the μ-synthesis (for structured uncertainty) or the H_∞ (for unstructured uncertainty) design formulation may be used.

Robust control design essentially consists of minimizing some form of a weighted sum of (1) the sensitivity of the closed-loop system to model error (i.e., variations in $G(s)$, $K(s)$, etc.), and (2) the sensitivity of the output y to measurement error ξ. The transfer function $S(s)$ in (1) also describes the dependence of the output y on the external disturbance d. Thus,

$$S(s) = \frac{1}{1 + G(s)K(s)H(s)}, \quad T(s) = \frac{G(s)K(s)H(s)}{1 + G(s)K(s)H(s)}. \tag{2.19}$$

Letting $W_1(s)$ and $W_2(s)$ denote the weights multiplying $S(s)$ and $T(s)$, respectively, $W_1(s)$, is often chosen to:

- penalize sensitivity $S(s)$ at low frequencies for good disturbance rejection (i.e., insensitivity) at low frequencies, and
- ensure that system performance (i.e., dynamic response) is maintained at low frequencies, despite parameter variations.

$W_2(s)$ is chosen to:

- ensure robust stability by "bounding" the closed-loop perturbation to an uncertainty $\Delta(s)$ in the singular-value sense [56] (where $G(s) = G_o(s) + \Delta(s)$, $G_o(s)$ being the nominal plant model) and
- attenuate the effect of high-frequency measurement noise by driving down $T(s)$ at high frequency.

Robust control design is often augmented by a linear quadratic regulator (LQR) so that the desired tracking performance is also obtained. The LQR approach is discussed in Box 2.9.

theory as applied to linear systems were first summarized using verbal descriptions in place of mathematical relations. Block diagrams were introduced as an effective way to capture all major interconnected, interacting components included within a system. Important mathematical concepts basic to systems and control analysis were included

Box 2.9 Optimal Control and Model Predictive Control, I

The original formulation of optimal control is based on the calculus of variations. Just the basics are discussed in Chapter 9 as part of our discussion of the latching control method. The main goal is to find a functional variation that leads to an extremal value (i.e., a maximum or a minimum) of a functional E_a or J (e.g., an integral of some expression containing the functional variation to be determined), typically subject to some constraints on the variables for which the functional variation is defined. A simple example may be to find a control force on a mobile cart that carries the cart from point A to B in the shortest time, knowing the forces acting on the cart. The maximum velocity may be constrained to some specified value. The performance index here, that is, the total time to travel, may be augmented to minimize fuel consumption while moving the cart. The solution to the optimization problem here would be a force versus time variation that could form the reference variation for a closed-loop controller to track. The overall formulation can be further extended to incorporate the closed-loop control design, thereby allowing the tracking error also be minimized.

The overall idea leads to the linear quadratic regulator optimal control problem, formulated as

$$\min_u J = \int_0^{t^*} (e^T Q e + u^T R u) dt \tag{2.20}$$

where $e = r - y$ and Q and R are weighting matrices that define the trade-off between error minimization and reduction of the control energy used. For example, in high accuracy servomechanism problems, the choice of $Q \gg R$ is usually made, while in satellite control applications, where fuel is limited, a choice of $R \gg Q$ is more appropriate. Since there is an implicit relationship between Q, R, and the closed-loop poles, traditional optimal control can be seen as an alternative way of determining the closed-loop poles. The solution to the problem in (2.20), however, is not simple, requiring the solution to a "Ricatti equation" for systems described in state-space form (e.g., see [52]). In general, the formulation in (2.20) is for linear models and can be specified either as a finite-horizon (t^* is finite) problem or as an infinite-horizon ($t^* \to \infty$) problem.

A particular evolution of the finite-horizon optimal control problem that is formulated in discrete time and uses a future prediction of the system output is model-based predictive control (MPC) [57], which has seen significant adoption in the process, and other, industries, rivaling PID control in popularity.

in text boxes, as was a brief summary of state-space methods. Systems notions with particularly important practical consequences such as feedback, stability, and the ability of controllers to alter system dynamics were the principal highlights. The special challenges of transitioning this body of techniques to wave energy devices were next

Box 2.10 Optimal Control and Model Predictive Control, II

The basic problem formulation for MPC is:

$$\min_{u(0),\ldots,u(h_c)} \sum_{k=0}^{h_p} e(k)^T Q e(k) + \sum_{k=0}^{h_c} \Delta u(k)^T R \Delta u(k) \tag{2.21}$$

where $\Delta u(k) = u(k) - u(k-1)$, $e(k) = r(k) - y(k)$, and h_p and h_c are the prediction and control horizons, respectively. Using a model of the system, the future output is predicted up to h_p steps ahead and the optimal control sequence, $u(0), \ldots, u(h_c)$, calculated so that the system output follows a reference trajectory. The next control input, $u(0)$, is then applied to the system and the optimal future control sequence recalculated. This "receding horizon" procedure allows the system to effectively counteract output disturbances (d in Fig. 2.8) and gives the control paradigm good robustness properties.

MPC is flexible and can deal with a number of practical system characteristics. For example, if the system contains pure delay (of, say, n_d steps), then the performance function in (2.21) is recast with n_d as the lower limit on the first summation term. Inequality constraints (e.g., end-stops for oscillating primary energy converters) can also be included, by augmenting Equation (2.21) with a set of constraints:

$$y_{min} \le y \le y_{max}$$
$$u_{min} \le u \le u_{max}$$
$$|\Delta u| \le u_{max}$$
$$\text{etc.}$$

MPC can use both linear and nonlinear models. Finally, one very useful exploitation of the general MPC framework is the possible substitution of the quadratic regulation penalty (e.g., $e^T e$) in (2.21) with an energy-related term, such as the power absorbed by the power take-off, coupled with a change from minimization to maximization of the performance function. Such a possibility, of course, has particular relevance to wave energy conversion.

summarized. Two complementary approaches to controlling wave energy devices were next discussed, and it was pointed out that the reader will notice both approaches in this book. A brief introduction to trajectory tracking, robust control, optimal control, and model predictive control was also provided in the form of text boxes.

3 Bodies Oscillating in Air

Excellent discussions and examples pertaining to oscillating systems in air can be found in texts such as [58, 59]. As we have seen in Chapter 1 many wave energy conversion concepts essentially consist of moored bodies floating on the ocean surface and oscillating in response to waves. Many devices are designed (or otherwise constrained) such that motion in a particular direction dominates. We first consider such single degree of freedom devices here, as they enable us to focus just on the aspects fundamental to their dynamics and control. In calm water (absent waves, currents, or winds), such a body is in stable static equilibrium with its weight balanced by buoyancy. Any initial displacement off the equilibrium will disturb this balance, with the excess or deficit in buoyancy opposing the displacement and trying to restore equilibrium as a spring force would. Other forces would also try to oppose the body's motion; for instance, the force due to viscosity would act to damp the motion by dissipating some of its energy. Dynamic equilibrium would then require that the inertial force of the body, determined by its mass and acceleration, should balance the sum of the restoring (spring) and damping forces on the body.

3.1 Power Absorption from an Oscillatory Force

In the following, we first seek to understand the need and place for control in a hypothetical situation where the mass, spring, damping combination above are in vacuum or, for our purposes, in air. We thus ignore any forces that the air itself may apply on the mass. In the next chapter we shall consider the real situation where the body performs its motion on the water surface and begin the process of understanding of how this complicates its dynamics and any attempt to control it.

Because we have a mass and a spring, we have the basic elements necessary to set off oscillations. We suppose next that an arbitrary oscillatory force is now applied to the mass. To help develop a firm understanding, we proceed slowly and incrementally at first, and start by reviewing the dynamics underlying the transfer of power from this oscillatory force to our oscillator. Note that "unforced" oscillation requires both (and only) a spring and a mass. However, it is through damping that some or all of the energy in the oscillation could be dissipated or put to use. An "unforced," "free," or "natural" oscillation implies an oscillatory response to a momentary initial deflection or velocity in the absence of an external oscillatory force. As the easiest type of oscillation to think

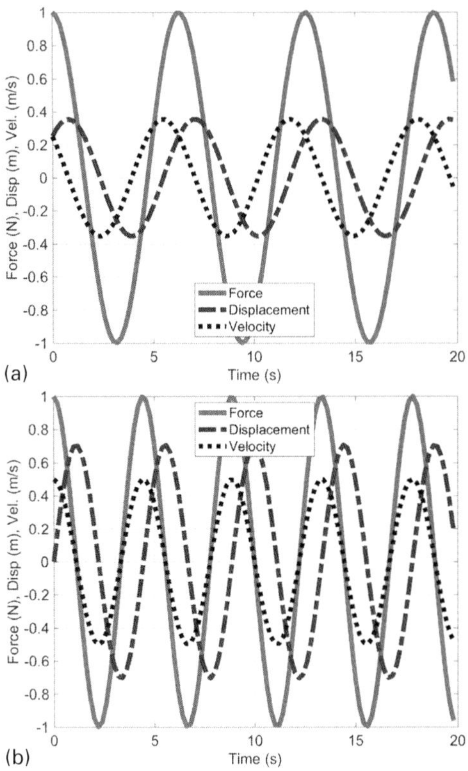

Figure 3.3 Behavior of an oscillator off (a) and at resonance (b), in regular forcing.

In other words the deflection must be 90 degrees out of phase with the force, or equivalently, the velocity must be in phase with the force. For a system with a nonzero mass and spring, this is possible only when the spring and the mass resonate at the forcing frequency and interact such that the force feels only the damper opposing it. This is when, as indicated by Equation (3.15),

$$k - \omega^2 m = 0 \text{ or } \omega = \omega_n. \tag{3.20}$$

Resonance, we thus see, is essential for maximum power absorption by the damper in the presence of a spring and a mass. We note in passing that with Equations (3.19) and (3.20) satisfied, the deflection $x(t)$ is identical to that for a pure damper, as stated in Equation (3.6). Fig. 3.3 illustrates the response of the damped system at and off resonance.

3.2 Control for Maximum Power Absorption

Let us proceed further with the single-degree-of-freedom oscillatory system in Fig. 3.1, with an oscillatory external force $f(t)$ applied in the oscillation direction. Suppose that we want to use our oscillating system to absorb power from this external force $f(t)$. Let k

denote the inherent stiffness in the system, and c the inherent damping constant. c could represent any viscous friction such as occurring in bearings. The natural frequency of the oscillatory system is $\omega_n = \sqrt{(k/m)}$. For a harmonic (i.e., sinusoidal) force at a frequency ω, power absorption is maximum when the forcing frequency $\omega = \omega_n$, as discussed in Section 3.1. Here we consider whether the motion of the body (mass m) can be controlled so as to mimic a resonance-like behavior at any given ω. Suppose we connect a linear motor or a dynamometer to the body that can apply a force $f_c(t)$ opposing the sense of $x(t)$ (the same way as the spring and damper do) directly to the mass; the equation of motion at any time instant t then can easily be found:

$$m\ddot{x} = -kx - c\dot{x} - f_c(t) + f(t), \tag{3.21}$$

where x, $\dot{x}$, and $\ddot{x}$ are all functions of t. If we sense $x(t)$ by means of a sensor such as an encoder embedded within the motor, then we can infer the instantaneous velocity $\dot{x}(t)$ simply by differentiating electronically the signal $x(t)$ once and the instantaneous acceleration $\ddot{x}(t)$ by differentiating the velocity signal once. These operations may be performed with analog or digital electronics, and may be affected by noise or digitization errors, but we assume here that they are precise enough for our purpose. Further, we suppose for the moment that the response of the electronics and the linear motor/actuator is very fast compared with the time period of the force $f(t)$, so that it does not significantly influence the dynamics of the system as captured by Equation (3.21). Let us suppose that we control the current flowing into the actuator such that the force applied by it is of the form

$$f_c(t) = Kx + C\dot{x} + M\ddot{x}. \tag{3.22}$$

In Equation (3.22), the term $C\dot{x}$ with $C > 0$ implies that a part of the force f_c opposes the body velocity and thus is dissipative or resistive (as seen in Section 3.1) with an "applied" (as opposed to what is inherent in the system, such as c), electronically adjustable damping constant C. The term Kx opposes displacement when $K > 0$ with K standing for an electronically set, applied, or "load" spring constant. Similarly, the term $M\ddot{x}$ opposes the acceleration by means of an electronically chosen applied or load inertia. The actuator used to apply force f_c is thus here being used to set the load damping, load spring, and load inertia components on the oscillating body m. In general, this actuator will be absorbing energy from the oscillating system, so that $C > 0$, corresponding to positive applied (or load) damping. It is recalled that physical spring or inertia do not absorb (supply) net power from (to) the external force, so that K or M can be either positive or negative, or zero. Inserting Equation (3.22) into Equation (3.21), we have

$$(m + M)\ddot{x} + (c + C)\dot{x} + (k + K)x = f(t). \tag{3.23}$$

Let us consider an external force $f(t)$ that varies as a sinusoidal function at an angular frequency ω. To be sufficiently general, we should use an expression of the form $f(t) = F_c \cos \omega t + F_s \sin \omega t$. It is easier for the subsequent developments to use an equivalent expression $f(t) = \Re(Fe^{i\omega t})$ instead, where $\Re$ denotes the real part of the complex quantity in the parentheses. F then is a complex quantity in general, written as $F(i\omega)$, whose real and imaginary parts together contain both the amplitude and phase information for the force. Since Equation (3.23) represents a linear system, the deflection $x(t)$ must have the same frequency ω and be of the form $x(t) = \Re[X(i\omega)e^{i\omega t}]$, where $X(i\omega)$ denotes the complex deflection amplitude containing both amplitude and phase information. For further simplification, we work directly with the complex quantities $F(i\omega), X(i\omega)$, and so on, until at the final step it is time to evaluate the temporal variation of the deflection $x(t) = \Re[X(i\omega)e^{i\omega t}]$. With these substitutions, Equation (3.23) now becomes an algebraic equation,

$$\left[(-\omega^2 m - \omega^2 M + K + k) + i\omega(C + c)\right] X(i\omega) = F(i\omega). \tag{3.24}$$

The complex amplitude $X(i\omega)$ can easily be evaluated using Equation (3.24) where the actual deflection amplitude $|X(i\omega)|$ and phase θ relative to the force are given by

$$|X(i\omega)| = \left| \frac{F(i\omega)}{(-\omega^2 m - \omega^2 M + K + k) + i\omega(C + c)} \right|,$$
$$\tan\theta = \frac{\omega(C + c)}{-\omega^2 m - \omega^2 M + (K + k)}. \tag{3.25}$$

Suppose for the moment that the control force f_c does not include a spring-like or inertia-like (i.e., "reactive") component, or that $K = 0$ and $M = 0$. In this case, an inspection of Equation (3.25) shows that the body displacement and velocity are maximized, and moreover, the phase difference between the force $f(t)$ and the deflection $x(t)$ is $\pi/2$ radians, when $\omega = \omega_n = \sqrt{k/m}$, that is, when the body resonates. At this frequency, the average power absorbed by the actuator over each cycle is

$$P_{abs} = \frac{1}{2}\omega^2 C\, |X(i\omega)|^2 \tag{3.26}$$

since at resonance, we have

$$|X(i\omega)| = \frac{|F(i\omega)|}{\omega(C + c)}. \tag{3.27}$$

Clearly, as Equation (3.26) implies, both C and $|X(i\omega)|$ affect the absorbed power. The least power is absorbed when $C = 0$. A large $|X(i\omega)|$ resulting from a small but nonzero C will not necessarily maximize P_{abs}, though C could be too large to allow a large enough $|X(i\omega)|$. Inserting Equation (3.27) into Equation (3.26),

$$P_{abs} = \frac{C\,|F(i\omega)|^2}{2(C + c)^2}. \tag{3.28}$$

In order to set a value for the load damping where most power can be absorbed, we differentiate both sides of Equation (3.28) with respect to C and use the condition that for P_{abs} to be maximum, $dP_{abs}/dC = 0$. Performing this differentiation, we find

$$\frac{dP_{abs}}{dC} = \frac{|F(i\omega)|^2}{(C + c)^2}(C^2 - c^2) = 0. \tag{3.29}$$

This implies that $P_{abs} = P_{max}$; that is, the actuator absorbs maximum power from a system in resonance when $C = c$, or when the load damping equals the damping inherent in the system. For $C = c$, it is easy to verify that the absorbed power P_{abs} equals the power dissipated in the inherent damping c. Hence at most only half of the power brought in by the force can be utilized by the applied damping C. If C were a linear generator, this would imply that at a maximum, 50% of the input power can be converted to electric power, while an equal amount is dissipated internally within c.

Consider now a more general situation where the forcing frequency ω does not equal the natural frequency ω_n. In this situation, $P_{abs} < P_{max}$. Using the linear motor or actuator providing the force f_c, we now apply a nonzero spring load K in addition to a damping load C so that

$$f_c(t) = C\dot{x} + Kx. \tag{3.30}$$

The denominator in Equation (3.25) now becomes

$$D(i\omega) = \left[-\omega^2 m + k + K + i\omega(C + c)\right], \tag{3.31}$$

with K in Equation 3.31 chosen such that

$$K = \omega^2 m - k. \tag{3.32}$$

The first term in $D(i\omega)$ now vanishes. In other words, use of K as given by Equation (3.32) for any particular forcing frequency ω enables a resonance-like behavior. Elimination of reactive terms in the denominator now sets the stage for the choice $C = c$, which leads to $P_{abs} = P_{max}$. Note that for $\omega < \omega_n$, $K < 0$. In other words, a negative spring load is required to force resonance below the natural frequency. On the other hand, for $\omega > \omega_n$, $K > 0$. Thus, at frequencies where the oscillation is spring dominated, the reactive load should work to reduce the effective spring, while the opposite will be true for inertia-dominated oscillation. At resonance, of course, there is no need for a reactive component.

Suppose we choose an inertial rather than spring load to augment the damping load, as

$$f_c(t) = C\dot{x} + M\ddot{x}. \tag{3.33}$$

To drive the reactive term in $D(i\omega)$ to zero we now need

$$M = \frac{k}{\omega^2} - m. \tag{3.34}$$

Again, with an inertia load augmenting the optimal damping load $C = c$, the absorbed power $P_{abs} = P_{max}$. It is easy to see that for forcing frequency $\omega < \omega_n$, $M > 0$, while for $\omega > \omega_n$, $M < 0$. In other words, for frequencies at which the oscillation is spring dominated, the load should increase the effective inertia of the body, while for inertia-dominated oscillation, a negative inertia load is required to force a resonance at that frequency.

We may also choose to apply both spring and inertia loads to supplement the optimal damping load $C = c$. In this case, we choose

$$\begin{aligned} M &= -m, \\ K &= -k. \end{aligned} \tag{3.35}$$

This choice enables $P_{abs} = P_{max}$ at any forcing frequency ω and requires no retuning from one ω to another. The use of reactive load components along with the resistive component C to enable resonance and optimal damping at a chosen frequency rather than just the system natural frequency is analogous to the condition of "tuned impedance matching" for electric and electronic circuits, where the most efficient power transfer between circuits or between circuit and load occurs.

3.3 Irregular Forcing

As we know, however, ocean waves are irregular and are frequently held up as one of nature's random phenomena. They are often best understood statistically via probability distributions, which in turn lead to their description in terms of "power spectra" (as we see later in Chapter 5). The forcing due to ocean waves is therefore not sinusoidal or harmonic. Indeed, this is the very reason for developing a control system that will attempt to tune the device to perform efficiently in real seas. It is as a preparation for further work on real ocean waves that, in this section, we will consider non harmonic forcing on our system in air.[1] We ask how we may build on our understanding of response to harmonic forcing and explore techniques that will enable a straightforward extension to non harmonic forcing.

We are helped in our goal by an important tool in engineering analysis, namely, that of Fourier decomposition and synthesis. It may be recalled that any periodic but nonsinusoidal function can be approximated by series of sine and cosine functions of different amplitudes and frequencies.[2] In other words, we can determine a set of sinusoids, of certain amplitudes, frequencies, and phases, which when added together lead to a function that very nearly resembles the original function. Our function does need to be piecewise continuous, or integrable over its period, for this approximation to make sense, but most functions that we will deal with are piecewise continuous. If we are faced with a nonsinusoidal but periodic odd function, we may approximate it by an infinite series of sines of different frequencies and amplitudes. If it is a nonsinusoidal periodic even function, we may approximate it with cosines instead. More generally, if this function is neither odd nor even but still periodic, both the cosine and sine series are required. This in turn means we can even use an infinite series of complex exponential functions (see Section 3.2) to approximate a general nonsinusoidal but periodic function.

Effectively, we can thus interpret any such function in nature as a collection of many sinusoids of different amplitudes and phases. Already it looks as if some of the control ideas in Section 3.2 will apply immediately to such an excitation. We know, however, that many types of forcing (including ocean waves) are not periodic in addition to being not sinusoidal. Fortunately, however, the Fourier series approach generalizes to accom-

[1] Note that we have chosen to work with a linear system which admits linear superposition; in other words, the system response to a sum of forces equals the sum of system responses to the individual force components.

[2] This is an approximation in part in the sense that the infinite series are usually truncated to a finite number of terms in practical applications; but see, e.g., [60].

modate nonperiodic functions, and the Fourier series become Fourier transforms, based on the argument that a nonperiodic function is simply a function that takes infinitely long to repeat itself, that is, has infinite period. Application of Fourier transforms also requires not only the use of complex exponential functions, but also the notion that even though time has gone on since $t \rightarrow -\infty$ through the present instant (but more generally, through $t \rightarrow \infty$), our functions $f(t)$ themselves are such that as $t \rightarrow \pm\infty$, $f(t) \rightarrow 0$ [51]. In addition, we also allow for mathematical negative frequencies. Thus, a nonperiodic integrable function of time $f(t)$ can be represented in terms of its frequency domain equivalent $F(i\omega)$, by means of (e.g., [60, 61])

$$F(i\omega) = \int_{-\infty}^{\infty} f(t)\mathrm{e}^{-i\omega t}\,dt. \tag{3.36}$$

Equation (3.36) represents the "forward Fourier transformation" process, and $F(i\omega)$ is the Fourier transform of $f(t)$. The reverse process, where we recover $f(t)$ from $F(i\omega)$, is defined as

$$f(t) = \frac{1}{2\pi}\int_{-\infty}^{\infty} F(i\omega)\mathrm{e}^{i\omega t}\,d\omega. \tag{3.37}$$

Intuitively, $F(i\omega)$ represents the contribution in terms of complex amplitude (i.e., magnitude and phase) of a particular frequency ω to the nonperiodic, nonsinusoidal function $f(t)$. There is an infinity of such $F(i\omega)$ needed, but the fact remains that each represents a unique frequency ω at a unique complex amplitude. This brings us back to the complex exponential-based treatment of harmonic inputs of Section 3.3, which seems applicable in this context. In Equation (3.21) if we temporarily let $f_c(t) = 0$, for $f(t) = F(i\omega)\exp(i\omega t)$, and $x(t) = X(i\omega)\exp(i\omega t)$,

$$\left(-\omega^2 m + k + i\omega c\right) X(i\omega) = F(i\omega). \tag{3.38}$$

We can interpret Equation (3.38) as a condition that describes the body's response to the complex force component $F(i\omega)$ at frequency ω, with ω representing one of the frequencies making up the nonperiodic excitation $f(t)$. $F(i\omega)$ at any ω would be given by a forward Fourier transformation. Note that despite the allusion to ocean waves above, we are still, for the remainder of this chapter, considering a spring-mass-damper system oscillating in air. Drawing further on the similarity between mechanical spring-mass-damping systems and electrical capacitance-inductance-resistance circuits,[3] we use here a complex quantity representing "mechanical impedance," which, for harmonic forcing, directly determines velocity through an algebraic relation to the external force. We define explicitly a body velocity $u(t) = \dot{x}(t)$. Realizing that $u(t)$ for harmonic excitation at frequency ω is of course also harmonic at frequency ω, we let $u(t) = U(i\omega)\exp(i\omega t)$, where a phase angle is implicit in each of $F(i\omega)$, $X(i\omega)$, and $U(i\omega)$. We rewrite Equation (3.38) as

$$Z(i\omega)U(i\omega) = F(i\omega) \tag{3.39}$$

[3] Recall that both are modeled with second-order differential equations, and hence are referred to as "second-order systems."

where

$$Z(i\omega) = c + i\left(\omega m - \frac{k}{\omega}\right). \tag{3.40}$$

Equations (3.38) and (3.40) are equivalent to each other. We define $Z(i\omega)$ as the mechanical impedance of our system. The damping constant c represents the dissipative or resistive real part $\Re[Z(i\omega)]$, while the inertia and spring terms together represent the reactive imaginary part $\Im[Z(i\omega)]$. At resonance, $\omega = \omega_n$, and $\Im[Z(i\omega)] = 0$.

For irregular external excitation, the load force will also be irregular. The load force $f_c(t)$ from Section 3.2, with M, C, and K included, transforms as:

$$F_c(i\omega) = \left[C + i\left(\omega M - \frac{K}{\omega}\right)\right] U(i\omega) \equiv \Lambda(i\omega)U(i\omega). \tag{3.41}$$

$F_c(i\omega)$ denotes the complex amplitude of the component at ω. We have defined above a complex load impedance $\Lambda(i\omega)$, whose real part equals the dissipative load C, and the imaginary part contains the spring and inertia loads K and M. Equation (3.39) with a nonzero control force becomes

$$[Z(i\omega) + \Lambda(i\omega)]\, U(i\omega) = F(i\omega), \tag{3.42}$$

Our effort in Equations (3.32)–(3.35) was to maximize power absorption at a chosen ω by setting the load such that the resistive part of the load (or load damping) equaled the inherent resistance in the system c, while the reactive part canceled the reactive effects arising from the system spring and inertia, effectively mimicking a system in resonance at any chosen forcing frequency ω. How may we express these steps by means of the complex load $\Lambda(i\omega)$? To ensure optimal load damping, the real part of load must equal the real part c of the system mechanical impedance. Further, we can see that the imaginary or reactive part of the load must cancel the reactive part of the system mechanical impedance. Thus, we need

$$\Re[\Lambda(i\omega)] = C = c = \Re[Z(i\omega)],$$
$$\Im[\Lambda(i\omega)] = -i\left(\omega m - \frac{k}{\omega}\right) = -\Im[Z(i\omega)]. \tag{3.43}$$

In other words, we need to apply a load such that the "load impedance" is the complex conjugate of the system's inherent impedance, that is,

$$\Lambda(i\omega) = Z^*(i\omega) \tag{3.44}$$

where * denotes complex conjugate. Any type of control (such as applied here) that achieves the condition (3.44) can be referred to as "complex-conjugate control." We note that since both $\Lambda(i\omega)$ and $Z(i\omega)$ are complex quantities, both amplitudes and phases need to be matched in Equation 3.44. Equation (3.44) applies to general non harmonic or irregular excitation that may be nonperiodic and comprised of an infinite number of frequencies at different amplitudes and phases, as represented by $F(i\omega)$.

If we seek to force a resonance at a chosen, dominant frequency ω in a nonperiodic, nonsinusoidal excitation, and further want to maximize the absorbed power $P_{abs} = P_{max}$ at ω, we simply choose a load spring (and/or inertia) and damping such that, per Equation (3.44), $\Lambda(i\omega) = Z^*(i\omega)$. As mentioned, this cancels, for the chosen ω, the reactive part, and matches the load resistance to the inherent resistance to enable maximum power transfer to the load resistance from that frequency component. In practice, this resistive load could be a simple resistance connected to the actuator $f_c(t)$ dissipating the power as heat, or a light bulb using the power to provide light and heat. We can also seek good performance when the frequency content in the excitation is changing over time (albeit much more slowly compared with the oscillation time-scale of our system). We can adjust the load impedance $\Lambda(i\omega)$, primarily just the reactive part here, so that it now equals the complex conjugate of the inherent impedance of the system at the new dominant frequency. This adjustment could be non-real time, in that it could be made every few minutes after the excitation is freshly measured and analyzed for frequency content. Or it could be done in real time as follows.

The first step is to express Equation (3.42) in the time domain so that we may watch events in real time. This requires an inverse Fourier transformation of both sides of Equation (3.42). We recall that the inverse Fourier transform of a product of two functions in the frequency domain produces a convolution of the two functions in the time domain (Faltung's theorem; e.g., [56]). Splitting the two terms on the left side of Equation (3.42) and writing their inverse Fourier transforms, we have, separately,

$$\frac{1}{2\pi}\int_{-\infty}^{\infty} Z(i\omega)U(i\omega)e^{i\omega t}d\omega = \int_{-\infty}^{\infty} z(\tau)u(t-\tau)d\tau,$$
$$\frac{1}{2\pi}\int_{-\infty}^{\infty} \Lambda(i\omega)U(i\omega)e^{i\omega t}d\omega = \int_{-\infty}^{\infty} h(\tau)u(t-\tau)d\tau. \tag{3.45}$$

Here t refers to the current instant, and τ represents a dummy time variable of integration. $u(t) = \dot{x}(t)$ denotes the instantaneous velocity. $z(t)$ and $h(t)$ are the "kernels" of the two convolutions and are, respectively, the impulse response functions modeling the dynamic effects of the inherent inertia, spring, and damping for the system and of the load applied on it (note here that we are working with a linear system). To arrive at $z(\tau)$ we begin by recalling the equation of motion, written with just the inherent system components, without the load $f_c(t)$, and without the external force $f(t)$:

$$m\ddot{x}(t) + c\dot{x}(t) + kx(t) = 0. \tag{3.46}$$

m, c, and k are all independent of frequency. They thus represent, respectively, real-valued functions of acceleration, velocity, and displacement whose value is the same at any frequency. Since m is constant with respect to acceleration at any frequency, and since uniform distributions in frequency domain "inverse Fourier transform" as impulses (modeled by the Dirac delta function) in the time domain, the first term in Equation (3.46) can be expressed as

$$m\ddot{x}(t) = \frac{1}{2\pi}\int_{-\infty}^{\infty} ma(i\omega)e^{i\omega t}d\omega = \int_{-\infty}^{\infty} m\delta(\tau)\ddot{x}(t-\tau)d\tau \tag{3.47}$$

where $a(i\omega)$ is the Fourier transform of acceleration. The sifting property of the Dirac delta function $\delta(\tau)$ means that the second integral in Equation (3.47) equals the value of $m\ddot{x}(t-\tau)$ at $\tau = 0$, which is just $m\ddot{x}(t)$. The kernel operating on acceleration in the convolution in Equation (3.47) is simply $m\delta(\tau)$. In a similar manner, we can argue that the damping-related term in the inherent response is

$$c\dot{x}(t) = \frac{1}{2\pi}\int_{-\infty}^{\infty} cU(i\omega)e^{i\omega t}d\omega = \int_{-\infty}^{\infty} c\delta(\tau)\dot{x}(t-\tau)d\tau. \tag{3.48}$$

Here $U(i\omega)$ equals the Fourier transform of velocity $u(t) = \dot{x}(t)$, and the kernel operating on the velocity $\dot{x}$ in the convolution to the right is $c\delta(\tau)$. It follows, via a similar argument, that the kernel operating on displacement $x(t)$ in the convolution leading to the third term in Equation (3.46) is $k\delta(\tau)$.

The kernel $z(\tau)$ in Equations (3.45), however, operates on the velocity $u = \dot{x}$. Therefore, without changing the value of the individual convolutions implied in Equation (3.46), if we express the respective kernels such that they all operate on velocity and add the results, we should arrive at $z(\tau)$. To that end, we find that (see also Section 3.3)

$$\int_{-\infty}^{\infty} m\frac{d\delta}{dt}(\tau)u(t-\tau)d\tau = m\dot{u}(t) = m\ddot{x}(t). \tag{3.49}$$

The kernel $c\delta(\tau)$ already operates on velocity. The kernel $k\delta(\tau)$ operates on deflection, which is an integral of velocity. In frequency domain,

$$kX(i\omega) = \left[\pi\delta(\omega) + \frac{1}{i\omega}\right]kU(i\omega). \tag{3.50}$$

The inverse Fourier transform of $[\pi\delta(\omega) + 1/i\omega]\,k$ is $kH(t)$, where $H(t)$ is the unit step function, or the Heaviside step function. Thus we can write

$$\frac{1}{2\pi}\int_{-\infty}^{\infty}\left[\pi\delta(\omega) + \frac{1}{i\omega}\right]kU(i\omega)e^{i\omega t}d\omega = \int_{-\infty}^{\infty} kH(\tau)u(t-\tau)d\tau = k\int_{0}^{\infty}\dot{x}(t-\tau)d\tau. \tag{3.51}$$

The last integral in Equation (3.51) is simply the integral of velocity from $t \to -\infty$ to the present instant, which is just the current displacement $x(t)$. The desired kernel $z(\tau)$ is thus

$$z(\tau) = m\frac{d\delta}{dt}(\tau) + c\delta(\tau) + kH(\tau). \tag{3.52}$$

With the effect of the load added, we can now rewrite the complete equation of motion with the left side as a convolution integral,

$$\int_{-\infty}^{\infty}[z(\tau) + h(\tau)]\,v(t-\tau)d\tau. \tag{3.53}$$

To apply an equivalent of complex-conjugate control in real time on this body oscillating in air, we choose $h(\tau)$ to match the conclusions of Equations (3.35). Thus, if we use

$$h(\tau) = -m\dot{\delta}(\tau) + c\delta(\tau) - kH(\tau), \tag{3.54}$$

we would find that this will lead to an instantaneous velocity $v(t)$ as determined by

$$2\int_{-\infty}^{\infty} c\delta(\tau)v(t-\tau)d\tau = f(t). \tag{3.55}$$

In other words, $v(t) = f(t)/(2c)$ at each instant, for forcing that may be irregular in nature. This control is possible in real time, because only measurements of displacement, velocity, and acceleration (the latter two could derived from the $x(t)$ measurements) at the present instant are needed for computing the instantaneous control force. Such control is possible only because the body oscillates in air. For bodies oscillating in a general, six-degree-of-freedom oscillation, the treatment above can be generalized with the aid of matrix algebra. Matrix methods are also helpful in control of systems of connected bodies oscillating in air, where control of one body may depend on a linear combination of motions of other bodies. However, the fundamental arguments above remain largely unaltered.

Finally, as we will see shortly, in these several operations from Equation (3.45) to (3.55) is embedded the potential difficulty we face when attempting to control bodies in water that are oscillating close to the water surface.

3.4 Conclusion

Starting with a very basic consideration of power transfer from an oscillatory force to a body whose motion was opposed by a restoring force produced by a spring and a resistive or dissipative force produced by a damper, in this chapter, we discussed how the correct phase relationship between the force and the body oscillation can dictate the average power transferred to the body. It was seen further that not only resonance but also a damping match between internal and external damping (as applied by a power-absorbing load) was essential for maximum power transfer averaged over a cycle. At forcing frequencies not equal to the natural frequency of body oscillation, we saw that an external control force based on motion measurements (displacement, velocity, acceleration) could be applied to enable such an impedance match for maximum power capture. We found that, since the system parameters themselves had no frequency dependence, implementing such control in the time domain and generalizing it to a range of frequencies (at different amplitudes and phases) would not be a serious challenge. Finally, we stress that all discussion in this chapter applied to bodies oscillating in air, with the consequence that the equation of motion was a differential equation with constant coefficients, that is, non-frequency-dependent coefficients.

4 Bodies Oscillating in Water

We noted in Chapter 3 that for bodies oscillating in air, control of motion in response to an irregular excitation requires only current motion information. We have assumed a control hardware with an infinitesimally small response time, as we will in the following discussion. In this chapter we see that, when a body oscillates in response to waves on the water surface, motion information from the future is required in the control of the motion at the current instant – even if all control hardware were to respond instantaneously. We find that control to produce optimal absorption from a wave to which the body is currently responding requires knowledge of the body response to the waves it will encounter up to several seconds into the future. While this poses no serious challenge in the regular wave idealization, for real irregular seas modeled as random phenomena, such knowledge is difficult to obtain. In the sections that follow, we attempt to understand the fundamental physics underlying this need for future information. We begin to learn here why impedance matching control such as considered in Chapter 3 is not possible for a floating body in irregular waves, unless one could find some way to predict or estimate either the body's velocity or the incident wave profile. It is useful to note that our understanding here is based on linear theory, and if the waves are too large in height or body oscillations are too large, we would need to amend our thinking. However, it is also useful to realize that in such situations, our attention would be more on surviving the situation and less on energy conversion.

4.1 Oscillation Near Free Surface

A more direct analogue to a body oscillating in air is an oscillating body that is submerged deep in water, such that the water surrounding it in all directions can, for all practical purposes, be thought to be unbounded or infinite in extent. A body oscillating in a fluid must lose some energy due to viscous friction in the boundary layer surrounding it, and therefore encounters an opposing damping force. In addition, as the body accelerates and decelerates, so must a quantity of fluid surrounding it. In air, both effects are small enough to ignore in many applications (including our system in Chapter 3) excepting some specialized systems such as small-scale sensitive microphones and speakers. For a body oscillating deep in water these effects can be appreciable. For a body oscillating in a single mode in an unbounded fluid, the equation of motion can be expressed as

$$(m + m_a)\ddot{x} + (C + c)\dot{x} + kx = f(t). \quad (4.1)$$

Here m_a denotes the equivalent mass of the surrounding fluid accelerated by the oscillating body, while c represents the damping effect of the viscous losses in the boundary layer between the oscillating body and the surrounding fluid. For air, m_a and c are much smaller than body mass m and applied damping C, respectively, and may be ignored in many cases, while in water, these quantities may be the same order of magnitude as m and C, and hence cannot be ignored. m_a is the added mass in unbounded fluid and is independent of the frequency of oscillation. c is the damping constant due to viscous damping assuming linearity. Considerable work on these parameters has been reported over several decades. Equation (4.1) and its generalization to multiple modes of motion are often adequate when describing the motion of deeply submerged underwater vehicles. As long as the appropriate expressions (e.g., [16, 62]) are used to determine m_a and c for bodies oscillating in an unbounded fluid (i.e., deeply submerged in deep water), the control of such bodies is fundamentally similar to their control in air and will follow the same thought process as used in Chapter 3 (note that we are ignoring density stratifications and internal waves here).

For a body oscillating in water, therefore, it is not the oscillation in a denser fluid that causes difficulty in control, as long as the surrounding fluid can be thought to be uniform and unbounded. If the body is deeply submerged and not too close to the sea floor, constant values for m_a and c may be used. Proximity to the sea floor could alter the values of the coefficients somewhat from their values in unbounded fluid. Control of underwater vehicles may present unique problems in its own right, but our attention here is on the unique problems of controlling wave energy devices. For in this case, it is proximity to the water surface ("air–water interface" or the "free surface") that leads to complications. Our difficulties arise directly from the fact that any body oscillating in water but close enough to the free surface will impart an oscillatory perturbation on the free surface, some of which will propagate away as a wave. The effect of the body's oscillation on the free surface decreases exponentially with the depth to which the body is submerged. An interesting reciprocity exists between body oscillation at a depth and free-surface perturbation. Thus, the effect of a free surface perturbation on the body also decreases exponentially with the depth of submergence of the body. As waves are after all a free-surface perturbation, a body that is designed to oscillate in response to surface waves must not be so deeply submerged that it feels negligible wave action at its depth. It follows then that as long as a body is close enough to the free surface to respond to waves, it must also be close enough to produce waves due to its oscillation. It is these two mutually reciprocal effects associated with body oscillation close to the free surface that lead to interesting dynamics, but complicate control.

Briefly, then, a wave energy device utilizing mechanical oscillation of a body must force the water around it and make waves of its own as it responds to incoming waves, which it must, in order to absorb energy from the incoming waves. The forcing of the water by the body is oscillatory and manifests an exchange of energy that contains both a reactive and a nonreactive or resistive part. The reactive part is associated with the oscillatory acceleration of the water at the free surface, while the nonreactive part is associated with propagation of energy along the free surface away from the body. The former represents a net zero exchange of energy over an oscillation period, while the latter represents an outward propagation of energy carried away from the body by waves radiating from it as it oscillates. The former is interpreted as an added mass or inertia for the body oscillating near the free surface, while the latter as "radiation damping."[1] As we will see here and in Chapter 6, there are three main facts surrounding these effects for bodies oscillating near the free surface:

1. Both added mass and radiation damping are frequency dependent, with the added mass being expressible as a sum of a purely frequency-dependent component and a frequency-independent component.
2. The purely frequency-dependent added mass and radiation damping are related to each other.
3. The radiation damping is related to the force an incoming wave would apply on the body if it were somehow held fixed in the wave.

The goal of the following sections is to gain a deeper understanding of the first two effects above, with particular attention to the physical relationships that quantify them.

4.2 Regular Waves

We suppose at first that the forcing is harmonic, in other words, that the wave incident on the body is of a single frequency and amplitude, and its form is that of a traveling sinusoid. As harmonic excitation of the type seen in Chapter 3 is the basic building block of irregular forcing, so are harmonic waves the building blocks of the more realistic irregular waves (as alluded to briefly in Chapter 3). The body may be either floating on the free surface or submerged such that the wave can still force it into oscillation. We suppose for convenience that the body is spherical and moored tightly enough, so that its primary motion is in the up-down heave mode. We begin with an equation of motion,

$$m\ddot{x} + c\dot{x} + kx = f(t). \tag{4.2}$$

m denotes the mass of the body, and c any viscous damping, which we shall ignore in what follows as small in comparison with the radiation damping mentioned in Section 4.1. k is the hydrostatic stiffness due to buoyancy and is thought to be a constant here for small enough deflections of the body. $f(t)$ is the total force experienced by the body in the presence of the wave while it is in motion. All other forces are ignored, and the body is in static equilibrium (i.e., weight = buoyancy) in the absence of waves. For a harmonic wave, the force can be expressed as $f(t) = F(i\omega)e^{i\omega t}$ with $F(i\omega)$ denoting the complex force amplitude as before. The solution $x(t)$ can be found by looking for a form

[1] This is only an effect of radiated wave energy and understood completely in terms of ideal fluid dynamics, often considerably exceeding boundary-layer effects such as viscous damping, which can sometimes be ignored in comparison.

$x(t) = X(i\omega)e^{i\omega t}$, where the complex deflection amplitude $X(i\omega)$ can be determined algebraically using

$$\left[-\omega^2 m + k\right] X(i\omega) = F(i\omega). \tag{4.3}$$

For waves and deflections small enough to be treated using linear differential equations, linear superposition holds. Then it can be shown [6, 8, 20] that the force term on the right side of Equation (4.3) is a linear superposition of two effects:

1. the force acting on the body with the incident wave striking it and the body held fixed, and
2. the force acting on the body when there is no incident wave, and the body oscillates at a velocity amplitude $U(i\omega)$.

As discussed further in Chapter 6, when an incident wave field approaches the body, just by virtue of its presence, the body alters the wave field. In addition, wave-induced motion of the body creates its own waves, further altering the wave field. Linear superposition allows the two effects to be treated separately and added together. The former is the so-called diffracted wave field (which in turn is a superposition of the incident wave field and a "disturbance" wave field produced by the body if it were stationary), and the latter is the radiated wave field. Both fields contain near-field-only "evanescent" components and propagating components that reach out to the far field. The propagating components of just the body-disturbed wave field and radiated wave field can travel long distances. In an open ocean in deep water for an isolated body, the propagating components are more amenable to approximate but insightful interpretation (as discussed in Chapter 6).

The diffracted and radiated wave fields make for an interesting fundamental study in potential theory and the hydrodynamics of floating body–wave interaction in their own right (Chapter 6). From an applications standpoint, very useful relationships exist between the diffracted and radiated fields, which we utilize extensively in both analysis and control. As mentioned above, the force that the body feels due to these two wave fields can be found by linear superposition, and the combined force represents the total force a body feels in waves. We denote the force component due to the diffracted wave field as $F_f(i\omega)$ and the force due to the radiated wave field as $\overline{F}_R(i\omega)$. $F_f(i\omega)$ is referred to as the diffraction wave force or excitation force (often "exciting" force), and typically will be a six-component vector, with the latter three being moments. Here we suppose it is just a complex scalar representing the force magnitude and phase in the heave direction. $\overline{F}_R(i\omega)$ is the radiation force and is again a six-component vector for a general, six-degree-of-freedom oscillation of the body, though here we consider only the force in heave direction. In summary, the entire hydrodynamic interaction of the body with an incident wave field is modeled using the relation

$$F(i\omega) = F_f(i\omega) + \overline{F}_R(i\omega). \tag{4.4}$$

The exciting force $F_f(i\omega)$ depends on the geometry of the body and its depth of submergence. For slender bodies or bodies with a characteristic diameter (circumscribing the larger body dimension) over an order of magnitude smaller than a wavelength, the body-disturbance wave field components are small enough to be ignored and the force can be obtained simply via an integration of the incident wave velocity potential over the submerged surface. This is often referred to as the Froude–Krylov force component. When the characteristic diameter is comparable to or greater than the incident wavelength, the body-disturbance component is substantial, and the diffraction problem needs to be solved, which in most cases requires numerical computation (e.g., [6, 8]). The force $\overline{F}_R(i\omega)$ also depends on the body geometry and submergence, and in most cases also requires numerical computation, though known relations between the diffraction and radiation forces can be used to provide values or their verification at chosen frequencies. Because this force is caused by body motion, it is helpful to consider it as a function of body velocity and acceleration. For small amplitude motions, the force is simply expressed as

$$\overline{F}_R(i\omega) = -\left\{b(\omega) + i\omega\left[\overline{a}(\omega)\right]\right\} U(i\omega). \tag{4.5}$$

Both $\overline{a}(\omega)$ and $b(\omega)$ are functions of frequency ω. In particular, $b(\omega) \to 0$ for large ω. However, for large ω, $\overline{a}(\omega) \to \overline{a}(\infty)$, which is a constant. We define $a(\omega) = \overline{a}(\omega) - \overline{a}(\infty)$, so that $a(\omega)$ denotes just the purely frequency-dependent part. We should add that, because $\overline{a}(\omega)$ tends to a constant value (typically nonzero) as $\omega \to \infty$, its inverse Fourier transform does not exist. However, since $a(\omega) \to 0$ as $\omega \to \infty$, its inverse Fourier transform does exist.

The frequency dependence of $b(\omega)$ and $\overline{a}(\omega)$ follows from the fact that the body produces a different radiated field at different frequencies. $b(\omega)$ is a real-valued function of frequency and has the effect of a linear damping constant for a harmonic wave at frequency ω. It represents the rate at which energy radiates away from the oscillating body via the radiated wave field. $\overline{a}(\omega)$ has the effect of an inertia or mass. This quantity represents the reactive exchange of energy between the body and the evanescent or nonpropagating wave field the body produces (see also Chapter 6). As mentioned, $\overline{a}(\omega)$ can be written as

$$\overline{a}(\omega) = a(\omega) + \overline{a}(\infty) \tag{4.6}$$

where the broken-off frequency-independent (or infinite-frequency) part is a typically nonzero constant. $a(\omega)$ is a real valued function of frequency. As an example, Fig. 4.1 shows the radiation damping and added mass results for a circular disc in heave that is submerged just below the free surface.

If we express Equation (4.3) in terms of velocity $U(i\omega)$ and insert Equations (4.4) and (4.6) into it, we have

$$\left[i\left(\omega\left(m + \overline{a}(\infty) + a(\omega)\right) - \frac{1}{\omega}k\right) + b(\omega)\right] U(i\omega) = F_f(i\omega). \tag{4.7}$$

Under the assumption that the body interacts with regular waves at frequency ω only, we may express the dynamics above by means of a differential equation,

$$\left[m + \overline{a}(\infty) + a(\omega)\right]\ddot{x} + b(\omega)\dot{x} + kx = F_f(i\omega)\mathrm{e}^{i\omega t}. \tag{4.8}$$

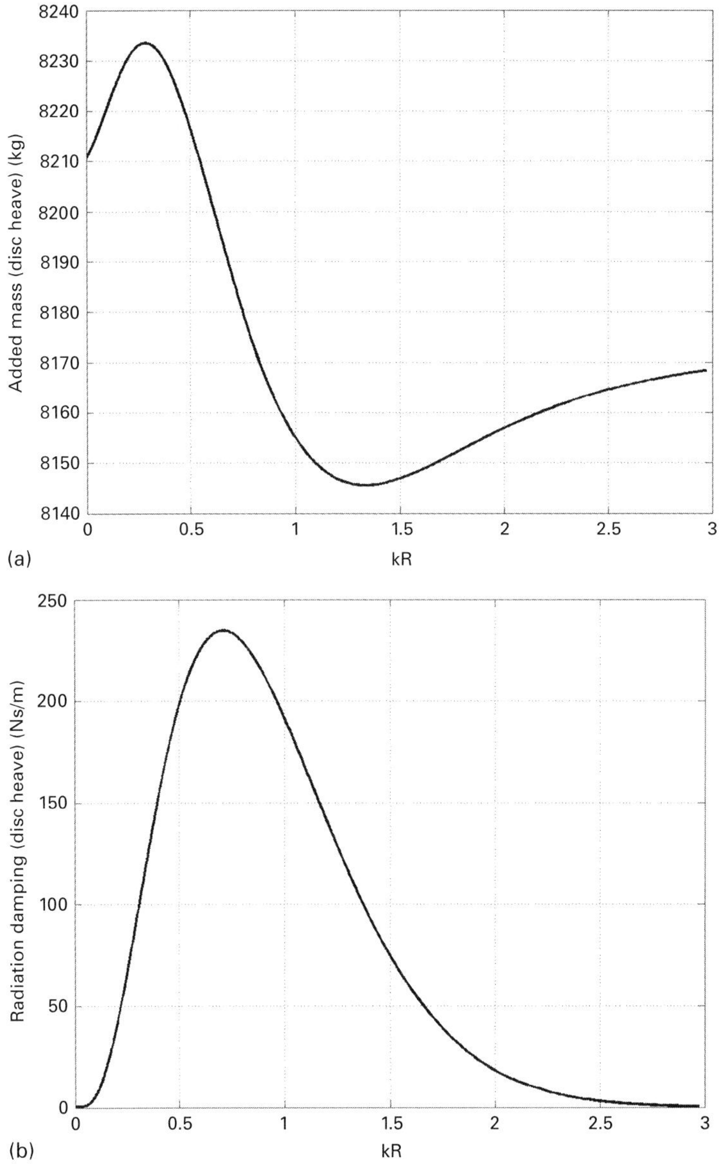

Figure 4.1 (a) The added mass (including frequency-independent component) of a neutrally buoyant circular disc floating just below the water surface and (b) radiation damping for the same disc. Heave is assumed to be the predominant motion.

This is a second-order ordinary differential equation with frequency-dependent coefficients that can be solved using techniques applicable to ordinary differential equations with constant coefficients only when the excitation is harmonic as shown on the right side at the frequency matching with that for the parameters on the left side. The solution procedure for Equation (4.8) then essentially just results in the algebraic relation in Equation (4.7). The quantity in brackets multiplying $U(i\omega)$ in Equation (4.7) can again be understood as a complex impedance where the parameters are themselves frequency dependent. Thus we have

$$Z(i\omega)U(i\omega) = F_f(i\omega) \tag{4.9}$$

where $Z(i\omega) = Z_r(\omega) + iZ_i(\omega)$ and

$$\begin{aligned} Z_r(\omega) &= \Re\left(Z(i\omega)\right) = b(\omega), \\ Z_i(\omega) &= \Im\left(Z(i\omega)\right) = \omega\left[m + \overline{a}(\infty) + a(\omega)\right] - \frac{1}{\omega}k. \end{aligned} \tag{4.10}$$

We may apply an external force on the body either from the sea floor itself or from another fixed reference (such as a platform fixed to the sea floor). This could be accomplished with hydraulic pistons or actuators, or in some cases with permanent magnet linear generators. A component of this force may serve as the power take-off for energy absorption, while another component could be used to apply a reactive load much as in the case of the body oscillating in air in Chapter 3. The external force $f_e(t) = F_e(i\omega)\exp(i\omega t)$ thus comprised could be expressed using

$$F_f(i\omega) = \left[\Lambda_r(\omega) + i\Lambda_i(\omega)\right] U(i\omega). \tag{4.11}$$

With the addition of this force, Equation (4.7) modifies to

$$\left[i\left(\omega\left(m + \overline{a}(\infty) + a(\omega)\right) - \frac{1}{\omega}k + \Lambda_i(\omega)\right) + b(\omega) + \Lambda_r(\omega)\right] U(i\omega) = F_f(i\omega). \tag{4.12}$$

More briefly,

$$\left[(Z(i\omega) + \Lambda(i\omega)\right] U(i\omega) = F_f(i\omega). \tag{4.13}$$

For harmonic excitation at a given frequency, we can argue that maximum power absorption occurs when that frequency is the natural frequency of the body, or $\omega = \omega_n$ where

$$\omega_n = \sqrt{\frac{k}{m + \overline{a}(\infty) + a(\omega_n)}}. \tag{4.14}$$

In addition, the damping constant provided by the power absorption mechanism must match the internal damping constant, the radiation damping being the dominant internal damping component and viscous damping being ignored in this case. We need

$$\Lambda_r(\omega_n) = b(\omega_n). \tag{4.15}$$

Thanks to resonance, in this case, the reactive part of the body's mechanical impedance vanishes naturally. In the more common situation when resonance is not possible at the incident frequency ω, that is, $\omega \neq \omega_n$, we could set the reactive component in the load (as in Chapter 3) such that the reactive component value in the overall mechanical impedance is forced to vanish. In either case, of course, the damping component of the load needs to match the internal damping. Summarizing these statements mathematically, as before, we need

$$\Lambda(i\omega) = Z^*(i\omega) \tag{4.16}$$

where we choose

$$\Lambda_r(\omega) = b(\omega) \tag{4.17}$$

and

$$\Lambda_i = -\left[\omega\left(m + \overline{a}(\infty) + a(\omega)\right) - \frac{1}{\omega}k\right]. \tag{4.18}$$

This form of complex-conjugate control is strictly only applicable in regular waves. It may be applied on real wave energy devices, where the load is chosen such that complex-conjugate impedance match occurs at the dominant frequency in a wave spectrum. More discussion on the effect of complex-conjugate control in regular waves on power absorption and the efficiency of power absorption is included in Chapter 6. In the following section, we examine non harmonic excitation as applied by irregular waves, with particular attention given to the increased complexity of the model that now becomes necessary, and the impact of this on motion control. Here we note once again that the procedure above could be adopted only because the excitation was harmonic and because the differential equation model was valid (at that frequency). In a sense, this restricts the applicability of this approach to the case of steady-state harmonic excitation and steady-state response at one frequency only, where full knowledge of the time dependence of the harmonic forcing at all times is available.

4.3 Irregular Waves

In Chapter 3 we saw how one could build on the relations derived for harmonic forcing to develop a control approach for irregular forcing, for bodies oscillating in air. Small deflections were assumed, so that, if the irregular input forcing is thought to be a superposition of a number of sinusoids (at different frequencies, amplitudes, and phases), then the response of the system to the irregular forcing is simply a linear superposition of the individual functions representing the system's response to each individual sinusoid. Thus, Fourier transformation could be used on both sides of the equation of motion to predict the system's response to any general forcing, including an irregular forcing, such as encountered in irregular waves.

We attempt here to extend the approach of Chapter 3 to bodies close to the free surface and responding to waves in an open sea and in deep water. It is in this effort that the difficulty of controlling the response of such bodies in waves becomes apparent. Note that the difficulty arises in both regular and irregular waves, though in strictly regular waves, the easier approach of Section 4.2 is possible. For irregular waves, the answer here is much more complicated. We will essentially only pose the question here, leaving possible answers to subsequent chapters.

As before, we begin with a body such as a sphere producing a single-mode response to incident waves. The equation of motion can, as before, be written as

$$m\ddot{x}(t) + kx(t) = f(t) \tag{4.19}$$

where we have ignored viscous damping or other similar loss mechanisms. $f(t)$ is now an irregular excitation due to waves, and as in Chapter 3, we take this to be a linear superposition of a number of harmonic waves at different frequencies, phases, and amplitudes. The decomposition of the total wave force $f(t)$ into the diffraction force and radiation force thus extends to irregular wave action, leading to

$$m\ddot{x}(t) + kx(t) = f_f(t) + f_r(t) \tag{4.20}$$

where $f_f(t)$ and $f_r(t)$ denote the diffraction and radiation force terms, respectively. As we know, $f_f(t)$ is independent of body motion, being the force on the body due to the wave field that would result if the body were held fixed in its equilibrium position. $f_r(t)$, on the other hand, results from body oscillation, and therefore closely depends on the geometry, draft, or depth of submergence, oscillation velocity, and frequency. If a body were to float at static equilibrium in calm water with no incident waves approaching, both $f_f(t)$ and $f_r(t)$ would be zero. If the body were to oscillate at a single frequency in otherwise calm water, $f_r(t)$ would no longer be zero, but $f_f(t)$ would be. What if, as the body is resting in calm water in its equilibrium position, it is somehow given a sudden impulsive, instantaneous movement? Clearly, through the short duration of that impulse, the body feels a force f_r. However, with its instantaneous movement near the free surface, the body has already created a wave field around it. This wave field will cause a force on the body, and will likely cause the body to move. So even if the impulsive movement lasted only an instant, its memory lingers. The motion of the water and the force it applies on the body will decrease as the wave field propagates away, slowly approaching zero as the wave field draws farther and farther away. If the body was forced into or allowed another instantaneous movement immediately following the previous movement, the force it feels is now the sum of the force that has lasted from the previous impulse and the force due to the new impulse. An entire irregular motion sequence of the body can be considered a succession of such impulses, one after the other, each new impulse adding a new force to what remained from the previous impulses. As the process continues, there comes a point when the force due to the first impulse has asymptotically gone to zero, and next the force due to the second impulse, and the third, will approach zero, and so on. The length of this "memory" is generally a function of body geometry and its draft or depth of submergence. The total force $f_r(t)$ at any time t can, thanks to linear superposition as just discussed, be written in the form of a convolution integral as

$$f_r(t) = -\int_{-\infty}^{\infty} z_r(\tau)u(t-\tau)d\tau \tag{4.21}$$

where $z_r(t)$ is the force per unit velocity acting on the body as a result of a unit impulsive velocity and contains the "radiation impulse response" function, or "radiation kernel," as discussed below. Mathematically speaking, here u, z_r, and f_r are all real-valued functions. We recall from Section 4.2 that one of the contributors to the radiation force is the "infinite-frequency or frequency-independent added mass" experienced by all bodies oscillating in water. A mathematically consistent way to include this in Equation (4.21) is to express $z_r(t)$ as

$$z_r(t) = h_r(t) + \overline{a}(\infty)\frac{d\delta}{dt}(t). \tag{4.22}$$

Then, by means of integration by parts,

$$f_r(t) = \int_{-\infty}^{\infty} h_r(\tau)u(t-\tau)d\tau + \overline{a}(\infty)\dot{u}(t) \tag{4.23}$$

where the second term is simply the infinite-frequency added mass times the instantaneous acceleration, and h_r is the radiation impulse response function or radiation kernel. Three facts about f_r and h_r emerge from the discussion above:

1. Although the motion at the present instant has the greatest influence on the magnitude of f_r at that instant, the motion at the previous instants still contributes to the force at the present instant; this is the so-called memory effect.
2. The memory is "finite" for all practical purposes, and thus the function $h_r(t)$ above tends toward zero as t increases, thanks to the fact that the energy in the waves the body makes propagates away from it.
3. No force f_r existed before the first motion occurred. The motion causes the force. Prior to the start of the motion, no force was present. Moreover, the force at the present instant depends only on the motion up to the present instant. It is not influenced by motion that has not yet occurred.

The last item above is simply a collection of statements implying that the radiation force $f_r(t)$ is a causal function of $u(t)$, and the impulse response function $h_r(t)$ is such that this condition holds; often, h_r is then referred to as a causal impulse response function. As an illustration, Fig. 4.2 plots the radiation impulse response function for the submerged circular disc in heave. The extent of h_r along the time axis shows the length of memory in this case. In other words, for the case shown in Fig. 4.2, velocities that occurred more than 18 seconds or so ago do not affect the force value at the present instant.

In light of item 3 above, we see that h_r is zero until $t = 0$, where it has a nonzero value, reflecting the sudden impulse applied on the body at $t = 0$. Equation (4.23) shows that for $\tau > 0$ seconds, $h_r(\tau)$ multiplies $u(t-\tau)$, or u that occurred τ seconds ago. If, on the other hand, $\tau < 0 = -\epsilon$ where ϵ is a positive quantity, then $h_r(\tau)$ would multiply $u(t-\tau) = u(t+\epsilon)$ (where $\epsilon > 0$), which has not occurred yet, being the velocity from ϵ seconds into the future. In such a case, $f_r(t)$ at the current instant t would be influenced by motion from the future. But we know that the force cannot exist before motion begins. Hence, $h_r(\tau)$ must be zero for $\tau < 0$, or more generally, $h_r(t) = 0$ for $t < 0$, which in effect is a mathematical summary statement of item 3 above. This implies that

$$\int_{-\infty}^{0} h_r(\tau)u(t-\tau)d\tau = 0, \tag{4.24}$$

which results in

$$f_r(t) = \int_{0}^{\infty} h_r(\tau)u(t-\tau)d\tau. \tag{4.25}$$

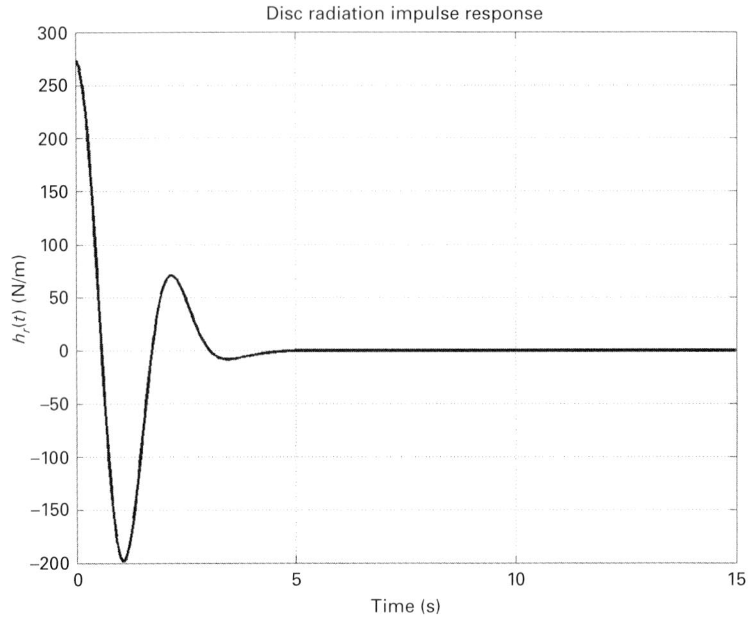

Figure 4.2 The impulse response function $h_r(t)$ for the disc in heave for which the added mass and radiation damping variations are shown in Fig. 4.1.

Based on Fig. 4.2 and item 2 above, we realize that τ does not really need to go all the way to infinity ($\tau \to \infty$), as h_r will have become zero for all practical purposes long before that. We leave the upper limit in Equation (4.25) as ∞ with the understanding that the ∞ in $\tau \to \infty$ can be 5 seconds or 35 seconds (or some similar value), depending on the geometry and submergence in a problem. The equation of motion with this background can now be written as

$$[m + \bar{a}(\infty)]\ddot{x}(t) + \int_0^{\infty} h_r(\tau)u(t-\tau)d\tau + kx(t) = f_f(t). \tag{4.26}$$

Rewriting this equation entirely in terms of the velocity $u(t)$,

$$[m + \bar{a}(\infty)]\dot{u}(t) + \int_0^{\infty} h_r(\tau)u(t-\tau)d\tau + k\int_{-\infty}^{t} u(\tau)d\tau = f_f(t). \tag{4.27}$$

Equations (4.26) and (4.27) are forms of the so-called Cummins' equation [63], which provides a way to avoid restrictive assumptions and awkward interpretations such as were found necessary in working with Equation (4.8) in Section 4.2. We note that Equations (4.26) and (4.27) are not differential equations, but rather linear integro-differential equations, and will provide much of the foundation for the model-based control algorithms developed in Chapters 7–10.

$f_f(t)$ is discussed briefly near the end of this chapter. Equations (4.26) and (4.27) are the linear time domain model for single mode oscillation on or closely below the free surface. As discussed earlier, for small-amplitude oscillation, forward Fourier transformation of Equation (4.27) must result in Equation (4.7), for a frequency $\omega > 0$, and inverse Fourier transformation of a generalized form of Equation (4.7) should result in Equation (4.27). Thus,

$$\frac{1}{2\pi}\int_{-\infty}^{\infty}\left[i\omega\left(m+\overline{a}(\infty)\right)+\left(\pi\delta(\omega)+\frac{1}{i\omega}\right)k\right]U(i\omega)\mathrm{e}^{i\omega t}d\omega$$
$$=\left[m+\overline{a}(\infty)\right]\dot{u}(t)+k\int_{-\infty}^{t}u(\tau)d\tau \tag{4.28}$$

where $H(t)$ is the Heaviside (unit) step function. The inverse Fourier transform of the right side of Equation (4.7) is

$$\frac{1}{2\pi}\int_{-\infty}^{\infty}F_f(i\omega)\mathrm{e}^{i\omega t}d\omega=f_f(t). \tag{4.29}$$

As we see later in Fig. 4.3, the magnitude of $F_f(i\omega)$ approaches zero for large ω. Therefore, its inverse Fourier transform does exist. We combine the remaining terms on the left side using a single frequency domain function $H_R(i\omega)$ as

$$\left[i\omega a(\omega)+b(\omega)\right]U(i\omega)=H_R(i\omega)U(i\omega). \tag{4.30}$$

$H_R(i\omega)$ represents just the frequency-dependent part associated with the radiation force, and the inverse Fourier transformation of Equation (4.30) results in

$$\frac{1}{2\pi}\int_{-\infty}^{\infty}H_R(i\omega)U(i\omega)\mathrm{e}^{i\omega t}d\omega=\int_{-\infty}^{\infty}h_r(\tau)u(t-\tau)d\tau \tag{4.31}$$

where

$$\frac{1}{2\pi}\int_{-\infty}^{\infty}H_R(i\omega)\mathrm{e}^{i\omega t}d\omega=h_r(t). \tag{4.32}$$

This is a point where the discussion of the time domain relations leading up to Equation (4.27) again becomes relevant. We argued above that $h_r(t)$ is a real-valued function. This implies, based on the properties of Fourier transforms [56, 61], that

$$H_R(-i\omega)=H_R^*(i\omega). \tag{4.33}$$

In terms of $a(\omega)$ and $b(\omega)$ this implies that

$$-i\omega a(-\omega)+b(-\omega)=-i\omega a(\omega)+b(\omega) \tag{4.34}$$

which in turn means that

- $b(\omega)$ is an even function
- $\omega a(\omega)$ is an odd function, and $a(\omega)$ itself is an even function.

Realizing that $\exp(i\omega t)=\cos\omega t+i\sin\omega t$,

$$h_r(t)=\frac{1}{2\pi}\int_{-\infty}^{\infty}H_R(i\omega)\mathrm{e}^{i\omega t}d\omega=\frac{1}{2\pi}\int_{-\infty}^{\infty}\left[b(\omega)+i\omega a(\omega)\right]\left(\cos\omega t+i\sin\omega t\right)d\omega. \tag{4.35}$$

Here we can again use the knowledge that $h_r(t)$ is real valued and, further, the inferences above that $b(\omega)$ is an even function, and $\omega a(\omega)$ is an odd function. We recall that the cosine function is an even function and the sine function is an odd function. Furthermore, since the product of two even functions is even and the product of two odd functions is also even, from Equation (4.35) we can deduce that

$$h_r(t) = \frac{1}{\pi}\int_0^{\infty} b(\omega)\cos\omega t d\omega - \frac{1}{\pi}\int_0^{\infty} \omega a(\omega)\sin\omega t d\omega. \tag{4.36}$$

It can be verified (not surprisingly, since $h_r(t)$ on the left side is real) that the terms in the integrals attached to i automatically vanish because $b(\omega)$ is an even function and $\omega a(\omega)$ is an odd function, and because the product of an even function with an odd function is an odd function.

Finally, because we found $h_r(t)$ to be a causal function, with $h_r(t) = 0,\ t < 0$, we may use this condition now with Equation (4.36), where we find that a negative value of t will alter the sign on the second term because of the sine function being an odd function. However, it will not change the sign of the first term. Because $h_r(t)$ must be zero for a negative value of t, we conclude that

$$\frac{1}{\pi}\int_0^{\infty} b(\omega)\cos\omega t d\omega = -\frac{1}{\pi}\int_0^{\infty} \omega a(\omega)\sin\omega t d\omega. \tag{4.37}$$

Therefore,

$$h_r(t) = \frac{2}{\pi}\int_0^{\infty} b(\omega)\cos\omega t d\omega = -\frac{2}{\pi}\int_0^{\infty} \omega a(\omega)\sin\omega t d\omega. \tag{4.38}$$

We may present the above discussion in a more formal light, by arguing as follows. Because $h_r(t)$ is causal, it follows that its Fourier transform is an analytic function in the upper half of the complex-frequency plane (i.e., its derivative exists everywhere in the upper half plane) [64]. Using the symbol ϖ to denote the general complex frequency variable with the understanding that the angular frequency ω defines a specific point on the real axis in the ϖ plane, $H_R(i\varpi)$ defines the Fourier transform of $h_r(t)$ over the entire complex frequency plane. Further, as Equation (4.33) asserts, since $h_r(t)$ is real valued, $H_R(-i\varpi) = H_R^*(i\varpi)$. It can thus be argued that the following contour integral equals zero provided the singularity at ω is excluded, as shown in Fig. 4.3:

$$\oint_C \frac{H_R(i\varpi)}{\varpi - \omega} d\varpi = 0. \tag{4.39}$$

The large curved portion $\overline{C}$ has a radius $R \to \infty$, while the small curved section c around ω has a radius $\varepsilon \to 0$. Expanding the contour integral in Equation (4.39) over the different sections comprising the contour,

$$\lim_{\varepsilon\to 0}\left[\int_{-\infty}^{\omega-\varepsilon} \frac{H_R(i\varpi)}{\varpi - \omega} d\varpi + \int_c \frac{H_R(i\varpi)}{\varpi - \omega} d\varpi + \int_{\omega+\varepsilon}^{\infty} \frac{H_R(i\varpi)}{\varpi - \omega} d\varpi + \int_{\overline{C}} \frac{H_R(i\varpi)}{\varpi - \omega} d\varpi\right] = 0. \tag{4.40}$$

The integral over the small semi circular portion can be expressed as half of what the integral value would be over a full contour encircling the singularity ω. The residue

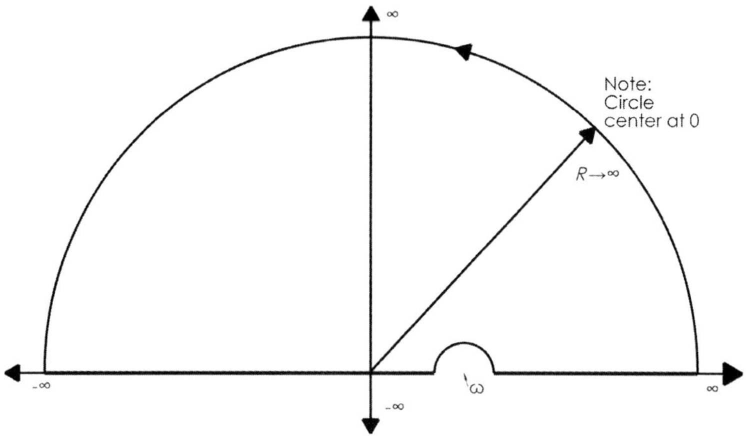

Figure 4.3 The integration path for the contour integral in Equation (4.39). Note that the radius of the small semicircular portion around ω is infinitesimal, while the radius of the large semicircle tends to infinity.

at ω is simply $H_R(i\omega)$. Based on the Cauchy residue theorem, the integral over the "full version" of the semi circle c would be $2\pi i H_R(i\omega)$. Therefore,

$$\int_c \frac{H_R(i\varpi)}{\varpi - \omega} d\varpi = \pi i H_R(i\omega). \tag{4.41}$$

The first and the third integrals together define the Cauchy principal value of the integral from $-\infty$ to ∞. Finally, we recall that $H_R(i\varpi) = b(\varpi) + i\varpi a(\varpi)$. Since both $b(\varpi), a(\varpi) \to 0$ more rapidly than $\varpi \to \infty$, the integral over $\overline{C}$ tends to zero. Thus we are left with

$$PV \int_{-\infty}^{\infty} \frac{H_R(i\varpi)}{\varpi - \omega} d\varpi = \pi i H_R(i\omega) \tag{4.42}$$

where PV denotes principal value. Separating the real and imaginary parts of Equation (4.42),

$$\begin{aligned} b(\omega) &= \frac{1}{\pi} PV \int_{-\infty}^{\infty} \frac{\varpi a(\varpi)}{\varpi - \omega} d\varpi, \\ \omega a(\omega) &= -\frac{1}{\pi} PV \int_{-\infty}^{\infty} \frac{b(\varpi)}{\varpi - \omega} d\varpi. \end{aligned} \tag{4.43}$$

The two parts of Equation (4.43) are known as the Kramers–Kronig relations, which arise in other phenomena of physics such as optics. They are frequently written as one-sided integrals on the positive real axis. By multiplying and dividing the right side of the first of Equations (4.43) by $\varpi + \omega$,

$$b(\omega) = \frac{1}{\pi} PV \int_{-\infty}^{\infty} \frac{\varpi^2 a(\varpi)}{\varpi^2 - \omega^2} d\varpi + \frac{\omega}{\pi} PV \int_{-\infty}^{\infty} \frac{\varpi a(\varpi)}{\varpi^2 - \omega^2} d\varpi. \tag{4.44}$$

Because $\varpi a(\varpi)$ is an odd function of ϖ, the positive and negative sides of the second integral in Equation (4.44) are equal and opposite, while the two sides of the first integral are equal and additive, leaving

$$b(\omega) = \frac{2}{\pi} PV \int_0^\infty \frac{\varpi^2 a(\varpi)}{\varpi^2 - \omega^2} d\varpi. \tag{4.45}$$

Similarly, performing the same operations with $\varpi + \omega$ on the second of Equations (4.43),

$$\omega a(\omega) = -\frac{1}{\pi} PV \int_{-\infty}^\infty \frac{\varpi b(\varpi)}{\varpi^2 - \omega^2} d\varpi - \frac{1}{\pi} PV \int_{-\infty}^\infty \frac{\omega b(\varpi)}{\varpi^2 - \omega^2} d\varpi. \tag{4.46}$$

Since $b(\varpi)$ is an even function of ϖ, $\varpi b(\varpi)$ is an odd function. The positive and negative sides of the first integral in Equation (4.46) are equal and opposite, while the two sides of the second integral are equal and additive. Therefore,

$$a(\omega) = -\frac{2}{\pi} PV \int_0^\infty \frac{b(\varpi)}{\varpi^2 - \omega^2} d\varpi. \tag{4.47}$$

We have essentially arrived at the conclusion in this section that the real-valued nature and the causality of $h_r(t)$ place some important constraints on the two quantities forming the heart of the radiation force. Specifically, $b(\omega)$ and $\omega a(\omega)$ are mutually related, and must respectively be even and odd functions of ω.

Using inferences on the physics of the bodies oscillating near the free surface, and the mathematical relations accompanying Fourier transformation and properties of functions, we have arrived at the defining condition (Equation (4.38)) that relates the time domain radiation impulse response function to the frequency domain, frequency-dependent radiation damping and added mass functions. However, in the process, we have also discovered that the frequency-dependent radiation damping and added mass are not independent functions. Rather, the dynamics of bodies oscillating near the free surface requires them to be related to one another through Equations (4.37) and (4.43)–(4.47). Equations (4.37), (4.38), and/or (4.43)–(4.47) arise entirely because $h_r(t)$ is real valued and causal with respect to t. Paradoxically, at first glance, this fact is one main reason why control of the oscillations of bodies near the free surface requires knowledge of the velocities of the future. We will find in Chapter 7 that for most efficient absorption using impedance matching such as we have seen here requires that the "time-reversed" version of h_r be synthesized, which is fully noncausal (or "anti causal"). As explained in Chapter 7, such a synthesis could be achieved by treating h_r as a sum of two impulse response functions, one odd and the other even, such that they add to zero for $t < 0$ and to h_r for $t \geq 0$. One of these and the negative of the other need to be synthesized for impedance matching control, if it is to be applied wave by wave. Because both are non-causal, we need velocity information from the future.

The exciting force $F_f(i\omega)$ in Equation (4.29) is often defined in terms of the incident complex wave amplitude $A(i\omega)$ (which includes both amplitude and phase) at the body centroid as

$$F_f(i\omega) = H_f(i\omega) A(i\omega). \tag{4.48}$$

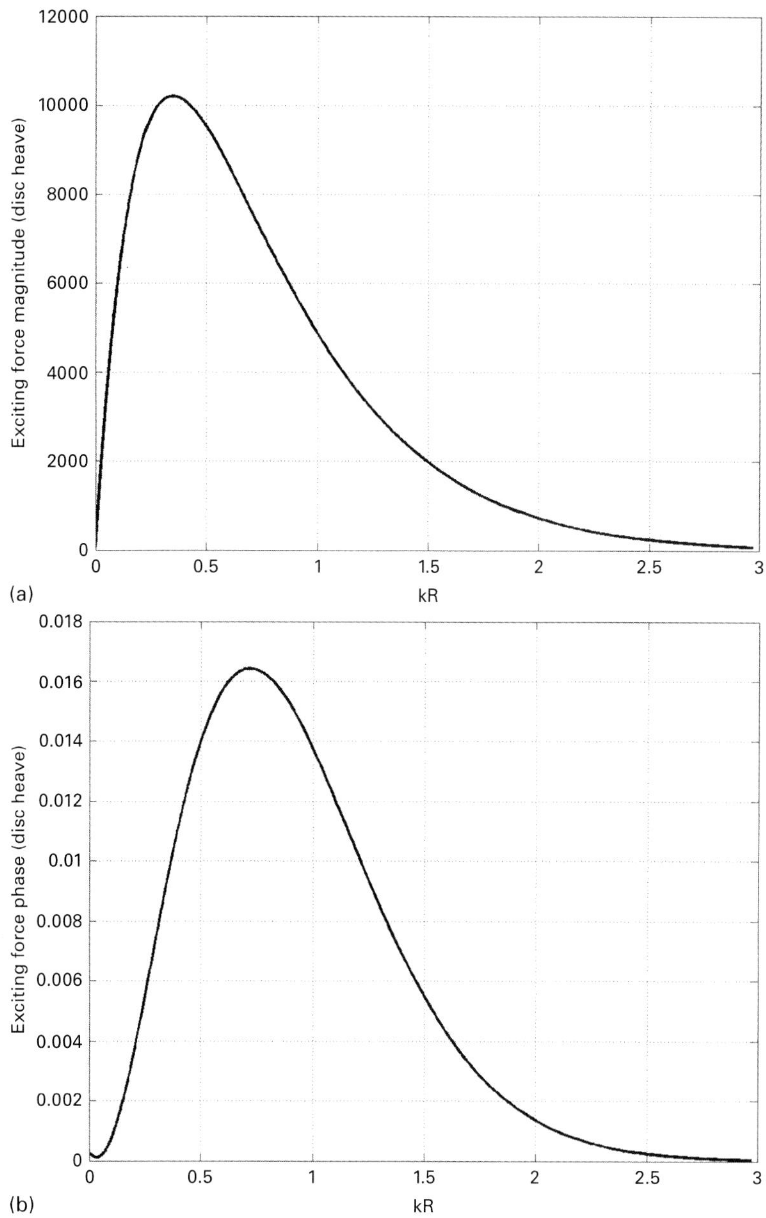

Figure 4.4 Exciting force magnitude (a) and phase angle (b) for heave oscillation of a circular disc submerged close to the surface for unit wave amplitude ($|A(i\omega)| = 1$ m).

Fig. 4.4 shows an example of exciting force magnitude and phase variation for unit wave amplitude (i.e., the amplitude and phase, respectively, of the function $H_f(i\omega)$). In the time domain, an inverse Fourier transformation of Equation (4.48) leads to

$$F_f(t) = \int_{-\infty}^{\infty} h_f(\tau)\eta(x_B, t - \tau)d\tau, \tag{4.49}$$

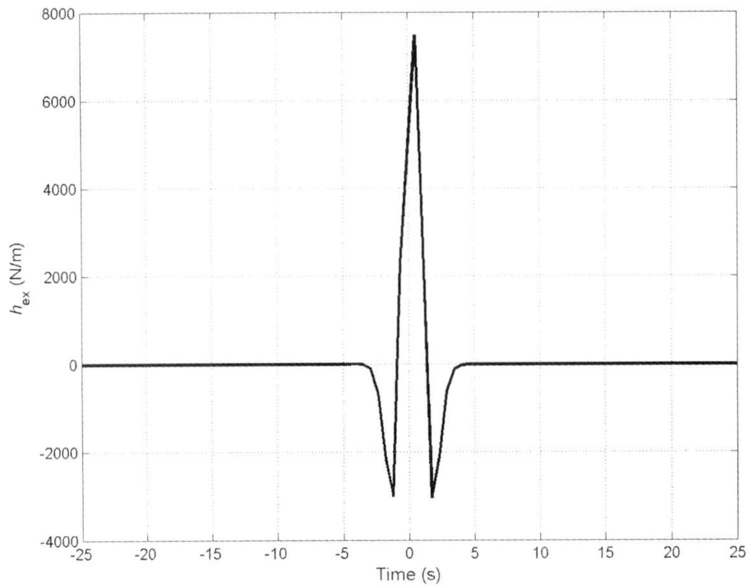

Figure 4.5 Impulse response function for exciting force in heave for a heaving near-surface submerged circular disc. This is expressed with respect to an impulsive wave excitation at the body location. Note that $h_f(t) \neq 0,\ t < 0$ here, so that h_f is noncausal relative to the surface elevation at the body.

where $h_f(t)$ is the exciting force impulse response function, and $\eta(x_B, t)$ is the instantaneous incident wave surface elevation at the body centroid. Fig. 4.5 plots the impulse response function h_f for the force variation in Fig. 4.4.

While the definition of the exciting force frequency response function $H_f(i\omega)$ as in Equation (4.48) is convenient in control, it masks important features of the true physics. In particular, the exciting force is the net result of a pressure distribution acting over the entire submerged surface. Further, because water under the free surface is a continuum, the body begins to feel the particle velocities and therefore pressure due to an approaching wave considerably before the wave actually arrives at the body surface. For these reasons, even though the actual process leading to the exciting force is causal when expressed fully accounting for the two features just mentioned, when expressed in terms of the incident wave field at the body centroid alone, the process appears noncausal. Hence, evaluation of the exciting force at the present instant requires not only the present and past values but also the future values of the surface elevation. Thus, even though $h_f(t)$ is real valued, it is non causal in that $h_f(t) \neq 0, t < 0$. This is the other potential difficulty with real-time control methods that require exciting force determination for the purpose of evaluating the control force.

The discussion above should shed some light on why control such as discussed in Section 4.2 has no straightforward extension to any excitation–response combination other than harmonic (regular, sinusoidal) waves. Strictly speaking, it works only in regular waves because the future knowledge is implicitly available in the assumption of steady-state harmonic excitation and response. Finally, we note that it is the presence of the free surface that thus complicates control to levels not quite tractable using the treatment of Section 3.3 in Chapter 3.

4.4 Conclusion

In summary, the difficulties surrounding the radiation and the exciting forces arise purely from the nature of the hydrodynamic interaction of the body with the free surface and the approaching waves. A good understanding of what lies beneath the tale that is constantly unfolding in the process where energy from the waves is mechanically converted is essential for effective control of this process. A number of diverse phenomena play a part: the waves that bring in the energy, the device that interacts with them hydrodynamically to capture the energy, the mechanism that we use to drive electricity generation (which often is also used to control the device), and the goals and algorithms that dictate the actions of the mechanism(s).

A few points are important to summarize, within the linear theory descriptions we use: a differential equation description of a wave energy device is valid only for regular waves, that is, sinusoidal waves at a single frequency, single amplitude, and single phase angle. This equation still contains frequency-dependent parameters, which take on different values at different frequencies. The more reliable description includes a convolution integral that captures more accurately the "memory effect" and the causal nature of the force felt by a body due to its own "wave-making." The consequences of this description and the noncausal nature of the exciting force as expressed here also apply to regular waves. However, we then have complete information about the wave, from past through present to future because its amplitude, phase, and frequency are all known. A complete description such as this is not instantaneously available for irregular waves. This is why the control approach of Equations (4.14) and (4.15) does not work in irregular waves. In the following chapters, we attempt to understand the physics behind this, and what might be done to provide solutions with varying degrees of practical applicability and hydrodynamic accuracy. Indeed, such solutions form the bulk of the wave energy control literature today.

In the following chapters, we try to provide the correct framework under which we can articulate the problems just outlined, and formulate possible solutions. Thus, our goal is not only to provide a context and a perspective on the problem of hydrodynamic control, but also to develop an understanding of the logical structure that needs to be employed for effective application of control. This will involve close contact with some of the mathematical machinery that has accelerated and directed progress in wave energy conversion over the last few decades. We begin with a brief study of the waves themselves.

Part III

The Hydrodynamics

Here we at last step into the trenches of wave energy conversion. Our first goal is to understand, at least insofar as it pertains to its immediate "livelihood," the environment to which a wave energy converter is to be designed to respond. We begin our study with a mathematical description of waves on the ocean surface, both regular and irregular (Chapter 5). Next, in Chapter 6, we probe into the heart of wave energy conversion and review that mathematical structure that provides succinct and far-reaching results and insights that can be used to design "good" wave energy converters that a control system makes better.

5 Nature of the Wave Input

A sound understanding of the type of energy resource a wave energy device must convert can be developed if we confine ourselves to "small amplitude" waves, that is, waves for which the crest-to-trough height is small enough compared with the length of the wave (i.e., the distance separating like points (such as crests/troughs) on the wave) so that the height-to-length ratio or, in a sense, the "wave slope," can be assumed to be very small. A quantitative description of a seemingly random or irregular wave field is then possible to build using a quantitative description of harmonic (i.e., monochromatic or single-frequency) waves. Next, if the response of the wave energy device to such a wave is understood, and if the device is thought to respond like a linear system, then the response to an irregular wave can be reproduced via linear superposition. The focus of this chapter is on developing an understanding of complex, irregular wave patterns, preparatory to understanding the response and performance of a wave energy device in such waves.

5.1 Description of a Harmonic Wave

Our study begins with a description of a harmonic wave of single frequency and single amplitude. Excellent coverage of this topic can be found in [1, 3, 4, 6, 8, 65, 66, etc.]. We recall from before that in deep water the water particles under a regular wave move in circular orbits whose radii decrease rapidly with depth. Our immediate goal below is to obtain a quantitative understanding of regular waves. Later in this chapter, we will extend this to irregular waves more representative of what we find in the real ocean. Fig. 5.1 shows an illustration of a regular wave.

Wave action on the water surface is largely the combined effect of external forcing such as wind and a restoring force generated by the water body. For waves relevant to energy conversion, the restoring action is provided by gravity. Furthermore, the waves we really want to utilize have small amplitudes or, more precisely, have small slopes, in that their height-to-length ratio is less than an order of magnitude smaller than unity. This enables an analysis based on linear differential equations, allowing for the use of linear superposition throughout. It has been well observed, however, that this assumption of linearity does in general lead to meaningful results of practical applicability.

We see below that, beneath the surface, fluid motion due to a wave can be described using the Laplace equation, which follows from an application of the conservation of

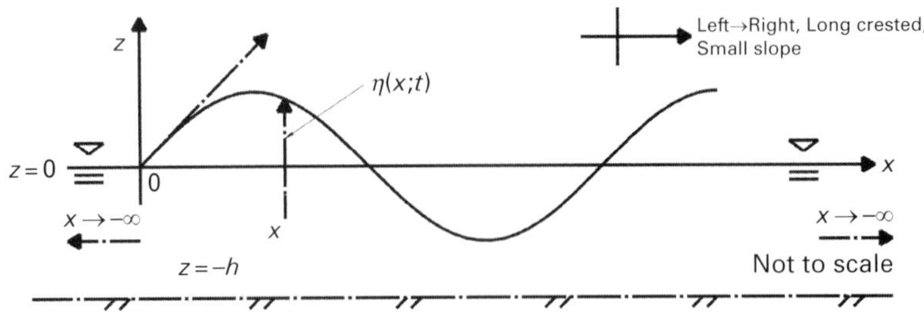

Figure 5.1 The most basic description of an ocean wave: a sinusoid with a single frequency and a single wavelength; that is, successive crests or troughs are separated by a constant length along the propagation direction, while at any single location, successive crests or troughs occur at the same time interval.

mass principle (i.e., the equation of continuity) throughout a fluid wherein motion is irrotational, inviscid, and incompressible in the large. Thus, a small fluid element cannot deform in shear except when inside a very thin layer immediately around an immersed body and the sea floor, and hence viscous effects over the bulk of the fluid affected by wave motion are negligible. The assumption of incompressibility holds well for water at speeds relevant to the waves of interest to us here. Motion of the fluid in the large can therefore be reasonably accurately described as irrotational flow, where the scalar velocity potential function $\phi(x, y, z; t)$ is related to the local velocity vector as

$$\mathbf{v} = \nabla\phi = \frac{\partial\phi}{\partial x}\mathbf{i} + \frac{\partial\phi}{\partial y}\mathbf{j} + \frac{\partial\phi}{\partial z}\mathbf{k} \tag{5.1}$$

in Cartesian coordinates. The coordinate system is located at the undisturbed free surface as shown in Fig. 5.1, with the z axis pointing upward. We assume a flat sea floor at a finite depth $z = -h$ except when the waters run deep, that is, where h is much greater than half the length of the wave being studied, in which case, $h \to \infty$ is used. Thanks to incompressibility, the continuity equation takes the form

$$\nabla\phi \cdot \mathbf{v} = 0. \tag{5.2}$$

Combining (5.1) and (5.2),

$$\nabla^2\phi = 0. \tag{5.3}$$

Equation (5.3) is the Laplace equation describing the fluid flow underneath the free surface and over the sea floor. Since the fluid particles are physically unable to move through the solid sea floor,

$$\frac{\partial\phi}{\partial z}(x, y, z; t) = 0, \quad \text{at } z = -h. \tag{5.4}$$

While the sea surface undulates up and down at any point $(x, y, 0)$, the fluid particles underneath the surface move in circular orbits. To assist description, a function $\eta(x, y; t)$ denoting just the free-surface undulation ("wave profile" or "wave surface elevation") is used, which at any point by definition can go only up or down. Absent any spraying particles breaking off the surface, the rate of change of $\eta(x, y; t)$ must equal the vertical component of velocity of the particle at $(x, y, 0; t)$, or,

$$\frac{\partial \eta}{\partial t}(x, y; t) = \frac{\partial \phi}{\partial z}(x, y, z; t); \quad \text{at } z = 0. \tag{5.5}$$

Strictly speaking, we should apply both conditions (5.5) and (5.6) below at $z = \eta$, the disturbed surface. However, since the waves of interest here have small slopes, η should be more than an order of magnitude too small compared with the overall spatial range spanned by the wave. For this reason, a reasonably correct description of the dynamics may be obtained by applying these conditions at $z = 0$ instead. This brings us to the Bernoulli equation, which interrelates pressure–velocity conditions at two suitably selected points, as in this case, at two points on the free surface. One point is at the chosen (x, y, z), and one a distance ∞ away where wave motion has not yet reached. We have

$$\frac{\partial \phi}{\partial t} + g\eta + \frac{1}{2}|\mathbf{v}|^2 + \frac{P_{atm}}{\rho} = \frac{P_{atm}}{\rho}, \quad \text{at } z = 0. \tag{5.6}$$

Note that P_{atm} on both sides is at $z = 0$, the undisturbed free surface. The use of P_{atm} (rather than $P_{atm} + p_f$, for instance) limits us to studying waves created by a local disturbance such as p_f elsewhere on the surface. This treatment gives us insights into the nature of the waves once they have moved away from the disturbance that created them, but does not allow study of the actual generation process, for which reference may be made to works such as [3] and [67] (this is an interesting area of study in its own right).

Next, we assume the water particle velocity to be small enough (with wave amplitudes being small) that $|\mathbf{v}|^2$ is an order of magnitude smaller than $|\mathbf{v}|$ as well as the first two terms in (5.6), so that we are left with

$$\frac{\partial \phi}{\partial t} + g\eta = 0, \quad \text{at } z = 0. \tag{5.7}$$

Differentiating with respect to t,

$$\frac{\partial^2 \phi}{\partial t^2} + g\frac{\partial \eta}{\partial t} = 0. \tag{5.8}$$

Combining Equations (5.5) and (5.8) we have

$$\frac{\partial^2 \phi}{\partial t^2} + g\frac{\partial \phi}{\partial z} = 0, \quad \text{at } z = 0. \tag{5.9}$$

This is the so-called free-surface condition, combining both the kinematic (5.5) and the kinetic (5.8) conditions. We remind ourselves that thanks to linearity, an irregular wave profile (i.e., free surface elevation) at any point can be expressed as a superposition of a large number of sinusoidal waves, as would result from a Fourier decomposition. Thus, here we look for a consistent description of a harmonic or regular sinusoidal wave that would fit within a Fourier decomposition. Using complex exponential notation for generality and for later convenience in handling amplitudes and phases, we separate the spatial and temporal parts of the velocity potential as

$$\phi(x, y, z; t) = \varphi(x, y, z)\mathrm{e}^{i\omega t}. \tag{5.10}$$

To help with the rest of the treatment in this text, we assume here waves propagating from left to right. From Equation (5.10),

$$\frac{\partial^2 \phi}{\partial t^2} = -\omega^2 \varphi(x, y, z)\mathrm{e}^{i\omega t} \quad \text{and} \quad \frac{\partial \phi}{\partial z} = \frac{\partial \varphi}{\partial z}\mathrm{e}^{i\omega t}. \tag{5.11}$$

Thus, for harmonic waves, the free-surface condition reduces to

$$\frac{\partial \varphi}{\partial z} - \frac{\omega^2}{g}\varphi = 0, \quad \text{at } z = 0, \tag{5.12}$$

while the sea-floor condition becomes

$$\frac{\partial \varphi}{\partial z} = 0, \quad \text{at } z = -h. \tag{5.13}$$

Equations (5.12) and (5.13) are thus the two boundary conditions for φ in the z direction. We consider next a long crested wave propagating from $x \to -\infty$ to $x \to \infty$, and align our y axis parallel to the crest. For long-crested waves, thus, variations in φ occur only in the x and z directions. Now we just need boundary conditions at the two ends of the domain in the x direction. In a sense, these are determined by the propagation direction of the wave being considered. For a rightward propagating wave, however far we look in the positive x direction, the wave travels away from us, the observer, that is, is outgoing at $x \to \infty$. Specifically, very informally, over the half-domain $x \geq 0$, $-h \leq z \leq 0$, rightward propagation implies that for a given ω, as t increases, so must x. For the harmonic wave with the temporal variation $\mathrm{e}^{i\omega t}$, a wave of the form $\sim \mathrm{e}^{-i(kx-\omega t)}$ satisfies this condition. For φ, this variation implies that

$$\frac{\partial \varphi}{\partial x} + ik\varphi \to 0, \quad x \to \infty. \tag{5.14}$$

On the other hand, for the half-domain $x < 0$, $-h \leq z \leq 0$, for a right-propagating harmonic wave, as t increases, x must become less negative. For the temporal variation $\mathrm{e}^{i\omega t}$, this condition is again satisfied by the same wave form $\sim \mathrm{e}^{-i(kx-\omega t)}$. Thus, as we look far into the negative x direction, the wave travels towards us, the observe, or is incoming at $x \to -\infty$. Hence,

$$\frac{\partial \varphi}{\partial x} + ik\varphi \to 0, \quad x \to -\infty. \tag{5.15}$$

Equations (5.14) and (5.15) represent the conditions at $x \to \infty$ and $x \to -\infty$, respectively, effectively describing a right-propagating harmonic wave that is incoming at $x \to -\infty$ and outgoing at $x \to \infty$. If we digress for a moment and consider a wedge at $x = 0$ extending from $y \to -\infty$ to $y \to \infty$ and forced into oscillation on a calm surface, the waves emanating from it are "outgoing" at infinity in either direction, $x \to \pm\infty$. This character is represented by the "radiation condition," which is expressed in a manner similar to Equations (5.14) and (5.15), but with appropriate modifications to represent outgoing behavior at both ends.

To return to the present problem, we have enough information now to solve the Laplace equation. Of the three second derivatives,

$$\frac{\partial^2 \varphi}{\partial y^2} = 0 \tag{5.16}$$

for long-crested waves. Restricting our wave behavior of interest to solutions that satisfy the form $\varphi(x, z) = X(x)Z(z)$,

$$\nabla^2 \phi = \nabla^2 \varphi = \frac{\partial^2 \varphi}{\partial x^2} + \frac{\partial^2 \varphi}{\partial z^2} = Z\frac{d^2 X}{dx^2} + X\frac{d^2 Z}{dz^2} = 0. \tag{5.17}$$

For a nontrivial solution $\varphi(x, z) \neq 0$, division by $\varphi(x, z) = X(x)Z(z)$ leads to

$$\frac{1}{X}\frac{d^2 X}{dx^2} + \frac{1}{Z}\frac{d^2 Z}{dz^2} = 0, \tag{5.18}$$

which implies

$$\frac{1}{X}\frac{d^2 X}{dx^2} = -\frac{1}{Z}\frac{d^2 Z}{dz^2} = -k^2. \tag{5.19}$$

For the left side to be a function of x alone and the right side to be a function z of alone, and for them to be equal to each other everywhere, they both must equal a constant throughout. We note that the constant $-k^2$ on the right embodies some forethought and an expectation of the type of solution we are looking for. Two ordinary differential equations thus result:

$$\begin{aligned} &\frac{d^2 X}{dx^2} + k^2 X = 0, \\ &\frac{d^2 Z}{dz^2} - k^2 Z = 0. \end{aligned} \tag{5.20}$$

The first of Equations (5.20) admits solutions of the form $X(t) = C_{x1}\mathrm{e}^{ikx} + C_{x2}\mathrm{e}^{-ikx}$. Taking each of the two parts and applying our two conditions (5.14) and (5.15), it is seen that only the second part represents a right-propagating wave from $x \to -\infty$ to $x \to \infty$. To retain just this part and reject the first part as inadmissible for a right-propagating wave, $C_{x1} = 0$ is chosen. The second of Equations (5.20) admits solutions of the form $Z(z) = C_{z1}\mathrm{e}^{kz} + C_{z2}\mathrm{e}^{-kz}$. In deep water, $h \to -\infty$, since

$$\frac{\partial Z}{\partial z} = kC_{z1}\mathrm{e}^{kz} - kC_{z2}\mathrm{e}^{-kz}. \tag{5.21}$$

For $\partial Z/\partial z = 0$, at $z = h \to -\infty$, the first part $\to 0$ while the second part $\to \infty$. Thus, only the first part is physically meaningful, and $C_{z2} = 0$ is chosen. Now turning to the free surface condition,

$$\frac{\partial \varphi}{\partial z} - \frac{\omega^2}{g}\varphi = 0, \Rightarrow \left(k - \frac{\omega^2}{g}\right) C_{z1}\mathrm{e}^{kz} = 0, \tag{5.22}$$

for a nontrivial solution, we arrive at

$$\omega^2 = gk, \tag{5.23}$$

which is the so-called deep-water dispersion relation. For finite depth, it is convenient to use

$$Z(z) = C_1 \sinh kz + C_2 \cosh kz, \tag{5.24}$$

so that application of the boundary condition at $z = -h$ provides a condition relating C_1 and C_2 as

$$C_1 = C_2 \frac{\sinh kh}{\cosh kh}. \tag{5.25}$$

This leads to

$$Z_2(z) = C_2 \left[\frac{\sinh kh \sinh kz}{\cosh kh} + \frac{\cosh kh \cosh kz}{\cosh kh} \right] = C_2 \frac{\cosh k(z+h)}{\cosh kh}. \tag{5.26}$$

Inserting this into the free-surface condition at $z = 0$,

$$k \tanh kh - \frac{\omega^2}{g} = 0; \Rightarrow \omega^2 = gk \tanh kh. \tag{5.27}$$

The latter part is the finite-depth dispersion relation for the waves under study, and is the more general form wherein no assumptions as to the depth (more precisely, the kh value) are implied. We can now write the entire solution for ϕ for the right-propagating wave:

$$\phi(x, z; t) = C \frac{\cosh k(z+h)}{\cosh kh} \mathrm{e}^{-i(kx - \omega t)}, \tag{5.28}$$

for finite depth h. For infinite depth $h \to -\infty$,

$$\phi(x, z, ; t) = C\mathrm{e}^{kz} \mathrm{e}^{-i(kx - \omega t)}. \tag{5.29}$$

Recall that z is positive upward. It is seen that ϕ is periodic in x and t. The period in time units is $T = 2\pi/\omega$, while the wavelength is $\lambda = 2\pi/k$. The constant k from Equation (5.19) thus ends up becoming a physically meaningful quantity, namely, the wave number. The dispersion relations connect the "spatial frequency" or wave number k with the "temporal frequency" ω.

The wave speed, that is, the speed with which this harmonic wave propagates (note that it is just the points on the wave in the same phase of motion, most visibly the wave crests, that are seen to move at this speed), is simply

$$v_c = \frac{\omega}{k}. \tag{5.30}$$

Given the dispersion relations (5.23) and (5.27), we see that v_c, the "phase velocity," is not independent of ω and k. In fact, the lower frequencies (or longer periods) corresponding to lower k (and longer wavelengths) propagate faster and overtake the higher frequencies (shorter waves) in a group, so that a wave group comprised of many frequencies propagates at a "group velocity" v_g,

$$v_g = \frac{d\omega}{dk} = \frac{1}{2}\frac{\omega}{k}, \tag{5.31}$$

in deep water. A similar expression can be derived for finite depth, by differentiating both sides of the dispersion relation in Equation (5.27). After a little algebra, this is found to be

$$v_g = \frac{1}{2}v_c\left(1 + \frac{2kh}{\sinh 2kh}\right). \tag{5.32}$$

It can be verified, by computing the limit ($\lim_{kh\to\infty} v_g$) in Equation (5.32) (applying L'Hôpital's rule in the process), that in deep water,

$$v_g = \frac{1}{2}v_c. \tag{5.33}$$

In other words, the individual waves in a group travel two times faster than the group as a whole. At the other extreme on the depth scale, as $kh \to 0$, it can be observed (by directly plotting the function, for instance) that $\tanh kh \to kh$. Substitution of this approximation in Equation (5.32) in the limit $kh \to 0$ leads to

$$v_g = v_c. \tag{5.34}$$

That is, individual waves in a group travel at the same speed as the entire group. The speed at which they propagate can be found by letting $\tanh kh \to kh$ in Equation (5.27) and using Equation (5.30):

$$v_g = v_c = \sqrt{gh}. \tag{5.35}$$

We note that this speed is independent of wave frequency and wave number. This means that a short pulse-like disturbance on the water surface will travel without losing its shape, that is, without "dispersing" into a group of crests and troughs. This property is characteristic of nondispersive waves such as acoustic waves in air, electromagnetic waves, and wave patterns on ideal stretched strings. The relation (5.34) shows that waves satisfying the assumption $kh \to 0$ are nondispersive. The following heuristic criterion enabling the use of deep- or shallow-water relations above has been used [67]: if $\lambda < h/2$, that is, $kh > 4\pi$, the deep-water assumption can be used. If $\lambda > 10h$, that is, $kh < \pi/5$, then the shallow-water relations are applicable. Anywhere in between, that is, $h/2 < \lambda < 10h$, or $4\pi < kh < \pi/5$, the full dispersion relation should be applied. A number of other criteria have also been examined in the literature [16]. The discussion above is relevant to control, as some of this understanding will guide a practical choice of the up-wave distance(s) at which wave profile measurements should be made in order to approach full impedance matching control in the time domain. The dispersive nature of waves has consequences for propagation of irregular waves in the ocean. We examine this briefly in Section 5.3 later in this chapter.

Meanwhile, also of relevance to control, the wave surface elevation $\eta(x;t)$ can be expressed as

$$\eta(x;t) = A\mathrm{e}^{-i(kx-\omega t)} \tag{5.36}$$

where A is the wave amplitude. In a study of irregular waves, it will be helpful to associate a phase directly with the amplitude A in a description such as Equation (5.36). Then the amplitude A will be interpreted as a complex quantity dependent on frequency and inclusive of the phase of that frequency component, as in $A(i\omega)$. While the focus is on a regular wave here, it should be noted that the actual surface elevation is $\zeta(x;t) = \Re\left[\eta(x;t)\right]$. It is useful to continue with the complex representation through the next steps, however, as long as the real parts of the final results are used. We return to the kinetic free surface condition

$$\frac{\partial\phi}{\partial t} + g\eta = i\omega\phi + g\eta = 0 \tag{5.37}$$

to express

$$\eta = \frac{-i\omega}{g}C\mathrm{e}^{-i(kx-\omega t)} = A\mathrm{e}^{-i(kx-\omega t)} \tag{5.38}$$

leading to

$$C = \frac{-g}{i\omega}A = \frac{igA}{\omega} \tag{5.39}$$

and

$$\phi(x,z;t) = \frac{igA}{\omega}\mathrm{e}^{kz}\mathrm{e}^{-i(kx-\omega t)} \tag{5.40}$$

for deep water. For finite depth, this solution takes the form

$$\phi(x,z;t) = \frac{igA}{\omega}\frac{\cosh k(z+h)}{\cosh kh}\mathrm{e}^{-i(kx-\omega t)}. \tag{5.41}$$

The particle velocities v_x and v_z in deep water can be found as

$$\begin{aligned} v_x &= \frac{\partial\phi}{\partial x} = \frac{gkA}{\omega}\mathrm{e}^{kz}\mathrm{e}^{-i(kx-\omega t)}, \\ v_z &= \frac{\partial\phi}{\partial z} = \frac{igkA}{\omega}\mathrm{e}^{kz}\mathrm{e}^{-i(kx-\omega t)}, \end{aligned} \tag{5.42}$$

indicating that the two components v_x and v_z are $\pi/2$ radians out of phase but have the same amplitudes. This essentially illustrates that in deep water the particle motion is on a circular trajectory. In addition, we can now see that particle velocities drop exponentially as water depth increases. Thus, surface wave influence becomes negligible below a certain depth at which e^{kz} reasonably approaches zero.

The energy propagated by a wave is often expressed in units of power (rate of energy propagation). As shown by a number of authors (e.g., [6, 8, 16]), the average energy density inside a regular wave of amplitude $A(\omega)$ at frequency ω is

$$\overline{e} = \frac{1}{2}\rho g A^2(\omega). \tag{5.43}$$

The energy content per unit crest length for a single wave of wavelength λ is simply $\bar{e}\lambda$. However, the rate at which the energy in a wave propagates is in units of power, and the mean rate across unit crest length is given by [6]

$$P_w = \bar{e} v_g \tag{5.44}$$

where v_g is just the group velocity. In deep water, P_w can be expressed in units such as W/m or kW/m, as

$$P_w = \frac{\rho g^2 A^2(\omega)}{4\omega}. \tag{5.45}$$

This concept generalizes to irregular waves, and the rate of energy propagation can be expressed using statistical forms representative of the quantities $A(\omega)$ and ω in Equation (5.45).

5.2 Description of Irregular Waves

The treatment in Section 5.1 assumes small wave slopes and results in linear equations for the overall dynamics. One advantage of this treatment is that any linear superposition of different regular waves, each as derived in Section 5.1, is also a solution to the hydrodynamic equations described therein.This is the mathematical basis for using a linear sum of a large number of wave components of different frequencies, amplitudes, and phases to represent irregular waves as occurring in the ocean. Reference may be made to the considerable published work in this area, notably, [68, 69, etc.]. An excellent discussion of the methods used to understand irregular waves (for the purpose of determining the response of structures operating in such waves) is provided in [67]. Following [67], if we consider just the surface undulations in long-crested waves at a single arbitrary location $(x_0, 0)$, and define the real part of $\eta_i(x_0, t)$ from Equation (5.36) as $\zeta_i(t)$, the surface profile at the chosen point may be represented as

$$\zeta(t) = \sum_{i=1}^{N} \zeta_i(t) \tag{5.46}$$

where $N \to \infty$, and each ζ_i can be expressed as

$$\zeta_i(t) = A_i \cos(\omega_i t + \theta_i) \tag{5.47}$$

where A_i is the amplitude, ω_i is the frequency of the wave component, and θ_i is the phase. One condition on the various ω_i is that none of them be an integer multiple of some fundamental frequency. Otherwise the waveforms may not be statistically independent. When they are statistically independent, the time average over a long enough duration of the product $\zeta_i \zeta_j$ summed over all N:

$$\begin{aligned} \overline{\zeta_i \zeta_j} &= 0, \quad i \neq j \\ &= \frac{1}{2} A_i^2, \quad i = j. \end{aligned} \tag{5.48}$$

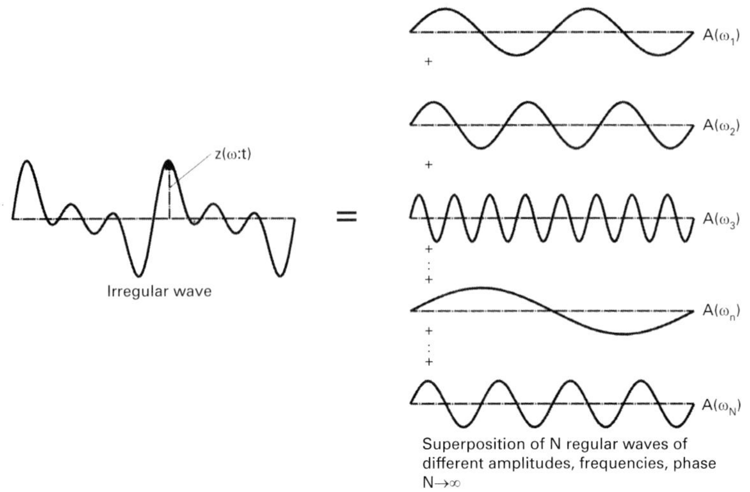

Figure 5.2 An illustration showing how an irregular wave may be considered as a superposition of a large number of regular wave components of different amplitudes and frequencies combining at different phases. Note that such a description draws on the idea of the "power spectral density," discussed in Section (5.2.3).

We suppose thus that the wave profile in Equation (5.46) is a net result of wind blowing over different regions. Further, we note that the various components are being added linearly to form the $\zeta(t)$, so that neither nonlinear interactions among waves nor Stokes-type steeper waves are considered here. As an example, Fig. 5.2 illustrates how different regular wave components may be added together to represent an irregular wave input.

Under the conditions outlined above, as $N \to \infty$, the probability density function for ζ at the chosen point x_0 approaches a Gaussian normal distribution defined as

$$p(\zeta) = \frac{1}{\sqrt{2\pi}\sigma} \exp\left[-\frac{1}{2}\left(\frac{\zeta}{\sigma}\right)^2\right] \tag{5.49}$$

where the time averages over a long enough time

$$\begin{aligned} \frac{1}{N}\sum_{i=1}^{N} \overline{\zeta_i} &= 0, \\ \frac{1}{N}\sum_{i=1}^{N} \overline{\zeta_i \zeta_j} &= \frac{1}{N}\sum_{i=1}^{N} A_i^2 \equiv \sigma^2. \end{aligned} \tag{5.50}$$

The first of Equations (5.50) implies that the mean value is zero, while the second defines the variance. Then, the probability that the circular particle motion under a wave will have an amplitude between A and $A + dA$ is

$$p(A) = \frac{A}{\sigma^2} \exp\left[-\frac{A^2}{2\sigma^2}\right]. \tag{5.51}$$

This is the Rayleigh distribution, describing the probability density of occurrence of a certain wave amplitude A at the point of interest x_0.

5.2.1 Probability Density Functions

Equation (5.51) provides useful information, but further progress is needed before we have a technique to relate statistical descriptions of ocean waves to statistical descriptions of the response of floating bodies and wave energy devices. Such progress is enabled by a further study of probability density functions (e.g., see [52]). Briefly, at the chosen point $x = x_0$, if the probability of finding a surface elevation ζ at time t is $p(\zeta; t)$, then a joint probability density function (or distribution) giving the probability of finding a wave profile ζ_1 at t_1 and ζ_2 at t_2 can also be defined, as $p(\zeta_1, \zeta_2; t_1, t_2)$. Similarly, and more generally, the probability that the surface elevation may be $\zeta_1, \zeta_2, \ldots, \zeta_n$ at times $t_1, t_2, \ldots, t_n$ is given by the joint distribution $p(\zeta_1, \zeta_2, \ldots, \zeta_n; t_1, t_2, \ldots, t_n)$. Now, the mean or "expectation" or "expected value" of wave elevation at the point x_0 and time t can be found as

$$\mu(t) = E\{\zeta(t)\} = \int_{-\infty}^{\infty} \zeta p(\zeta; t) d\zeta \tag{5.52}$$

with the mean square given by

$$E\{\zeta^2(t)\} = \int_{-\infty}^{\infty} \zeta^2(t) p(\zeta; t) d\zeta. \tag{5.53}$$

The variance can be defined as

$$\sigma^2(t) = E\{[\zeta(t) - \mu(t)]^2\} = \int_{-\infty}^{\infty} [\zeta(t) - \mu(t)]^2 p(\zeta; t) d\zeta. \tag{5.54}$$

We can derive an idea of how "correlated" the different ζ_i are by setting up a correlation function, such as

$$\varrho(t_1, t_2) = E\{\zeta(t_1)\zeta(t_2)\} = \int_{-\infty}^{\infty} \zeta_1 \zeta_2 p(\zeta_1, \zeta_2) d\zeta_1 d\zeta_2. \tag{5.55}$$

For a total of N ζ_i values obtained at N different times t_i, an $N \times N$ correlation matrix can be formed with elements $\varrho(t_i, t_j)$. If the correlation matrix is diagonal, the different ζ_i, ζ_j, $i \neq j$, are uncorrelated. Recall that correlation functions are often used in "template matching" algorithms to identify true objects or signals embedded within noisy image or measurement data.

5.2.2 Stationarity and Ergodicity

Our goal here is to derive convenient and compact but physically meaningful statistical descriptions. Hence, we need to narrow our field of view somewhat. In particular, we are here helped by a frequent practice in work related to ocean waves. This consists of making some assumptions regarding the variability (or absence thereof) of the statistical measures that describe them. For instance, it is customary to assume that the probability density functions above are invariant to translation in time. Thus, the waves themselves are of course changing, but the probability density functions that describe their changes within individual records are not changing from one record to another. In other words,

$$\begin{aligned} p(\zeta_i; t) &= p(\zeta_i; t+T), \\ p(\zeta_i, \zeta_j; t_i, t_i, t_j) &= p(\zeta_i, \zeta_j; t_i + T, t_j + T). \end{aligned} \tag{5.56}$$

In this circumstance, ζ is said to be a stationary process. Strictly speaking, this needs to be true for any T. However, it is reasonable to suppose that these relations will hold for records acquired over about 10–30 min at most sites. Nearby wind patterns can challenge or restrict such an assumption to shorter durations, however. The relation in Equation (5.56) implies that the quantities μ, σ, and ρ obtained for N samples starting at $t = 0$ will be the same as those evaluated for another N samples starting at $t = T$. Conceivably, we could then record a single sample continuously over a long time duration $T \to \infty$ and compute the mean, variance, and so on over that sample using

$$\begin{aligned} \hat{\mu} &= \lim_{T\to\infty} \int_{-T/2}^{T/2} \zeta(t)dt, \\ \hat{\sigma}^2 &= \lim_{T\to\infty} \int_{-T/2}^{T/2} [\zeta(t) - \mu]^2 dt, \\ \hat{\varrho}(\tau) &= \lim_{T\to\infty} \int_{-T/2}^{T/2} \zeta(t)\zeta(t+\tau)dt. \end{aligned} \tag{5.57}$$

A stationary process is also ergodic if the "statistics" in Equation (5.57) are no different from those found using Equations (5.52), (5.54), and (5.55), or when

$$\hat{\mu} = \mu; \quad \hat{\sigma}^2 = \sigma^2; \quad \hat{\varrho}(t, t+\tau) = \varrho(\tau). \tag{5.58}$$

Our understanding below is based on the assumption that over the time-scales of interest (10–30 min), ocean waves are both stationary and ergodic.

5.2.3 Power Spectral Density

The correlation function $\varrho(\tau)$ for such processes is the product of two values at different times (different by τ) of the same signal, and hence is often referred to as "autocorrelation." The autocorrelation is a function defined in the time domain. A special significance is attached to its Fourier transform $S(\omega)$, defined as

$$S(\omega) = \int_{-\infty}^{\infty} \rho(\tau) e^{i\omega\tau} d\tau. \tag{5.59}$$

The reverse relation can be expressed as an inverse Fourier transform,

$$\varrho(\tau) = \frac{1}{2\pi} \int_{-\infty}^{\infty} S(\omega) e^{i\omega\tau} d\omega. \tag{5.60}$$

$S(\omega)$ provides a measure of the "power" associated with each frequency ω, bringing us much closer to the property of waves of most interest to us. It reflects the contribution of each frequency to the wave record $\zeta(x_0, t)$ used to find the autocorrelation $\varrho(\tau)$ and then to $S(\omega)$.

We should note that, for a random process such as $\zeta(t)$ defined over $-\infty < t < \infty$, for which the mean-square value is nonzero, the Fourier transform of the surface elevation $\zeta(t)$ itself does not exist. However, the correlation function $\varrho(t)$ is Fourier transformable [51]. $S(\omega)$ is real and positive and is referred to as the power spectral density, or simply spectral density, and has dimensions $\sim \mathrm{m}^2\mathrm{s}$. The time series $\zeta(x_0, t)$ can thus be considered a superposition of an infinite number of harmonic waves of amplitude $A(\omega)$ and frequency ω such that

$$A^2(\omega_0) = \lim_{\Delta\omega \to 0} \int_{-\omega_0 - \Delta\omega/2}^{\omega_0 + \Delta\omega/2} S(\omega) d\omega \tag{5.61}$$

for a chosen frequency ω_0.

We note, however, that the phase information associated with each $A(\omega)$ is lost in this process. A common procedure when using a power spectrum $S(\omega)$ as a starting point for time domain simulations is to assume the phase $\theta(\omega)$ to be a random number between 0 and 2π. The complex amplitude at each wave frequency is then expressed as

$$A(i\omega) = A(\omega) e^{i\theta(\omega)}. \tag{5.62}$$

For a range of frequencies $\omega_i, i = 1, \ldots N$, therefore, the amplitudes A_i and phases θ_i from Equation (5.62) can be used in conjunction with Equations (5.46) and (5.47) to construct an irregular wave record (time series) for a particular power spectrum $S(\omega)$. We note, however, that because θ_i is chosen randomly (but $\in [0, 2\pi]$), each time the time series $\zeta(t)$ is generated. This discussion thus provides a perspective on the statistical usefulness of harmonic wave analysis and, by extension, of the theories developed for linear regular waves only. This treatment is of particular value for devices with a response that can be approximated as linear, because in this case, the extension of regular wave results to irregular waves (as described by the spectral densities at particular sites) is somewhat straightforward.

It is interesting to see that for $\tau = 0$, Equation (5.60) becomes

$$\varrho(0) = \frac{1}{2\pi} \int_{-\infty}^{\infty} S(\omega) d\omega = \lim_{T \to \infty} \int_{-T/2}^{T/2} \zeta^2(t) dt \tag{5.63}$$

where the quantity to the right of the second equality is just the mean-square surface elevation. Thus, the area under the spectral density curve is 2π times $\varrho(0)$ or 2π times the mean-square elevation. On a related note, sometimes a one-sided rather than two-sided spectral density is used, as follows from

$$\varrho(0) = \frac{1}{2\pi} \int_{-\infty}^{\infty} S(\omega) d\omega = \frac{1}{2\pi} \int_{0}^{\infty} 2S(\omega) d\omega = \frac{1}{2\pi} \int_{0}^{\infty} \overline{S}(\omega) d\omega \tag{5.64}$$

where $\overline{S}(\omega) = 2S(\omega)$. Care needs to be taken in simulations or laboratory experiments to ensure that a consistent spectral density representation is used. For example, in the spectra examined below, a one-sided spectral density is used and the bar over $S(\omega)$ is dropped altogether. Fig. 5.3 shows a typical spectral density plot obtained via wave elevation

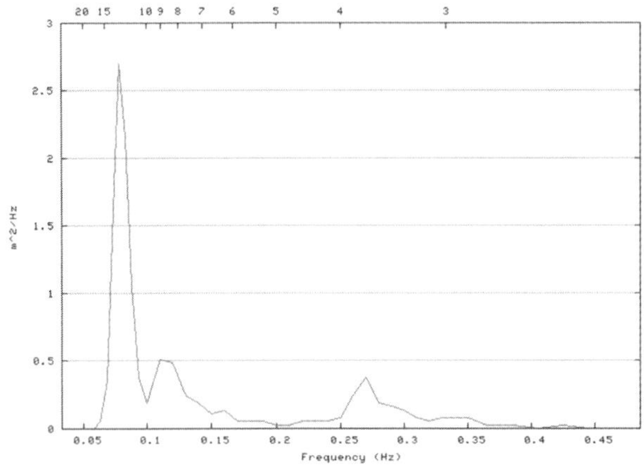

Figure 5.3 An example spectrum measured by a wave rider buoy deployed at station 46013 off the coast of Northern California. This buoy is part of a wider, worldwide network, and data such as this spectrum can be obtained from a website maintained by the National Data Buoy Center.

measurements with a wave rider buoy in the ocean. Often, however, for convenience, speed, and generality in analysis and simulation, it is customary to assume certain "standard" power spectral variations, with the particular choice guided by available approximate knowledge of the general area where the device is to be located. Examples include the Bretschneider spectrum, JONSWAP spectrum, and Pierson–Moskowitz spectrum [70]. A spectral representation frequently used is the so-called two-parameter spectrum, expressed as [6]

$$S(\omega) = \frac{\alpha g^2}{\omega^5} \exp\left[-\beta (g/V_w \omega)^4\right]. \tag{5.65}$$

Here ω denotes the wave frequency and V_w the wind speed. The spectral representation here is of a fully developed spectrum in that it is independent of the duration and distance over which the wind has been blowing. For particular parameter values $\alpha = 8.1 \times 10^{-3}$, and $\beta = 0.74$, Equation (5.65) represents the Pierson–Moskowitz spectrum.

In the treatment of irregular waves, it has been a practice to define a number of spectral moments, generally expressed as [67]

$$M_n = \int_0^\infty \omega^n \overline{S}(\omega) d\omega. \tag{5.66}$$

Many of the commonly referenced parameters used to describe irregular waves for design purposes can be related to the spectral density $S(\omega)$ via the spectral moments M_n, regardless of the particular $S(\omega)$ variation assumed or encountered at the device site.

The average time duration between successive zero crossings of the wave profile, or the "average zero-crossing period" is given by

$$T_z = 2\pi \sqrt{\frac{M_0}{M_2}}. \tag{5.67}$$

The average time period between successive crests, or the "average crest period" is

$$T_c = 2\pi\sqrt{\frac{M_2}{M_4}}. \tag{5.68}$$

The average wave height (crest to trough) can be found as

$$H_a = \sqrt{2M_0}. \tag{5.69}$$

A commonly used statistical measure of wave spectra is the significant wave height, or the average of the one-third highest waves observed over a length of time long enough to allow a spectral representation. This is simply

$$H_s = 4\sqrt{M_0}. \tag{5.70}$$

Also commonly used in wave energy is the "energy period" defined as the average period of energy propagation in a spectrum. This can be computed as

$$T_e = 2\pi\left(\frac{M_{-1}}{M_0}\right). \tag{5.71}$$

It is possible to express standard spectral representations such as the Pierson–Moskowitz spectrum directly in terms of the significant wave height H_s and energy period T_e (e.g., [70, 71]) as

$$S(\omega) = 131.5\frac{H_s^2}{T_e^4\omega^5}\exp\left[-1054/(T_e\omega)^4\right]. \tag{5.72}$$

The Pierson–Moskowitz spectrum represents unimodal spectra, that is, with single peaks. Actual spectra at the converter sites may be multimodal, with two or more peaks. Typically, in such spectra, the peaks at greater peak frequencies ($\sim$ 1 rad/s) represent the shorter, so-called wind seas more local in origin, while the peaks at lower frequencies ($\sim$ 0.6 rad/s) represent the longer swells, generated by distant high-wind/storm activity. In design, the so-called Ochi–Hubble spectra [70] are frequently used to model wave climates at sites where multiple peaks are common.

So far, our discussion has considered only unidirectional, or long-crested waves. The effects of directionality can be considered by means of an angular distribution, such as [7]

$$S(\omega, \theta) = \kappa S(\omega)\cos^n\theta \tag{5.73}$$

where $\theta \in (-\pi/2, \pi/2)$ represents the approach directions of the wave components. θ is generally thought to be independent of ω. $\kappa = 2/\pi$ when $n = 2$. $\cos^2\theta$ is the so-called Mitsuyasu spread function. The appropriate angular spread functions may be site dependent, but the $\cos^2\theta$ spread has frequently been used in laboratory testing and simulations.

5.3 Group Behavior of Waves

A fascinating aspect of the actual propagation of waves described using power spectra as above is the way they seem to propagate in "groups." This is largely a manifestation of the dispersive property of waves we briefly considered earlier. Specifically, seen from a structure floating or fixed in the ocean, several large waves may appear to approach in succession as a group, to be followed by several smaller waves. Wave grouping occurs on time scales relevant to energy conversion. This has consequences for a wave energy converter (WEC), in that there may be periods of successive large waves where energy conversion must shut down as the device goes into "survival mode", after which there would be periods of several successive smaller waves in which energy conversion can proceed and be maximized [72, 73].

The term "wave groupiness" has been used in the literature to describe the tendency of propagating waves to form groups. In general, the narrower a spectrum, the stronger the tendency to propagate in groups, or the groupier the waves. This is particularly relevant when we wish to answer a question such as: What is the probability that the next surface elevation will exceed a certain value? If we can convince ourselves of the groupiness in an approaching wave field, the answer to this question could reflect greater certainty than when all we can say is that the surface elevations follow a Gaussian probability distribution.

We consider the approach of Longuet-Higgins [69] here, which assumes that wave components are uncorrelated, as touched on in Section (5.2), with the mean surface elevation being zero (which is just the mean water level). Further, stationarity is also assumed over time-scales of about 30 min. A spectral width parameter ν may be defined as

$$\nu^2 = \frac{M_0 M_2}{M_1^2} - 1. \tag{5.74}$$

The mean square elevation $\overline{\zeta^2}$ and "mean frequency" may be expressed as,

$$\begin{aligned} \overline{\zeta^2} &= M_0, \\ \overline{\omega} &= M_1/M_0. \end{aligned} \tag{5.75}$$

The rth spectral moment about the mean frequency may be found using

$$\mu_r = \int_0^\infty (\omega - \overline{\omega})^r S(\omega) d\omega. \tag{5.76}$$

Then $\mu_0 = M_0$, $\mu_1 = 0$, and $\nu^2 = \mu_2/(\mu_0 \overline{\omega}^2)$. Further, a spectrum is thought to be narrow when $\nu \ll 1$, although the treatment below proves accurate up to $\nu \to 0.35$. In general, the "narrower" the spectra, the more accurate it is to utilize a complex envelope function to capture group behavior, and to arrive at suitable group statistics (since the envelope is meant to "trace" a group by simply connecting successive peaks or troughs

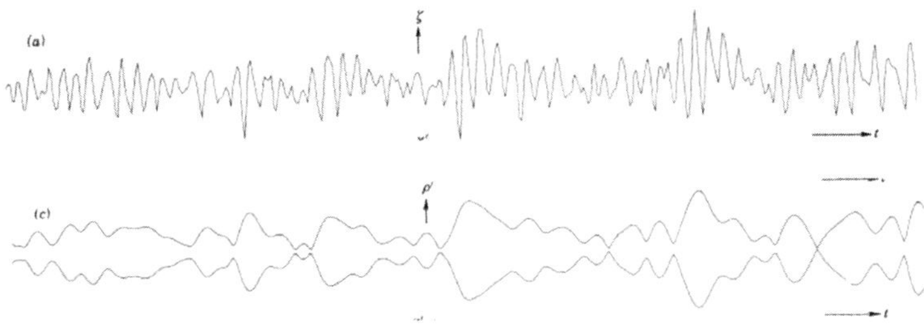

Figure 5.4 Example of an unfiltered wave surface elevation record and an envelope fitted to it to capture the group behavior [69]. Copyright: The Royal Society of London. Reproduced by permission.

in a record of elevation $\zeta(t)$; see Fig. 5.4). The wave–envelope complex amplitude is represented as $A(t) = \rho_e e^{i\phi_e}$, where

$$\rho_e \cos\phi_e = \sum_{i=1}^{\infty} A_i \cos\left[(\omega_i - \overline{\omega})t + \theta_i\right],$$

$$\rho_e \sin\phi_e = \sum_{i=1}^{\infty} A_i \sin\left[(\omega_i - \overline{\omega})t + \theta_i\right]. \tag{5.77}$$

The real and imaginary parts of the envelope thus form a Hilbert-transform pair $[\zeta(t), \eta(t)]$ [69]. Realizing that the previously used σ is (Equations (5.49) and (5.51)) is just $\sigma = \sqrt{M_0}$, the probability distribution function for the envelope function is

$$p(\rho_e) = \frac{\rho_e}{\mu_0} \exp\left[-\left(\frac{\rho_e^2}{2\mu_0}\right)\right] \tag{5.78}$$

which is also a Rayleigh distribution. Note that $\mu_0 = \sigma^2$. We can define a group as waves contained within a time interval that is bounded by successive up-crossings of the envelope of a certain level ρ_e. Now, the number N' of up-crossings of a level ρ_e can be found as

$$N' = \sqrt{\frac{\mu_2}{2\pi}}\, p(\rho_e). \tag{5.79}$$

The length of a group l can be defined as the time interval between two successive up-crossings of a level ρ_e. The mean value $\bar{l}$ is

$$\bar{l} = \frac{1}{N'}. \tag{5.80}$$

By suitably choosing the threshold level ρ_e, one can easily define a "quiescent period," during which "calmer" conditions can be expected to prevail. The number of up-crossings N' is maximum when $\rho_e = \sqrt{\mu_0}$. For narrow spectra with $\nu < 0.35$, the mean number of waves $\overline{G}$ in a group defined by up-crossings of level ρ_e of the envelope can be estimated as

$$\overline{G} = \frac{1}{\sqrt{2\pi}} \sqrt{\frac{M_2}{\mu_2}} \left(\frac{\sqrt{\mu_0}}{\rho}\right) \exp\left(\frac{\rho_e^2}{2\mu_0}\right). \tag{5.81}$$

This number reaches a minimum when there are most up-crossings, that is, when N' is maximum, which, as mentioned above, occurs when $\rho_e = \sqrt{\mu_0}$. In terms of the spectral width parameter, this minimum group length (in units of number of waves) is

$$\overline{G}_{mn} = \sqrt{\frac{\mathrm{e}}{2\pi}}\frac{\sqrt{1+\nu^2}}{\nu}. \tag{5.82}$$

Above, l and $\bar{l}$ are in units of time, while N' are in the units of "occurrences per unit time" (i.e., frequency in Hz), and therefore $\overline{G}$ and $\overline{G}_{mn}$ are in the units of "number of waves" (up-crossing a prescribed level) and e = 2.7182.

The same approach can be used to quantify the number of waves exceeding a high threshold, for example, at which the converter is to enter survival mode. The length of a "high run" in units of time $\bar{l}'$ can be found as

$$\bar{l}' = q\bar{l} \tag{5.83}$$

where

$$q(\rho_e) = \int_{\rho_e}^{\infty} p(\rho_e)d\rho_e = \exp\left(\frac{-\rho_e^2}{2\mu_0}\right). \tag{5.84}$$

To estimate the average number $\overline{H}$ of waves in a high run,

$$\overline{H} = \frac{1}{\sqrt{2\pi}}\sqrt{\left(\frac{M_2}{\mu_2}\right)}\frac{\sqrt{\mu_0}}{\rho_e}. \tag{5.85}$$

Note that $\overline{H}$ is inversely proportional to ρ_e, so that the higher the threshold, the fewer the waves up-crossing it. If the threshold is defined as $\rho_e = 2\sqrt{\mu_0}$, that is, equal to the significant wave amplitude $H_s/2$, then $\overline{H}$ can be expressed as

$$\overline{H}_s = 0.1995\sqrt{1+\nu^2}/\nu. \tag{5.86}$$

With ρ_e defined suitably for a particular device (the power take-off [PTO], device dimensions, etc. would determine this threshold), survival mode could be activated for a length of time estimated as above (see also [73]). Note that the probability that ρ_e will be exceeded can also be estimated as part of this process. As noted earlier, these statistics are subject to the envelope function being a good approximation for the delineation of wave groups. Thus they may provide useful information when $\nu \leq 0.35$.

5.4 Wave Power as Rate of Energy Propagation

The rate of energy propagation in units of power per unit crest length is useful in understanding the wave energy potential of coastal or offshore sites. If we stay with long-crested spectra here to keep the discussion more straightforward, this be found using the ideas expressed in Equations (5.43)–(5.45). In particular, realizing that

$$A^2(\omega) = \lim_{\Delta\omega\to 0}\int_{\omega-\Delta\omega}^{\omega+\Delta\omega} S(\omega)d\omega \tag{5.87}$$

and that theoretically ω ranges from 0 to ∞, the average rate of energy propagation can be expressed in power per unit crest length as

$$P_W = \rho g \int_0^{\infty} 2S(\omega) v_g(\omega) d\omega, \tag{5.88}$$

Equation (5.88) illustrates the significance of the spectral density function $S(\omega)$ to WEC design. Based on [67], the average rate of energy propagation per unit crest length in deep water is

$$P_W = \frac{1}{2}\rho g^2 M_{-1}. \tag{5.89}$$

Using the definitions for the energy period T_e and the significant wave height H_s, this becomes

$$P_W = \frac{\rho g^2}{64\pi} H_s^2 T_e. \tag{5.90}$$

In units of kW/m, the power carried per unit crest length by a particular wave spectrum with significant wave height H_s and energy period T_e can be expressed by evaluating the fraction above as

$$P_W = 0.49 H_s^2 T_e \tag{5.91}$$

where ρ is the sea-water density (1.025×10^3 kg/m^3).

In practical design applications for WECs, it is often useful to show the power per unit crest length in the form of a "scatter diagram," where the incident power in kW/m is indicated within a grid pattern, with T_e forming the x-axis of the grid, and H_s forming the y-axis. "Bins" corresponding to particular combinations of T_e and H_s contain the corresponding power values averaged over a certain time period. The entries in the scatter diagrams thus change with location. Directly derived from the scatter diagrams are the "power matrices" depicting the energy capture performance of wave energy devices. These have the same horizontal and vertical axes as the scatter diagrams, but the entries in the bins contain the actual absorbed power, or the actual energy capture length. What lies between the scatter diagrams and the power matrices is the entire device, including the primary energy converter, the power conversion chain, and any control strategies. The power matrix provides the necessary information for an economic performance calculation and enables decisions on the relative merits of different sites, devices, or control techniques.

For devices whose dynamics can be assumed to be linear, the information from regular wave analyses can be used in a straightforward manner to obtain the required performance measures for irregular waves described by any spectra. The following approach could be used for that purpose.

5.5 Device Response in Irregular Waves

The notion of power spectral density of the wave input can easily be translated to a power spectral density describing an output of a linear system driven by the waves.

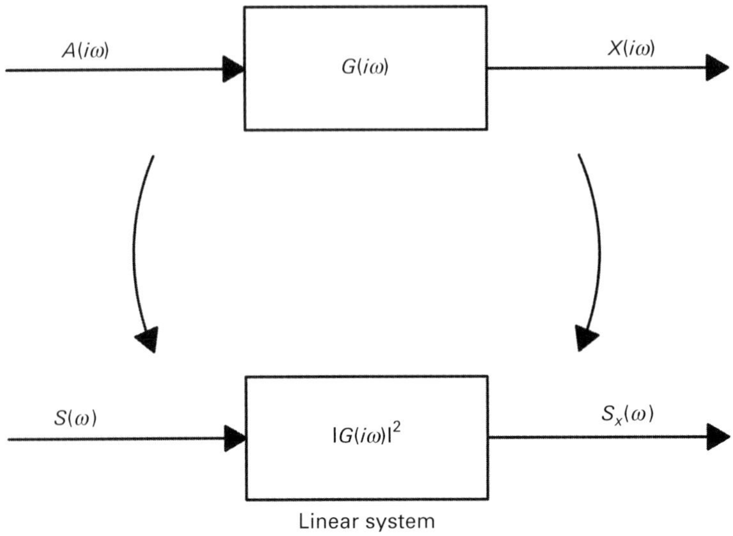

Figure 5.5 Block diagram illustration of the approach used to obtain the response spectral density for body oscillation in irregular waves.

If we use frequency domain analysis throughout and recognize that amplitude response is all we can find this way (the phase information being sacrificed to the convenience of the spectral treatment), then the output of a WEC (approximated as a linear system) can be related to any input wave spectrum using the following approach. Although illustrated below for a single mode converter, the approach can easily be generalized to multiple-mode devices using matrix techniques [52].

If $G_d(i\omega)$ denotes the frequency response function for the displacement response of a single-mode device such that (see Fig. 5.5)

$$G_d(i\omega) = \frac{1}{-\omega^2[m + \overline{a}(\infty) + a(\omega)] + i[b(\omega) + D + c_d] + k}, \tag{5.92}$$

c_d denotes the linearized damping constant due to viscosity, with other device parameters having been defined before. Thus,

$$X(i\omega) = G_d(i\omega)F_f(i\omega). \tag{5.93}$$

Note that no control is applied here and D is the load damping applied by the PTO. The exciting force is related to incident wave amplitude $A(i\omega)$ through another frequency response function, such that

$$F_f(i\omega) = H_f(i\omega)A(i\omega). \tag{5.94}$$

Combining Equations (5.92)–(5.94),

$$X(i\omega) = G(i\omega)A(i\omega). \tag{5.95}$$

where

$$G(i\omega) = G_d(i\omega)H_f(i\omega). \tag{5.96}$$

Up to this point, both amplitude and phase of the displacement $X(i\omega)$ are known. However, the treatment so far applies only to regular waves. When multiple regular wave components combine to form an irregular wave input, if we wish to continue in the frequency domain, the amplitude representation essentially gives way to power spectra with the frequency components appearing at random phases. Multiplying both sides of Equation (5.95) by their respective complex conjugates,

$$X^*(i\omega)X(i\omega) = A^*(i\omega)G^*(i\omega)G(i\omega)A(i\omega). \tag{5.97}$$

Though the reasoning through the following step is not straightforward (but is presented in texts such as [52]), for a single mode system Equation (5.97) can be extended to relate power spectral densities according to

$$S_X(\omega) = |G(i\omega)|^2 S(\omega) \tag{5.98}$$

where $S_X(\omega)$ denotes the spectral density of the displacement, and $S(\omega)$ is the spectral density of the incident irregular waves defined previously. Fig. 5.5 illustrates the process above leading to the spectral density for body oscillation in irregular waves. The function $|G(i\omega)|^2$ defines the square of the response amplitude operator for the body.

We point out again that since the phases associated with frequency components are thought to be random in the reasoning leading up to $S(\omega)$, the original phase information is not preserved, and the phases associated with the spectral density $S_X(\omega)$ are also random. Thus, while useful for generating power matrices, this treatment is of little use in real-time wave-by-wave control in irregular waves, the phase of the displacement relative to the exciting force being central to efficient absorption, as we argued in the earlier chapters. We examine the question of amplitudes and phases relative to the wave input further in the following chapter, with particular attention to how frequencies, device geometries, PTO loads, and control combine to determine the overall response and, ultimately, energy conversion. We will see how interesting the answers tend to be, and how dependent the overall performance of a device is, on the manner in which it responds to waves. As most of the treatment of the following chapter is in the frequency domain, the group behavior of waves is only indirectly accounted for. Real-time control techniques of subsequent chapters tacitly focus on wave groups comprised of small/moderate waves, and though this is not explicitly stated, we assume that a supervisory control system driven by wave group statistics is in place, which can potentially allow switching between conversion and survival modes.

5.6 Conclusion

In this chapter, we discussed the work environment in which a WEC operates. It was observed that the use of statistical quantities based on probability density functions enabled convenient descriptions such as power spectral densities that allowed a straightforward extension of the frequency domain analysis techniques. Thus, an understanding of the device operation in sinusoidal waves can be extended without too much effort to a clear assessment of their performance in real-life ocean waves. As a caveat, the entire analysis structure is based on linear theory, so nonlinear wave–wave interactions and nonlinear device dynamics are outside the scope of this discussion. We also examined statistical techniques that utilize the group behavior of waves to learn more about the waves than we could otherwise. Wave group statistics also allow short-term predictions of the wave input in statistical terms and are potentially important to wave energy device operation. As for control of device hydrodynamic response, either one could work with statistical measures all the way through the development of a controller or one could use a more deterministic approach. Either approach has its merits and limitations in terms of the resulting energy-conversion gains as well as the extra effort needed on different levels. In a very broad sense, the difference amounts to control on a wave group by wave group basis or control on wave-by-wave basis. In the following chapter, however, our interest is in understanding the hydrodynamic interactions driving the way a device responds to waves. In the subsequent chapters thereafter, we consider different ways to extend the hydrodynamic understanding so that devices perform efficiently in a range of wave climates.

6 A Closer Look at Wave Energy Hydrodynamics

It is the way the oscillating elements and their references respond to each approaching wave (however different from the one before) that determines how much of its energy the device will ultimately absorb. No matter how crucial to energy generation the power take-off (PTO) or the secondary converter may be, how important the generation equipment, how serious the difficulties from a defective mooring, how energy-rich the site of operation, how clever the control system design, it is often the action of the primary energy converter that determines the economic cost competitiveness of a wave energy device. Much of our discussion here therefore concerns the interaction between the primary converter and approaching waves. We cover the whole range, from the oscillation amplitude and the extent to which this oscillation lags behind or leads the surface undulations supplying the energy, to the forces and moments imparted on the device by the wave field. We are particularly interested in the role that the shape and size of the device play in making the best use of the ceaseless rhythmless dialog that describes the work life of a wave energy converter (WEC).[1]

We study what geometries and or motion combinations make devices better at energy absorption, and try to understand precisely where any contributions of control would fit within the "big picture," as defined by the larger wave-body dynamics. Our goal is to apply control such that the best energy absorption performance enabled by a particular geometry or a particular combination of device oscillations is achieved and sustained through changing wave fields. Our focus in this chapter is therefore on learning the insights provided to us by mathematicians, physicists, and engineers over the last four decades. We will see that, often, most of these insights are most clearly expressed in mathematical terms. While existing quantitative understanding of ship hydrodynamics and floating bodies serves as a starting point, important differences unique to wave energy devices are that the response must often be designed to be different for different elements of a jointed or articulated overall structure, and rather than moving forward steadily and/or minimizing motions through waves, the device should try to maintain its mean position and, when possible, respond in such a way that energy absorption

[1] Indeed, in the words of long-time wave energy researcher Jamie Taylor of Edinburgh University, "if it was possible to make a perfect wave energy converter, it might be compared to a graceful and clever dancer who only ever gets to work with a mad orchestra" [74].

from the surrounding and approaching wave field is maximized. It is the latter goal related to energy absorption that is our main concern here, and we will (1) quantify the importance of device geometry to its conversion efficiency with or without control, (2) understand the role hydrodynamic control can play for a chosen geometry, and (3) explain the hydrodynamic basis for some of the statements and inferences in Chapter 4.

Once again, we stress that our best understanding currently is given by linear theory with the associated applicability restrictions to small-amplitude waves and small-amplitude oscillations. Linear theory works reasonably well in most energy conversion situations, however, and can, at any rate, provide useful insights on which an interpretation of more complex nonlinear phenomena could be based. Finally, note that the discussion below is intended to provide a wider design/control perspective. It is not our intent to frustrate the reader by providing parameter definitions without giving a working recipe for actual numerical evaluation but, rather, to suggest that insights into the global behavior of some of these parameters can often "light the way" to successful designs. Closed-form analytical evaluations are possible for a few geometries under some approximations. Outside of such special cases, numerical evaluation using one of the standard software packages (referred to later in this chapter as well as in the later chapters of this book) or use of published results, if available, is recommended.

6.1 A Body in Waves

To understand the oscillatory motion of a body in waves, the forces and moments acting on it, and the energy absorbed by it, we begin by invoking the description of the small-amplitude propagating wave from Chapter 5. Specifically, we address (1) how the oscillating body influences the wave field and (2) how the wave field influences the body oscillations. Most wave energy devices "add" to the wave field near and far, and indeed, it is the nature of this very effect that holds the key to good energy absorption. As mentioned earlier, we employ a small-amplitude response approximation in addition to the small-amplitude wave approximation. These approximations enable extremely elegant and accurate descriptions of all of the hydrodynamic action occurring between a wave energy device and the wave field in which it works. The deeper power of linear theory-based arguments is in the way they lead to the device geometries and shapes that enable maximum absorption. Linear analysis also clarifies the way in which different oscillation modes can combine to enable maximum absorption. And importantly for us, our analysis here also provides fundamental insights into how device oscillations need to be controlled so that power absorption is maximized. To help with a basic-level understanding, we focus on a single-mode device for most of the discussion. Extensions to multiple-mode operation are often relatively straightforward with the use of matrix methods. On the understanding that the subscript i (where $i = 1, \ldots, 6$) denotes one of the six modes in which a floating body can oscillate (with $i = 1, \ldots, 3$ representing the three translational modes surge, sway, and heave, respectively, and $i = 4, \ldots, 6$ the three rotational modes roll, pitch, and yaw, respectively), our attention here is on a body

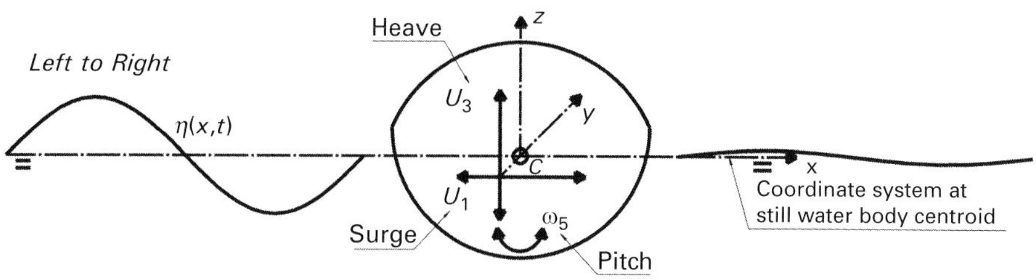

Figure 6.1 A body floating in regular waves. Waves are here thought to travel from left to right.

oscillating predominantly in just a particular translational mode $i = 1, 2, 3$. Fig. 6.1 illustrates a body floating in a regular wave:

$$m\ddot{x}_i + c_d\dot{x}_i + k_i x_i = f_i + f_L. \tag{6.1}$$

Here m denotes the in-air mass of the body. A secondary energy converter (PTO) based on a hydraulic cylinder or a linear motor/generator is assumed. The force f_L is applied on the oscillating body from a reference assumed immobile here, but in practice the reference could be a deeply submerged body, the sea floor, or a pier. f_L is chosen by the device operator and includes both the force responsible for power conversion and the force involved in controlling the device oscillations. The stiffness k_i can be assumed constant for a heaving body whose oscillation excursion spans a draft range over which the water-plane area is very nearly the same. k_i can be amended to include the stiffness effect of mooring lines, tension-leg supports, or any constant or linearized stiffness in the PTO mechanism. Note that for surge, sway, and yaw modes the hydrostatic stiffness will be zero, as no change results in the buoyancy force or moment during these oscillations. c_d is a constant here representing a linearized rate of energy loss through viscous friction.

The total wave force f_i is an important quantity. It is really the nature of this force, and the response of the body to this force, that we need to understand here. Excellent discussion on this force can be found in texts such as [5, 6, 8, 66], papers such as [20, 53], and others.

Briefly, a body large enough and close enough to the surface to be in a position to absorb a significant portion of energy from approaching waves can also significantly disturb the wave field. Further, as it oscillates in response to the waves, it continuously creates its own waves. These waves apply their own force on the body. This is where the small-amplitude wave/oscillation approximation adds value, in that linearity makes it possible for us to express the overall wave force as a linear superposition of the force as would be produced on a body if it were held fixed in the approaching wave field (the "diffraction force") and the force that would be produced on the body by the waves it would make if it were to oscillate on a calm surface (the "radiation force"). The diffraction force is frequently also called "exciting force." Thus

$$f_i(t) = f_{fi}(t) + f_{Ri}(t). \tag{6.2}$$

With $\mathbf{r}$ denoting the position vector defined by $[x, y, z]$ in the Cartesian (or $[r, \theta, z]$ in the cylindrical polar) coordinate system, the velocity potential $\phi(\mathbf{r}, t)$ describing the fluid action here can be considered a linear superposition of the diffraction potential $\phi_D(\mathbf{r}, t)$ and the radiation potential $\phi_R(\mathbf{r}, t)$. ϕ_D represents the effect of the original incident wave potential together with the disturbance to the wave field introduced by the body when held fixed in the fluid. ϕ_R represents the velocity potential produced by body oscillation in an otherwise calm fluid. Thus ϕ_D and ϕ_R must satisfy the Laplace equation and the free-surface condition, as well as the sea-floor condition. However, for ϕ_D to represent the velocity potential with the body held fixed, it must satisfy the condition that no fluid can flow through or across the immersed surface of the body, or that

$$\nabla \phi_D \cdot \mathbf{n} = \frac{\partial \phi_D}{\partial n} = 0 \tag{6.3}$$

where $\mathbf{n}$ is the outward-pointing unit vector normal to the immersed surface at some point (x, y, z) on the immersed surface. On the other hand, because ϕ_R is produced by body oscillation, the fluid particles adjacent to the body surface at some (x, y, z) must move at the same velocity at which that point on the body moves (we can relate the velocity at a point on the body surface to the translational velocity of the body centroid and the angular velocity of the body around the centroid if we have quantitative information on the body geometry). Because linear superposition holds throughout, we can now define the overall velocity potential $\phi(\mathbf{r}, t)$ as

$$\phi(\mathbf{r}, t) = \phi_D(\mathbf{r}, t) + \sum_i u_i \phi_i(\mathbf{r}, t) \tag{6.4}$$

where ϕ_R is represented as a summation over the number of degrees of freedom relevant to the problem, u_i is the velocity in each, and ϕ_i is the radiation potential due to unit velocity in the ith degree of freedom. For instance, the boundary condition to be satisfied by each $\phi_i (i = 1, \ldots, 3)$ is then

$$\frac{\partial \phi_i}{\partial n} = n_i \tag{6.5}$$

where n_i denotes the linear velocity component along the surface normal $\mathbf{n}$ at (x, y, z) due to unit velocity in the ith degree of freedom. For a body in single mode heave oscillation, $i = 3$, $\sum_i u_i \phi_i(\mathbf{r}, t) = u_3 \phi_3$, and $n_i = n_3$ is the z-direction cosine of $\mathbf{n}$ at (x, y, z).

Finally, in order to make further progress, we need to make it explicit that the wave field created by the body oscillation, that is, that described by ϕ_R, and just the disturbance wave field introduced by the body being held fixed, that is, $(\phi_D - \phi_I)$, must both propagate away from the body.

Focusing now on harmonic waves (regular or mono-frequency), the different velocity potentials may be written in the form

$$\phi(\mathbf{r}; t) = \varphi(\mathbf{r}) e^{i\omega t}. \tag{6.6}$$

The total diffraction force (exciting force) on the body in the ith degree of freedom can be written as

$$f_{fi} = \iint_{S_B} p(\mathbf{r}; t) n_i dS \tag{6.7}$$

where p denotes the pressure acting at a given point on the immersed surface S_B of the body. Within linear theory (using Bernoulli's theorem with the nonlinear term ignored) for harmonic waves,

$$F_{fi} = i\omega\rho \iint_{S_B} \varphi_D(\mathbf{r}) n_i dS \tag{6.8}$$

where F_{fi} denotes the complex amplitude of f_{fi}. The force due to $\phi_R(\mathbf{r}; t)$ is what is referred to as the "radiation force," and for harmonic oscillation of the body with a velocity $u_i(t) = U_i(i\omega)e^{i\omega t}$ (where $U_i(i\omega)$ is velocity complex amplitude, containing both magnitude and phase information) the complex amplitude of the radiation force can be expressed as

$$\overline{F}_{Ri} = -b_i(\omega)U_i(i\omega) - i\omega\overline{a}_i(\omega)U_i(i\omega). \tag{6.9}$$

The coefficients on $U_i(i\omega)$ can be understood as "impedances," and since $\overline{f}_{Ri}$ is the radiation force, they are really the two parts of the "radiation impedance" of the body in the ith mode. For single-mode oscillation in the ith mode, we have the relations connecting the radiation impedance terms with the radiation potential [8]:

$$\begin{aligned} \overline{a}_i(i\omega) &= \rho \iint_{S_B} \Re(\varphi_i) n_i dS \\ b_i(\omega) &= \rho\omega \iint_{S_B} \Im(\varphi_i) n_i dS \end{aligned} \tag{6.10}$$

where $\Re$ and $\Im$ denote the real part and imaginary part, respectively, of the complex quantity φ_i. We note that $b_i(\omega)$ is a real-valued quantity and multiplies velocity $U_i(i\omega)$, and is therefore damping-like. It is the radiation damping coefficient, different at different frequencies, as pointed out in Chapter 4. $b_i(\omega)$ represents the rate at which the body radiates energy away to the "far field" with the waves it creates with its oscillation in mode i. The quantity $\overline{a}_i(\omega)$ is mass- or inertia-like, being attached to the acceleration $\omega U_i(i\omega)$. It represents the rate at which the body exchanges energy with the evanescent, or nonpropagating, waves it creates with its oscillation. These waves remain near the body and hence do not permanently take energy away from the oscillation. $\overline{a}_i(\omega)$ approaches a constant value $\overline{a}_i(\infty)$ (see [6] for a discussion on the relationship between $\overline{a}_i(\infty)$ and the added mass (inertia) coefficients in unbounded fluid for different modes). Specifically, for the heave, pitch, and roll modes,

$$\overline{a}_i(\infty) \to \frac{1}{2} m_{a_i}, \; i = 3, \; 4, \; 5, \tag{6.11}$$

where m_{a_i} on the right side is the added mass (inertia) in an unbounded fluid of a double body comprised of the immersed volume and its reflection about the waterline. In any case,

$$\lim_{\omega \to \infty} \overline{a}_i(\omega) \to \overline{a}_i(\infty). \tag{6.12}$$

$\overline{a}_i(\omega)$ is frequently termed the "frequency-dependent added mass (inertia)," with $\overline{a}_i(\infty)$ referred to as the "infinite-frequency added mass (inertia)." As pointed out in Chapter 4, the reason for separating out the infinite-frequency added mass (inertia) also goes back to the fact that, because $\overline{a}_i(\infty)$ is typically a nonzero constant, the inverse-Fourier transform of $\overline{a}_i(\omega)$ does not exist. On the other hand, $\overline{a}_i(\omega) - \overline{a}_i(\infty)$ is inverse-Fourier transformable as it approaches a zero value as $\omega \to \infty$. We find it notationally convenient to introduce a symbol $a_i(\omega)$ defined as just the frequency-variable part of $\overline{a}_i(\omega)$, where

$$a_i(\omega) = \overline{a}_i(\omega) - \overline{a}_i(\infty). \tag{6.13}$$

Returning now to the equation of motion (6.1), we see why it is only a true differential equation as long as we understand that the excitation is harmonic. In this case only, we can write

$$[m + \overline{a}_i(\infty) + a_i(\omega)]\,\ddot{x}_i + [b_i(\omega) + c_d]\,\dot{x}_i + k_i x_i = F_{fi}(\omega)\mathrm{e}^{i\omega t} + F_L. \tag{6.14}$$

In general, for bodies of arbitrary immersed shapes, it is not easy to compute the added mass and radiation damping (hydrodynamic coefficients), and the exciting force amplitude plus phase. Numerical (e.g., boundary element analysis) computations may be required [75]. Examples of established computational packages include Wamit, Hydran, [76–78], and others. For some geometries, however, under some simplifying assumptions, closed-form analytical solutions may be available [66, 79].

For irregular excitation, recall from Chapter 4 that we no longer can work with a differential equation such as above (there being many frequencies embedded in the input approaching at any time t) and must use the convolution representation for the radiation force.

We note that we find it convenient to represent the total radiation force $\overline{F}_{Ri}$ in harmonic waves as

$$\overline{F}_{Ri}(i\omega) = -F_{Ri}(i\omega) - i\omega\overline{a}_i(\infty)U_i(i\omega), \tag{6.15}$$

and to let F_{Ri} represent just the frequency-variable effects, as

$$F_{Ri}(i\omega) = [i\omega a_i(\omega) + b_i(\omega)]\,U_i(\omega) \equiv H_{Ri}(i\omega)U_i(\omega). \tag{6.16}$$

As seen earlier, the equation of motion for the ith mode for irregular excitation can be written as

$$[m + \overline{a}_i(\infty)]\,\dot{u}_i + \int_0^{\infty} h_r(\tau)u_i(t-\tau)d\tau + c_d u_i + k_i \int_{-\infty}^{t} u_i(\tau)d\tau = f_{fi}(t) + f_L(t). \tag{6.17}$$

Alternatively,

$$[m + \overline{a}_i(\infty)]\,\ddot{x}_i + \int_0^{\infty} h_r(\tau)\dot{x}_i(t-\tau)d\tau + c_d\dot{x}_i + k_i x_i = f_{fi}(t) + f_L(t) \tag{6.18}$$

where $h_r(t)$ is the radiation impulse response function previously discussed and can be found via a cosine transform of $b_i(\omega)$ or a sine transform of $a_i(\omega)$, as discussed in Chapter 4. We recall also that the radiation damping $b_i(\omega)$ and the frequency-variable added mass $a_i(\omega)$ are interrelated via the Kramers–Kronig relations discussed in Chapter 4. As also discussed therein, the convolution form can also be used for the exciting force variation in terms of the surface elevation at the device location (typically the centroid). Our task here, however, is to try to gain some insight into, and appreciation of, device behavior, through a further discussion of the hydrodynamics underlying device response to regular waves. We should stress that, while the integro-differential equation model is essential for wave-by-wave control in irregular waves, the frequency domain representation does provide a clarity and understanding that could prove vital to the ultimate success of a device and any control strategy developed.

We briefly return to the velocity potentials to mention that the use of Green's second identity [60], which relates functions within a fluid domain to their variations over the fluid boundaries, leads to a number of findings of consequence. The functions have to be only twice-differentiable. Thus for two such functions f and g,

$$\iiint_V \left(f\nabla^2 g - g\nabla^2 f\right) dV = \iint_S \left(f\frac{\partial g}{\partial n} - g\frac{\partial f}{\partial n}\right) dS. \tag{6.19}$$

Since both φ_D and φ_R satisfy the Laplace equation, with the help of the boundary conditions (6.3) and (6.5) and the fact that $\varphi_D - \varphi_I$ and φ_R are outgoing far from the body, an application of Green's second identity leads to

$$F_{fi} = -i\omega\rho \iint_{S_\infty} \left(\varphi_I \frac{\partial \varphi_i}{\partial n} - \varphi_i \frac{\partial \varphi_I}{\partial n}\right) dS. \tag{6.20}$$

S_∞ represents a cylinder at infinity surrounding the body, where ∞ simply implies far from the body or in the far field. Equation (6.20) reveals to us the remarkable fact that the exciting force in the ith mode on the surface of the body is expressible entirely in terms of the radiation potential in the ith mode in the far field and the incident potential far from the body [80]. Equation (6.20) is the so-called Haskind (or Haskind–Hanaoka) relation [5].

6.2 Beam-Sea Devices

For two-dimensional bodies, we find that Equation (6.20) results in a particularly tidy expression for the exciting force in terms of the far-field amplitude of the radiated wave $\mathcal{A}_{UW}(i\omega)$ where the subscript UW denotes the "up-wave" direction, which is the negative x direction for waves traveling left to right in the positive x direction. Note that the up-wave and down-wave radiated amplitudes are for unit velocity of body oscillation. We assume deep-water conditions for most of the discussion below. Based on [21], for crest length equal to device length L in the crest direction, the complex amplitude of the exciting force is related to $\mathcal{A}_{UW}(i\omega)$ according to

$$F_{fi}(i\omega) = \rho g A \mathcal{A}_{UW}(i\omega) L \tag{6.21}$$

where A is the incident wave amplitude. Most of the discussion here assumes that viscous damping is small enough to be neglected, or that $c_d = 0$. Next we recall that the radiation damping, which determines a part of the force due to the radiation potential, relates to the propagating wave created by body oscillation (while added mass relates to the nonpropagating evanescent wave field around the body). It is possible to express the radiation damping $b_i(\omega)$ directly in terms of the far-field radiated wave amplitudes as [53]

$$b_i(\omega) = \frac{1}{2}\rho\omega\left(|\mathcal{A}_{UW}(i\omega)|^2 + |\mathcal{A}_{DW}(i\omega)|^2\right)L. \tag{6.22}$$

Now, we know that the incident wave power over crest length L is $P_w L$, where per unit width,

$$P_w(\omega) = \frac{\rho g^2 A^2}{4\omega}. \tag{6.23}$$

With the help of a little algebra, we can see that the radiation damping for single-mode oscillation in the i direction can be expressed as

$$b_i(\omega) = \frac{|F_{fi}(i\omega)|^2}{8P_w L\gamma}. \tag{6.24}$$

It is remarkable that the exciting force on the body as arising from the diffraction potential φ_D and the radiation damping as arising from the radiation potential φ_i should be so directly related to one another. Equation (6.24) is sometimes thought of as another form of the Haskind–Hanaoka relation. Here γ is a quantity of major significance to the overall absorption performance of the device, and is given by

$$\gamma = \frac{|\mathcal{A}_{UW}(i\omega)|^2}{|\mathcal{A}_{UW}(i\omega)|^2 + |\mathcal{A}_{DW}(i\omega)|^2}. \tag{6.25}$$

We note that for a beam-sea device spanning an approaching wave crest of length L, the diffraction potential can be expressed in terms of the potential for the reflected wave propagating away from the body in the up-wave direction, and the transmitted wave propagating away from the body in the down-wave direction. Far from the device, we can use the far-field complex reflection coefficient $R(i\omega)$ and the far-field complex transmission coefficient $T(i\omega)$ [20] to write ($A = 1$)

$$\varphi_D \sim (ig/\omega)\left(\mathrm{e}^{-ik(\omega)x} + R(i\omega)\mathrm{e}^{ik(\omega)x}\right)\mathrm{e}^{k(\omega)z}, \quad x \to -\infty,$$
$$\varphi_D \sim (ig/\omega)T(i\omega)\mathrm{e}^{-ik(\omega)x}\mathrm{e}^{k(\omega)z}, \quad x \to \infty. \tag{6.26}$$

In a similar manner, the potential associated with the radiated wave far from the device can also be written in terms of the far-field up-wave and down-wave potential amplitudes above, as

$$\varphi_i \sim (ig/\omega)\mathcal{A}_{UW}(i\omega)\mathrm{e}^{ik(\omega)x}\mathrm{e}^{kz}, \quad x \to -\infty,$$
$$\varphi_i \sim (ig/\omega)\mathcal{A}_{DW}(i\omega)\mathrm{e}^{-ik(\omega)x}\mathrm{e}^{kz}, \quad x \to \infty. \tag{6.27}$$

Fig. 6.2 shows the wave radiation properties of two sections in heave oscillation. The symmetric section is seen to radiate equally in the up-wave and down-wave directions, while the asymmetric section radiates more energy in the up-wave direction.

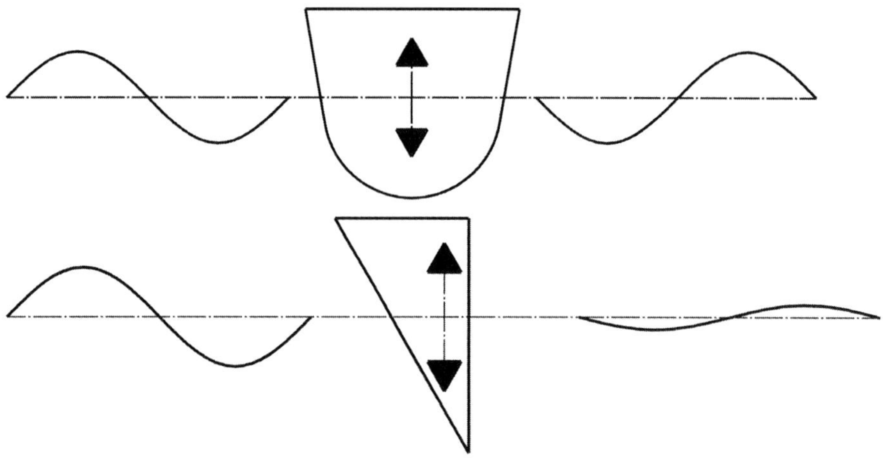

Figure 6.2 An illustration of the wave radiation properties of two crest-spanning sections in heave oscillation. The triangular section radiates larger waves into the direction from which the incident waves causing its oscillation approach.

Application of Green's second identity to the radiation potential and just the body-disturbed part of the diffraction potential (here we denote $\varphi_s = \varphi_D - \varphi_I$) and their complex conjugates leads to the remarkable conclusion that ([8, 21, 26, 80], etc.)

$$R(i\omega)\mathcal{A}^*_{UW}(i\omega) + T(i\omega)\mathcal{A}^*_{DW}(i\omega) + \mathcal{A}_{UW}(i\omega) = 0. \tag{6.28}$$

In other words, the radiation behavior of the device oscillating in calm water is quantitatively related to the reflection and transmission of incident waves by the same device when held fixed.

In general the device will experience both radiation as quantified by $\mathcal{A}_{UW}(i\omega)$ and $\mathcal{A}_{DW}(i\omega)$ and reflection/transmission as measured by $R(i\omega)$ and $T(i\omega)$. Now, it has been shown [53] that the energy conversion efficiency, defined as the ratio of the power absorbed by the device over the power incident on it, or $\eta_c = P_c(\omega)/(P_w(\omega)L)$, is given by

$$\eta_c = \gamma - \frac{\mathcal{A}_{UW}(i\omega)\alpha^* - \gamma}{\gamma}, \quad \gamma = \frac{|\mathcal{A}_{UW}(i\omega)|^2}{|\mathcal{A}_{UW}(i\omega)|^2 + |\mathcal{A}_{DW}(i\omega)|^2} \tag{6.29}$$

where for an oscillation velocity complex amplitude denoted by $U_i(i\omega)$

$$\alpha = [\omega U_i(i\omega)]/(gA). \tag{6.30}$$

This discussion is where we see why velocity control is fundamental to efficient energy absorption. Specifically, if $U_i(i\omega)$ can be controlled such that

$$\mathcal{A}_{UW}(i\omega)\alpha^* = \gamma, \tag{6.31}$$

then

$$\eta_c = \eta_{mx} = \gamma \tag{6.32}$$

where γ is as defined by Equations (6.25) and (7.10). This is another remarkable result in that γ is the best efficiency possible, achievable by controlling a device to oscillate at a complex amplitude determined by Equation (6.31). Of course, if the natural oscillation of the device were such that Equation (6.31) is automatically satisfied, no control would be needed. The velocity U_i that satisfies Equation (6.31) thus defines the "hydrodynamic velocity optimum" for the device.

Next we note that γ is completely defined by the "strengths" of the far-field radiation potential (or, simply, the respective wave-profile amplitudes) in the up-wave and down-wave directions. The manner in which the device radiates energy away from it thus determines how best it can absorb energy from an approaching wave. Specifically, if $\mathcal{A}_{UW}(i\omega) = \mathcal{A}_{DW}(i\omega)$, that is, an equal amount of energy is radiated away in both directions, then the best efficiency $\eta_{mx} = \gamma$ enabled by Equation (6.31) is 50%. Such symmetric radiation is characteristic of bodies symmetric about the yz plane (normal to the wave propagation direction along the x axis). If the shape of the device in the xz plane can then be designed such that $\mathcal{A}_{UW}(i\omega) > \mathcal{A}_{DW}(i\omega)$, it is possible to achieve $\gamma > 0.5$. Recall that Fig. 6.2 schematically illustrates these two situations.

Could a device be shaped then such that $\mathcal{A}_{UW}(i\omega) \gg \mathcal{A}_{DW}(i\omega)$, to a point where $\mathcal{A}_{DW}(i\omega) \approx 0$, so that $\gamma \to 1$ (and so $\eta_{mx} \to 1$)? One hundred percent of the incoming wave power would then be absorbed by the device. Such a shape was designed experimentally by Salter in the early 1970s (prior to the advent of much of the theory above). The resulting device was the Salter duck, which has sometimes been described as a "backless flap" pitching in response to incoming waves. As Fig. 6.3 shows, it is easy to see intuitively that a pitching motion of this section can generate a relatively large radiated wave in front (i.e., in the up-wave direction), while the cylindrical back surface generates a negligible wave in the down-wave direction. Fig. 6.4 shows an early design of the duck with a hydraulic pump using the pitch oscillation relative to the spine [17], and an early version of a floating crest-spanning oscillating water column device developed by the National Engineering Laboratory in the United Kingdom.

To gain a better understanding of the significance of the duck geometry in waves, we substitute $\mathcal{A}_{DW}(i\omega) = 0$ (i.e., negligible radiation in the down-wave direction) in Equation (6.28) and verify that then

$$|R(i\omega)| = 1, \text{ and consequently, } |T(i\omega)| = 0, \text{ since } |R(i\omega)| + |T(i\omega)| = 1. \quad (6.33)$$

Note that the present analysis is concerned with two-dimensional problems as applicable to crest-spanning beam-sea devices. The last part of Equation (6.33) simply states that

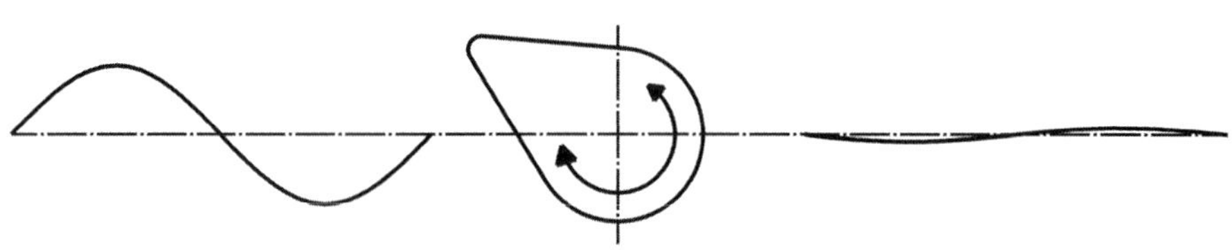

Figure 6.3 An experimentally designed shape for the Salter duck in single-mode pitching oscillation that radiates predominantly in the up-wave direction.

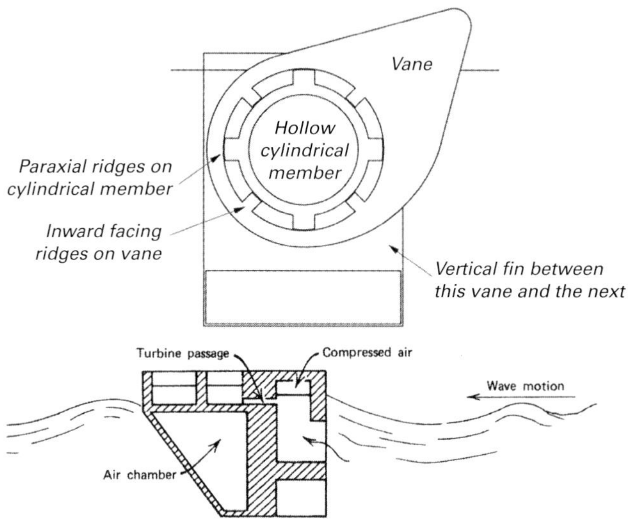

Figure 6.4 Crest-spanning WECs with preferred radiation in the up-wave direction. Incident waves in these diagrams are approaching from the right. Courtesy Edinburgh Wave Power Project; [16], Dover Publications Inc.; reproduced by permission.

for a fixed two-dimensional body, a part of the incident wave is reflected and the rest is transmitted. If for certain geometries, however, with the body oscillating in the desired mode, little or no energy is radiated in the down-wave direction, then a little or none is transmitted through "diffraction" effects. In such a situation, what control according to (6.31) accomplishes is that the waves that body oscillation creates in the up-wave direction cancel the reflected waves due the body (i.e., "diffraction" effects in the up-wave direction) according to $C\mathcal{A}_{UW}(i\omega) + R(i\omega) = 0$, where C is a constant. With no part of the approaching incident energy traveling away behind the device (Equation (6.33)), and none as just seen traveling away to the front of it, all of the incident energy is available for the device to absorb. We explore this point further in the following discussion. Recalling that $|\mathcal{A}_{UW}|^2 = \mathcal{A}_{UW}\mathcal{A}^*_{UW}(i\omega)$, it is easy to see using Equations (6.30) and (6.31) that, for energy best conversion,

$$U_i^*(i\omega) = \frac{gA\mathcal{A}^*_{UW}(i\omega)}{\omega\left(|\mathcal{A}_{UW}(i\omega)|^2 + |\mathcal{A}_{DW}(i\omega)|^2\right)}. \tag{6.34}$$

Using Equation (6.21),

$$U_i(i\omega) = \frac{F_{fi}(i\omega)}{\rho\omega\left(|\mathcal{A}_{UW}(i\omega)|^2 + |\mathcal{A}_{DW}(i\omega)|^2\right)L}. \tag{6.35}$$

Then using $b_i(\omega)$ as given by Equation (6.22) we learn that the optimum velocity condition in fact also implies

$$U_i(\omega) = \frac{F_{fi}(i\omega)}{2b_i(\omega)}. \tag{6.36}$$

Note that, since $b_i(\omega)$ is real, $U_i(i\omega)$ in Equation (6.36) is in phase with the exciting force $F_{fi}(i\omega)$. The striking similarity of this result to expressions derived earlier in Chapter 4 is not coincidental. $U_i(i\omega)$ as given by Equations (6.36) and/or (6.30) and (6.31) is the hydrodynamically optimum velocity enabling maximum power capture for a given shape. The conditions under which U_i is as given by Equation (6.36) are examined below in the context of the single-mode device as described by Equation (6.1).

6.3 Producing Optimum Velocity

Equation (6.1) is further developed below for harmonic waves. As we have seen before, one value of a harmonic analysis lies in being able to generalize the conclusions to dynamics in irregular waves through Fourier transformation for linear equations of motion. Thus, if we let $f_{Ri}(t) = F_{Ri}(i\omega)\mathrm{e}^{i\omega t}$, and $f_{fi}(t) = F_{fi}(i\omega)\mathrm{e}^{i\omega t}$, we expect that for this linear system, $x_i(t) = X_i(i\omega)\mathrm{e}^{i\omega t}$, and $\dot{x}_i(t) = U_i(i\omega)\mathrm{e}^{i\omega t}$. Then,

$$F_{Ri}(i\omega) = -\left[i\omega a_i(\omega) + b_i(\omega)\right] U_i(i\omega) = -\left[-\omega^2 a_i(\omega) + i\omega b_i(\omega)\right] X_i(i\omega). \quad (6.37)$$

Note that the other component of the radiation force, namely, $i\omega\bar{a}_i(\infty)U_i(i\omega) = -\omega^2\bar{a}_i(\infty)X_i(i\omega)$ due to the constant frequency-independent infinite-frequency added mass, also needs to be included. If we further let the load be harmonic for harmonic excitation, that is, $F_L(t) = F_L(i\omega)\mathrm{e}^{i\omega t}$, then, for harmonic waves alone, Equation (6.1) reduces to

$$\left[-\omega^2\left(m + \bar{a}_i(\infty) + a_i(\omega)\right) + i\omega\left(b_i(\omega) + c_d\right) + k_i\right] X_i(i\omega) = F_{fi}(i\omega) + F_L(i\omega). \quad (6.38)$$

Letting $F_L(i\omega) = -DU_i(i\omega) = -i\omega DX_i(i\omega)$ (where D represents a constant dissipative/resistive load, i.e., constant damping) for illustration, with $c_d \to 0$,

$$\left[-\omega^2\left(m + \bar{a}_i(\infty) + a_i(\omega)\right) + i\omega\left(b_i(\omega) + D\right) + k_i\right] X_i(i\omega) = F_f(i\omega). \quad (6.39)$$

It is straightforward to see that

$$X_i(i\omega) = \frac{F_{fi}(i\omega)}{\left[-\omega^2\left(m + \bar{a}_i(\infty) + a_i(\omega)\right) + i\omega\left(b_i(\omega) + D\right) + k_i\right]}. \quad (6.40)$$

The average power absorbed by the dissipative (resistive) load over a period can simply be written as

$$P_c(\omega) = \frac{1}{2}D|U_i(i\omega)|^2 = \frac{1}{2}\omega^2 D|X(i\omega)|^2. \quad (6.41)$$

From Equation (6.24),

$$|F_{fi}(i\omega)|^2 = 8P_w(\omega)b_i(\omega)L\gamma. \quad (6.42)$$

Substituting (6.42) and (6.40) into Equation (6.41),

$$P_c(i\omega) = \frac{4\omega^2 Db_i(\omega)P_w(\omega)L\gamma}{\left[-\omega^2\left(m + \bar{a}_i(\infty) + a_i(\omega)\right) + k_i\right]^2 + \left[\omega\left(b_i(\omega) + D\right)\right]^2}. \quad (6.43)$$

A power absorption efficiency can be defined as $\eta = P_c(\omega)/[P_w(\omega)L]$ for a crest spanning device:

$$\eta = \frac{4\omega^2 D b_i(\omega)\gamma}{\left[-\omega^2\left(m + \overline{a}_i(\infty) + a_i(\omega)\right) + k_i\right]^2 + \left[\omega\left(b_i(\omega) + D\right)\right]^2}. \tag{6.44}$$

Recalling the discussion in Chapter 4 and using a frequency variable ω_n to define the natural frequency in mode i, at resonance,

$$-\omega_n^2\left[m + \overline{a}_i(\infty) + a_i(\omega_n)\right] + k_i = 0. \tag{6.45}$$

It is easy to see that the first term in the denominator in Equation (6.44) $\rightarrow 0$ as wave frequency $\omega \rightarrow \omega_n$. In addition, it can further be seen that when $D = b_i(\omega)$, in addition to $\omega = \omega_n$, $\eta = \eta_{mx} = \gamma$. Note that we have assumed $c_d \rightarrow 0$. In other words, when an incident wave excites resonant oscillations of the body, and in addition, if the energy-absorbing load damping D applied on the body equals the radiation damping $b_i(\omega)$, the energy conversion efficiency is maximized, as informally concluded in Chapter 4. For a device with minimum down-wave radiation then, $\eta_{mx} \rightarrow 1$. Under the conditions $\omega = \omega_n$ and $D = b_i(\omega)$, Equation (6.40) reduces to

$$U_i(i\omega) = \frac{F_{fi}(i\omega)}{2b_i(\omega)} \tag{6.46}$$

where Equation (6.46) is just the condition for the hydrodynamic velocity optimum for mode i. As we saw in Chapter 4, when the approaching wave frequency $\omega \neq \omega_n$, a control force may be applied effectively to let a part of the load contribute to or take away from the inertia and stiffness, thereby enabling resonance of the body plus actuator system at the approaching wave frequency. Another part of the load may then apply energy-absorbing damping to match the radiation damping. As pointed out in Chapter 4, at least in theory, this can be achieved for chosen non resonant frequencies in regular waves.

6.4 Calculating the Average Absorbed Power

To calculate the actual average power absorbed by the mode U_i over a cycle, under the condition (6.46), we argue that, during each cycle, the exciting force provides a net power to the body given by (see, e.g., also [20, 66])

$$P_f(\omega) = \frac{1}{2}\Re\left[F_f(i\omega)U_i(i\omega)\right] = \frac{1}{4}\left[F_{fi}(i\omega)U_i^*(i\omega) + U_i(i\omega)F_{fi}^*(i\omega)\right]. \tag{6.47}$$

However, as the body oscillates, it radiates energy away from it at an average per-cycle rate related to the radiation damping $b_i(\omega)$ and velocity $U_i(i\omega)$ as

$$P_r(\omega) = \frac{1}{2}U_i(i\omega)b_i(\omega)U_i^*(i\omega). \tag{6.48}$$

The power that can be absorbed and converted by the PTO is thus

$$P_c(\omega) = P_f(\omega) - P_r(\omega) = \frac{1}{4}\left[F_{fi}(i\omega)U_i^*(i\omega) + U_i(i\omega)F_{fi}^*(i\omega)\right] - \frac{1}{2}U_i(i\omega)b_i(\omega)U_i^*(i\omega). \tag{6.49}$$

When the device oscillates at optimum velocity,

$$F_{fi}(i\omega) = 2U_i(i\omega)b_i(\omega). \tag{6.50}$$

Substituting Equation (6.50) into Equation (6.49), it is seen that

$$P_c(\omega) = \frac{1}{2}U_i(i\omega)b_i(\omega)U_i^*(i\omega). \tag{6.51}$$

Alternatively, using Equation (6.46), $P_c(\omega)$ is seen to be

$$P_c(\omega) = \frac{|F_{fi}(i\omega)|^2}{8b_i(\omega)}. \tag{6.52}$$

Equation (6.51) with $U_i(i\omega)$ given by Equation (6.46) and/or Equation (6.52) represent the maximum average power per cycle at a frequency ω that can be absorbed by the PTO when the damping it applies equals the radiation damping while the device is resonating at that frequency. Note that we have assumed $c_d \to 0$ and single-mode oscillation in the ith mode here.

6.5 Favorable Mode Combinations

We may find it useful to group the various oscillation modes for a floating body into two categories: symmetric and anti symmetric. For beam-sea devices, this understanding is based on the far-field radiation potential complex amplitudes. For a body with symmetry about the YZ plane, both up-wave and down-wave radiation potentials have the same magnitudes and phases when the body is in heave oscillation (Fig. 6.5(a)). On the other hand, when the same body performs surge oscillation, the magnitudes of the up-wave and down-wave radiation potentials are equal, but their phases are opposite, as can be verified intuitively by displacing the body back and forth (Fig. 6.5(b)). Heave motion is thus a symmetric mode and surge motion is anti symmetric. The oscillations in surge, sway, roll, pitch, and yaw are all anti symmetric. At least two ways to combine symmetric and anti symmetric modes for beam-sea devices have been investigated in the literature. The first, based on rectilinear translation of the body centroid, is more straightforward to understand (here referred to as "non orbital centroid motion"). The second is based on correctly utilizing the heave and surge modes, effectively producing controlled "orbital motion" of the centroid. The latter approach is manifested in the original versions of the Bristol submerged cylinder device.

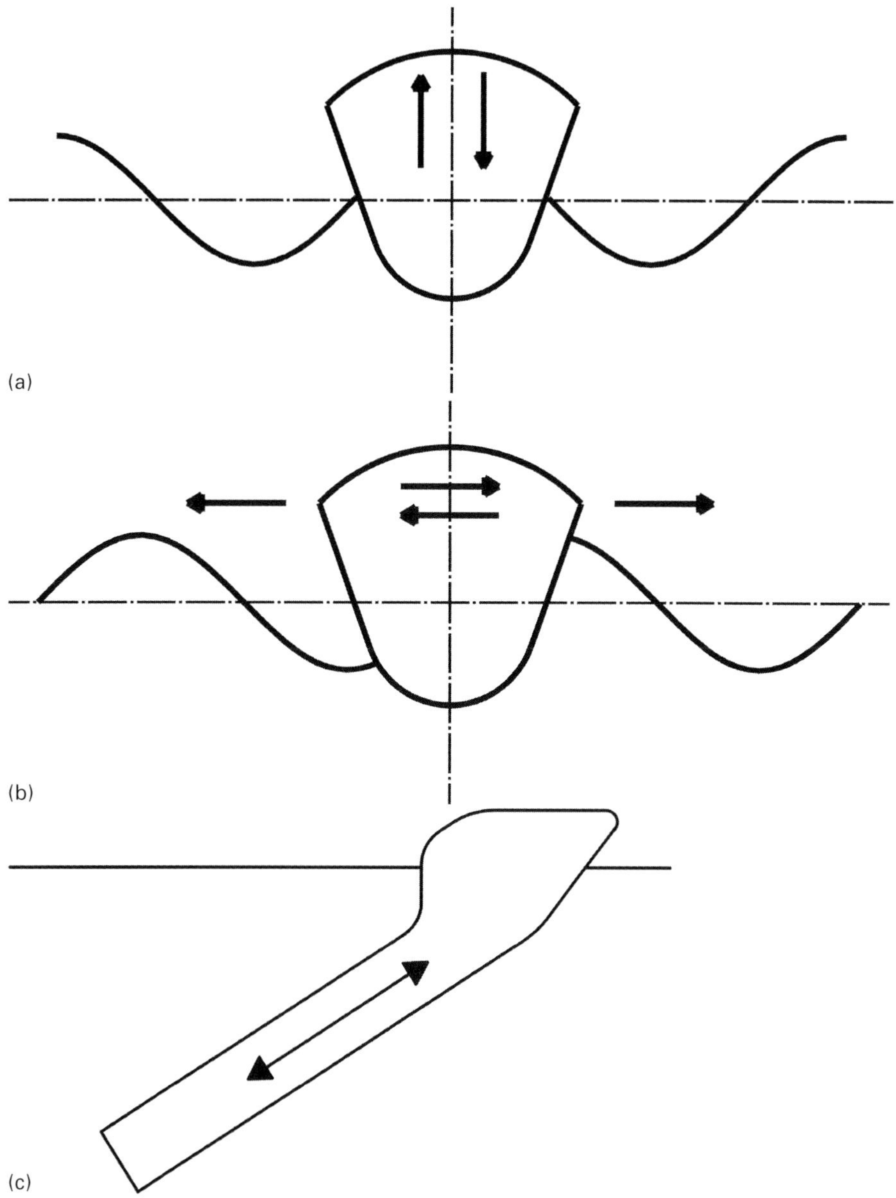

Figure 6.5 A body performing one symmetric (a) and one anti symmetric (b) oscillation (here heave and surge, respectively). The sloped IPS buoy profile (c) is an example of the use of two oscillation modes and geometric design for preferred radiation in the up-wave direction.

6.5.1 Non orbital Motion of Body Centroid

The linear theory world of small-amplitude oscillations permits linear superposition among different oscillation modes and the effects they produce. Writing Equation (6.28) for heave oscillation, with $i = 3$,

$$R(i\omega)\mathcal{A}^*_{3UW}(i\omega) + T(i\omega)\mathcal{A}^*_{3DW}(i\omega) + \mathcal{A}_{3UW}(i\omega) = 0. \tag{6.53}$$

Rewriting Equation (6.28) for surge oscillation, with $i = 1$,

$$R(i\omega)\mathcal{A}^*_{1UW}(i\omega) + T(i\omega)\mathcal{A}^*_{1DW}(i\omega) + \mathcal{A}_{1UW}(i\omega) = 0. \tag{6.54}$$

If a body oscillates such that heave and surge occur concurrently, then the combined action can be represented simply by adding Equations (6.53) and (6.54). Note that $R(i\omega)$ and $T(i\omega)$ represent the reflection and transmission coefficients (as seen in the far field) with the body held fixed. Thus,

$$R(i\omega)\mathcal{A}^*_{sUW}(i\omega) + T(i\omega)\mathcal{A}^*_{sDW}(i\omega) + \mathcal{A}_{sUW}(i\omega) = 0 \tag{6.55}$$

where

$$\mathcal{A}_{sUW}(i\omega) = \mathcal{A}_{1UW}(i\omega) + \mathcal{A}_{3UW}(i\omega), \text{ and } \mathcal{A}_{sDW}(i\omega) = \mathcal{A}_{1DW}(i\omega) + \mathcal{A}_{3DW}(i\omega). \tag{6.56}$$

The design objective in such a case would be to shape the geometry and to choose a combination of heave and surge amplitudes and phases, such that their two far-field up-wave radiated waves add constructively, with $|\mathcal{A}_{sUW}| \sim |\mathcal{A}_{1UW}| + |\mathcal{A}_{3UW}|$. At the same time, the two down-wave amplitudes should add destructively, so that $|\mathcal{A}_{sDW}| \sim |\mathcal{A}_{1DW}| - |\mathcal{A}_{3DW}|$. The combined down-wave radiation potential amplitude needs to be smaller than the combined up-wave amplitude, which would lead to $\gamma > 0.5$ and even $\gamma \rightarrow 1$ with correct design. This effect was one of the goals of the Sloped IPS buoy design (developed and tested by the Edinburgh Wave Power Group), with a two-dimensional profile, as shown in Fig. 6.5(c). The device was constrained to oscillate at angles ranging from 0 to 90 degrees relative to horizontal, thus representing a range of combinations of heave (90 degrees) and surge (0 degrees) modes at different amplitudes. Laboratory tests on a model with an approximately "semi-elliptical" profile indicated considerable improvement in bandwidth at angles such as 35 and 45 degrees [81].

6.5.2 Orbital Motion of Body Centroid

The Evans submerged cylinder device provides an especially elegant approach to realizing near-complete absorption (theoretically $\rightarrow$ 100%). Its development in the 1970s was prompted by a result obtained some years earlier that showed that the far-field radiation amplitudes generated in the heave and surge modes by a submerged cylinder in beam-sea configuration were 90 degrees out of phase but equal in magnitude [30]. As discussed briefly in Chapter 1, the foundations for this work may have been laid in the 1940s with the observation by Dean [27] that a submerged cylinder held fixed in beam seas produced no reflection but only transmission, though the transmitted wave had a phase different from that of the incident wave. This finding was formalized by Ursell [28], who also studied forced oscillation of such a cylinder. Ogilvie [29] found that the same effect (no reflection and transmission with a phase change) was also exhibited by a submerged cylinder simultaneously oscillating in heave and surge modes.

Beginning with [30]

$$\mathcal{A}_{3UW}(i\omega) = i\mathcal{A}_{1UW}(i\omega), \tag{6.57}$$

as we see below, this result can be used to show that when the cylinder is held fixed in approaching long-crested waves, there is no reflection, and all of the incident wave is transmitted, with a phase change. We recall that the amplitudes in Equation (6.57) are per unit velocity in the respective modes. Writing Equation (6.28) for the surge and heave modes separately [30],

$$\begin{aligned} R(i\omega)\mathcal{A}^*_{1UW}(i\omega) + T(i\omega)\mathcal{A}^*_{1DW}(i\omega) + \mathcal{A}_{1UW}(i\omega) &= 0, \\ R(i\omega)\mathcal{A}^*_{3UW}(i\omega) + T(i\omega)\mathcal{A}^*_{3DW}(i\omega) + \mathcal{A}_{3UW}(i\omega) &= 0. \end{aligned} \tag{6.58}$$

If the following relationships are used here

$$\begin{aligned} \mathcal{A}_{3UW}(i\omega) &= i\mathcal{A}_{1UW}(i\omega), \\ \mathcal{A}_{1DW}(i\omega) &= -\mathcal{A}_{1UW}(i\omega), \quad \text{surge (antisymmetric) mode,} \\ \mathcal{A}_{3DW}(i\omega) &= \mathcal{A}_{3UW}(i\omega), \quad \text{heave (symmetric) mode,} \end{aligned} \tag{6.59}$$

by utilizing all three of the relations in Equation (6.59) to replace the radiation amplitudes associated with the heave mode in the second of Equations (6.58) and dividing through by i,

$$-R(i\omega)\mathcal{A}^*_{1UW}(i\omega) + T(i\omega)\mathcal{A}^*_{1DW}(i\omega) + \mathcal{A}_{1UW}(i\omega) = 0. \tag{6.60}$$

Now adding the first of Equations (6.58) and Equation (6.60), and substituting the second of Equations (6.59) into the result,

$$T(i\omega) = \frac{\mathcal{A}_{1UW}(i\omega)}{\mathcal{A}^*_{1UW}(i\omega)}. \tag{6.61}$$

Equation (6.61) shows that $|T(i\omega)| = 1$, and therefore $\Rightarrow |R(i\omega)| = 0$. In addition, $T(i\omega) = \mathrm{e}^{2i\delta_1}$, where $2\delta_1$ is the phase shift suffered by the incident wave as it passes over the fixed submerged cylinder (δ_1 being the phase of the up wave far-field $\mathcal{A}_{1UW}(i\omega)$ for surge oscillation). All of the incident energy is thus transmitted and none is reflected when the cylinder is held fixed in place.

The radiation behavior expressed by Equations (6.57) suggests the following possibility. We note that the up-wave and down-wave far-field radiated-wave amplitudes due to surge and heave oscillations are 90 degrees out of phase. Then, if the surge and heave oscillations causing the up-wave and down-wave radiated amplitudes were themselves also 90 degrees out of phase, the radiated-wave amplitudes would reinforce each other in one direction and subtract from each other in the opposite direction. Whether the reinforcing effect occurs in the up-wave or down-wave directions would depend on the sign of the 90-degree phase difference between the two oscillations. Note that the cylinder centroid traces a circular orbit when surge and heave are 90 degrees out of phase. In practice, this would mean using the PTO or another means to constrain the oscillations of the submerged cylinder such that its centroid moved in a circular orbit.

We can deduce from Equation (6.61) and the paragraph following it that we need the two radiated wave fields resulting from surge and heave to subtract from each other in the up-wave direction and to reinforce each other in the down-wave direction. This is because we want the net radiated wave field to cancel the net diffracted/scattered wave field, which for a beam-sea crest-spanning device would normally just be the transmitted and reflected wave fields: transmitted field in the down-wave direction, and reflected field in the up-wave direction. Since there is no reflected wave in the up-wave direction for the present submerged cylinder to cancel, we need no radiated wave up-wave, and hence want the up-wave radiated wave-field contributions of surge and heave to subtract from each other. On the other hand, there is a strong transmitted wave behind the cylinder. The down wave radiated fields due to surge and heave should therefore combine additively. Then, we could potentially control the oscillations in such a manner that in addition to one mode leading the other by 90 degrees, the combined radiated wave down-wave cancels the transmitted wave.

Specifically, for waves approaching from $x \to -\infty$, the transmitted wave propagates to $x \to \infty$. The cylinder centroid must then trace a clockwise circle. Note that the radiated amplitudes in Equations (6.59) are for unit velocity oscillations. For clockwise rotation, $U_3 = iU_1$. It can also be seen that counterclockwise trajectory would be required for waves approaching from $x \to \infty$ (see [82] for a discussion of these points using Kochin functions). When the down-wave radiated wave fully cancels the transmitted wave, and there is no reflected wave and no radiated wave up-wave given the sense of the rotation, 100% of the incident energy is available for conversion by the mechanism causing the cylinder motion. Such complete conversion is possible when both modes resonate and when the power-absorbing load damping equals the radiation damping for each mode (ignoring viscous damping). This discussion is summarized below [30]. For instance, the response of the ith mode where $i = 1$ and/or 3,

$$X_i(i\omega) = \frac{F_{fi}}{-\omega^2\left(m + \overline{a}(\infty) + a(\omega)\right) + k + i\omega\left(b(\omega) + D_i\right)} \tag{6.62}$$

where $F_{fi}(i\omega) = \rho g A \mathcal{A}_{iUW}(i\omega)$. With $\mathcal{A}_{3UW}(i\omega) = i\mathcal{A}_{1UW}(i\omega)$ (Equation (6.57)), $b_1(\omega) = b_3(\omega) = b(\omega)$, and $a_1(\omega) = a_3(\omega) = a(\omega)$. The total absorbed power is given by

$$P_c(\omega) = \frac{1}{2}\sum_{i=1,\,3} \omega^2 D_i |X_i(i\omega)|^2. \tag{6.63}$$

Following arguments similar to those used for single-mode devices and applying the same load damping on both modes, it can be seen that the absorption efficiency (with viscous damping ignored) is

$$\eta_c = \frac{4b(\omega)D_i}{\left[-\omega^2\left(m + \overline{a}(\infty) + a(\omega)\right) + k\right]^2 + \left[b(\omega) + D_i\right]^2}. \tag{6.64}$$

With the cylinder centroid in circular motion in the correct sense (clockwise or counterclockwise depending on the incident-wave direction), when

$$-\,\omega^2\left(m + \overline{a}(\infty) + (\omega)\right) + k = 0, \text{ and } D_i = b(\omega),\ \eta_c = 1. \tag{6.65}$$

Thus, under correct orbital motion of the centroid, resonant oscillations in surge and heave combined with optimum damping in each mode (i.e., load damping equal to radiation damping, with viscous damping ignored) result in a radiated wave that completely cancels the transmitted wave for the submerged cylinder. Control of both the amplitude and phase of the forces applied in the two modes (perhaps via the PTO mechanism) would enable these conditions to be satisfied at frequencies different from the natural frequency. Finally, we should note that the effects discussed above are possible for non circular shapes, though then the orbital trajectory will also be non circular and an additional frequency dependence (due to inequality of added mass and radiation damping variations in the two modes) may also need to be considered.

6.6 Omni directional Devices

A three-dimensional treatment is required for devices that are not crest-spanning. Notable among three-dimensional devices are omni directional devices, which are usually defined by a vertical axis of symmetry. The response of omni directional devices is not dependent on the direction of incoming waves. When such devices in heave have very small dimensions compared with the wavelengths of interest, to such an extent that they can be considered "point sources" relative to the incoming wavelengths [83], they are referred to as "point absorbers." With such heaving point absorbers, wave radiation is symmetric about the vertical axis, so that they can be regarded as "monopoles" or point sources on the two dimensional free surface. We can make arguments similar to those used in the previous sections regarding the relevance of far-field radiation potential amplitudes for three-dimensional devices. In particular, the exciting force for a device requiring a three-dimensional treatment can be expressed, based on [84], as

$$F_{fi}(i\omega;\theta_I) = \frac{\rho g}{k} A\mathcal{A}(i\omega;\theta_I + \pi) \tag{6.66}$$

where A is the incident wave amplitude and $\mathcal{A}(i\omega;\theta_I + \pi)$ is given by

$$\mathcal{A}(i\omega;\theta_I + \pi) = \sqrt{2\pi}\mathrm{e}^{i\pi/4} f(\omega, \theta_I + \pi), \tag{6.67}$$

the far-field radiated amplitude against the incident wave direction θ_I [21, 84]. The function $f(\omega;\theta)$ for three-dimensional devices captures the far-field angular variation of the radiated wave field. $\mathcal{A}(i\omega;\theta)$ has length units and can be related to the Kochin functions used in the literature [21, 26, 66, 80, 85, 86, etc.]. In fact, note that, following the definitions of [26] and [80] for the Kochin functions, Equation (6.66) can be expressed as

$$F_{fi}(i\omega;\theta_I) = \rho g A \mathcal{K}_i(\theta_I + \pi) \tag{6.68}$$

where $\mathcal{K}_i(\theta_I + \pi)$ denotes the Kochin function for the radiated wave field against the incident wave direction θ_I. The Kochin function $\mathcal{K}_i$ in deep water can be expressed as

$$\mathcal{K}_i(\theta_I + \pi) = \iint_{S_B} \overline{\varphi}_D \frac{\partial \overline{\varphi}_i}{\partial n} dS \tag{6.69}$$

where $\overline{\varphi}_D$ and $\overline{\varphi}_i$ are defined as $\varphi_D \times [\omega/(iAg)]$ and φ_i, respectively, provided that φ_i is the velocity potential due to unit velocity in the ith mode [85]. Recall that φ_D represents the diffracted wave potential, while φ_i represents the radiation potential due to unit velocity in the ith mode.

Note that an alternative definition, that is, k times this function in Equation (6.69), is adopted in [66], so that the $\mathcal{A}(i\omega;\theta)$ here can be seen as the Kochin functions defined in [66].

Definitions similar to that of Equation (6.69) have been provided in [26] for the diffracted wave field for three-dimensional bodies, as well as for both radiation and diffraction potentials for two-dimensional, beam-sea type devices. In particular, it is interesting to consider how the far-field arguments used here and in the literature relate to the geometry of the body surface through the Kochin function.

Here we take the oscillation mode $i \leq 3$ as before. There is also a way to relate the radiation damping to the exciting force (via the Haskind–Hanaoka relation for three-dimensional floating bodies):

$$b_i(\omega) = \frac{1}{8\lambda P_w} \int_0^{2\pi} |F_{fi}(i\omega;\theta)|^2 d\theta. \tag{6.70}$$

Based on Equation (6.66) and Equation (6.23) for incident wave power per unit width,

$$b_i(\omega) = \frac{\rho\omega}{4\pi k} \int_0^{2\pi} |\mathcal{A}(i\omega;\theta+\pi)|^2 d\theta \tag{6.71}$$

where $\theta + \pi$ is the direction in which a wave is radiated. We note that in three-dimensional operation, waves can be radiated over a 360-degree arc around the device. For axi symmetric heaving omni directional devices, $F_f(i\omega)$ is independent of θ in Equation (6.70). The radiated wave field in this case is also independent of θ. Equation (6.70) can then more compactly be written as

$$b_i(\omega) = \frac{k|F_{fi}(i\omega)|^2}{8P_w}. \tag{6.72}$$

We can argue that, even though the device itself may be symmetric, its oscillations may be anti symmetric and directional. The radiated wave field may not then be uniform. For instance, for surge oscillation, radiation along the oscillation direction will be greater than radiation in any other direction. For heave, of course, the buoy radiates equally in any direction over the 360-degree arc.

An analysis similar to that used for beam-sea devices can be performed here to determine the best energy capture for oscillations in the ith mode for three-dimensional devices. Thus,

$$P_c(\omega) = \frac{1}{2}\omega^2 D|X_i(i\omega)|^2. \tag{6.73}$$

From a frequency domain equation of motion for a three dimensional device,

$$X_i(\omega) = \frac{F_{fi}(i\omega;\theta_I)}{\left[-\omega^2\left(m+\overline{a}_i(\infty)+a_i(\omega)\right)+k_i\right]+i\left[b_i(\omega)+D\right]}. \tag{6.74}$$

$P_c(\omega)$ can then be written, using Equation (6.70), as

$$P_c(\omega) = \frac{4\omega^2 Db_i(\omega)\lambda P_w}{\left[-\omega^2\left(m+\bar{a}_i(\infty)+a_i(\omega)\right)+k_i\right]^2+\left[\omega\left(b_i(\omega)+D\right)\right]^2} \frac{|F_{fi}(i\omega;\theta_I)|^2}{\int_0^{2\pi}|F_{fi}(i\omega;\theta)|^2 d\theta}. \tag{6.75}$$

For an axisymmetric device, the ratio of the two exciting force-related terms reduces to $\varepsilon/(2\pi)$:

$$\frac{P_c(\omega)}{P_w(\omega)} = \frac{4\omega^2 Db_i(\omega)}{\left[-\omega^2\left(m+\bar{a}_i(\infty)+a_i(\omega)\right)+k_i\right]^2+\left[\omega\left(b_i(\omega)+D\right)\right]^2} \frac{\varepsilon\lambda}{2\pi}. \tag{6.76}$$

The ratio above is the average power absorbed per cycle over the incident wave power per unit crest length. It thus represents the energy "capture length" for the device. As before, when (1) the device resonates at frequency ω_n, so that $-\omega_n^2\left(m+\bar{a}_i(\infty)+a_i(\omega_n)\right)+k_i=0$, and further, (2) absent viscous damping, when the load damping $D=b_i(\omega_n)$, that is, the rate of energy absorption equals the rate of energy radiation (leading to optimum damping), the capture length is maximized and can be expressed as

$$l_{max}(\theta_I) = \frac{\varepsilon\lambda}{2\pi}. \tag{6.77}$$

More generally, for three-dimensional devices,

$$l_{max}(\theta_I) = \frac{\lambda|F_{fi}(i\omega;\theta_I)|^2}{\int_0^{2\pi}|F_{fi}(i\omega;\theta)|^2 d\theta}. \tag{6.78}$$

l_{max} can also be expressed in terms of the function $f(\omega;\theta)$, as

$$l_{max}(\theta_I) = \frac{\lambda|f(\omega;\theta_I+\pi)|^2}{\int_0^{2\pi}|f(\omega;\theta)|^2 d\theta} = \frac{\varepsilon\lambda}{2\pi}. \tag{6.79}$$

Recall that the function $f(\omega;\theta)$ represents the angular variation of the radiated wave field. For an axisymmetric body in heave, the radiated wave field has no angular variation, and $\varepsilon = 1$. However, for an axisymmetric body in surge, there is greater radiation in the oscillation direction. In fact, $f(\omega;\theta)$ then varies as $\cos\theta$, where θ is measured from the line defining the incident wave direction. $\varepsilon = 2$ for surge oscillation of axisymmetric bodies [84]. It can be verified that the velocity leading to l_{max} is the hydrodynamic velocity optimum.

As discussed in [66], if multiple modes can be used for energy absorption, further gains in capture length are possible under hydrodynamically optimum velocity in each mode. For axisymmetric devices, if surge and heave are utilized together, it can be shown that $\varepsilon = 3$. Similarly, if heave and pitch are utilized together, $\varepsilon = 3$. Note that for axisymmetric geometries, heave and surge, and heave and pitch, are not hydrodynamically coupled. If surge, sway, and heave are utilized simultaneously, $\varepsilon = 3$ still, as also shown in [21]. These three modes are hydrodynamically uncoupled from each other. On the other hand, surge and pitch are mutually coupled hydrodynamically, in that oscillation in one mode causes a radiation force on the other. In fact, it is possible for the two modes to cancel each other's radiated wave fields, so no energy is radiated away from the body and no energy can be absorbed by the body. Mathematically, this situation illustrates that, when the axisymmetric body oscillates in surge and pitch, the radiation damping matrix is singular. Thus, if surge and pitch are used together, $\varepsilon = 2$ still. If surge, heave, and pitch are used together, $\varepsilon = 3$. As surge motion takes place along the incident wave fronts, greater wave radiation takes place up-wave and down-wave along this direction than in any other, leading to a greater capture length than with heave, where the radiated wave is equal in all directions over the free surface. Small axisymmetric devices in heave thus radiate waves as "monopoles," while surging devices radiate waves as "dipoles." A wave-making pattern similar to surge is seen for pitch. For surge and pitch, as discussed above, the directional dependence of the radiated wave amplitude in the far field is as $\cos\theta$, with θ denoting the radiation direction measured from the line along the incident-wave approach direction [53, 66]. For sway and roll, the dependence is as $\sin\theta$. For heave oscillation, there is no directional dependence of the radiated wave field, so that $\varepsilon = 1$. It should be noted that, in long waves, the radiation damping for small axisymmetric surging devices may still be smaller than that for the same geometry in heave, so that larger velocities (and excursions) in surge may be required to realize the greater capture lengths possible through the greater ε. As the device size increases in the surge direction, however, the counteracting effect of its two faces on wave radiation decreases, and smaller velocities would be required for optimum conversion. Similar considerations apply to pitch, roll, and sway modes.

It should be noted too that, for an axisymmetric geometry, the yaw motion does not produce any waves, nor does it couple to another oscillation mode. Therefore, if we visualize the radiation damping matrix for an axisymmetric body in general six-degree-of-freedom oscillation, the entire row corresponding to the yaw mode is zero. The general 6×6 radiation damping matrix for an axisymmetric body thus must be singular. Furthermore, as explained in [66], based on the interference between incident, diffracted, and radiated wave fields, the best capture length possible from an axisymmetric body in general six-degree-of-freedom oscillation corresponds to $\varepsilon = 3$, and equals $3\lambda/(2\pi)$.

For an axisymmetric geometry, this effect can be understood simply by redefining the coordinate frame so that the x axis aligns with the wave direction, so that the resulting 3×3 radiation damping has rank 2 when conversion is through heave, surge, and pitch modes (given that surge and pitch are coupled and can cancel each other's waves). If rotational modes are constrained and conversion is through surge, sway, and heave modes for an arbitrary wave incidence angle, then the radiation damping matrix is not singular and its rank equals 3. For a 6×6 radiation damping matrix, therefore, with the surge–pitch and sway–roll coupling in place, the rank can at most be 3 (similar considerations hold for crest-spanning devices, where two out of three modes are independent for an asymmetric shape [26]). An interesting consequence of the singularity of the radiation damping matrix is that at the hydrodynamically optimum operation, not only yaw, but one each from the surge–pitch and sway–roll pairs of coupled motions could be prescribed arbitrarily (i.e., as convenient). Often, simply locking one or the other may be the expedient approach, keeping in mind that there is no penalty to pay for this in terms of power capture.

To sum up, since l_{max} can potentially significantly exceed 1 in longer waves, the ability of an axisymmetric device to capture energy spanning a wave front much wider than its own dimensions is a considerable advantage. Since for point absorber devices, the device diameter $\ll \lambda$, much attention has been given in the literature and in practice to exploiting this "antenna effect" for cost-effective conversion. Small sizes generally imply shorter resonant periods, however, and control such as discussed here (i.e., forcing resonance away from natural resonant frequency and simultaneously matching load damping to radiation damping) then becomes an important aspect of the device design. It should be noted that small devices also require large oscillation amplitudes for best energy conversion.

6.7 Head-Sea Devices

It has been found that the ability to concentrate the radiated energy over a preferred direction is crucial for head-sea operation too. For an incident wave direction θ_I relative to the orientation of the long axis, the maximum energy capture length can be expressed as

$$l_{max}(\theta_I) = l_{max}(0)\frac{|F_f(i\omega;\theta_I)|^2}{|F_f(i\omega;0)|^2}, \tag{6.80}$$

showing that the maximum capture length is achieved when $\theta = 0$ or when head sea conditions occur; that is, when wave crests are aligned along a direction perpendicular to the long axis of the device, as intended. Further, $l_{max}(0)$ is given by

$$l_{max}(0) = q_1\frac{\lambda}{2\pi}. \tag{6.81}$$

q_1 represents the modification to the heaving axisymmetric device result due to elongation of the body, and it is found, for bodies with a rectangular plan area, that

$$l_{max}(0) > 0.4L \tag{6.82}$$

where L is the length of the device. As suggested by Equation (6.80), the directionality of the radiated wave field is important even here. Geometry choices for head-sea devices are more restricted, but multiple device elements such as three-hinged rafts could be driven to maximize capture length. The radiated wave field could be maximized in a preferred direction through mutual reinforcements and cancellations by correctly controlling the odd and even modes of oscillation. The maximum capture length can theoretically approach a value close to the length of the device when motions are unconstrained [53]. We could use control in this case to cause raft modes that enable the correct mode combinations, in addition to forcing resonance and optimum damping. More recent work on head-sea operation under oscillation constraints in the context of the Pelamis device appears in [87].

6.8 Energy Absorption under Displacement/Velocity Constraints

So far, we have given little regard to the practical limitations to optimum behavior. In practice, it is quite possible that when a device is operating at or close to optimum velocity as given by Equation (6.36), the velocity $U_i(i\omega)$ becomes so large that the device fully emerges or fully submerges during its excursion. In addition, the PTO may have only a certain oscillation range, and motions exceeding the PTO limits could damage the entire device. It is necessary therefore to understand what the best possible power absorption would be if the device oscillations are constrained to remain below a prescribed value. Note that Equations (6.51) and (6.52) assume that any required optimum velocity $U_i(i\omega)$ is practically realizable and no constraints need be applied. One way to impose an oscillation constraint while seeking the best absorption possible under that constraint is to adopt the constrained optimization approach of [30]. This approach is summarized below.

If the maximum permissible displacement amplitude is α, then the maximum allowable optimum velocity amplitude can be found as $\beta = \omega\alpha$ at frequency ω, and the constraint can be expressed as

$$|U_i(i\omega)| \leq \beta, \quad \text{or} \quad U_i(i\omega)U_i^*(i\omega) \leq \beta^2. \tag{6.83}$$

Whenever the actual optimum velocity exceeds the permissible value β, we reason that most power would be absorbed if $|U_i(\omega)|$ could at least equal β, with

$$U_i(i\omega)U_i^*(i\omega) = \beta^2. \tag{6.84}$$

It is then possible to adjoin $P_c(\omega)$ as

$$\begin{aligned}\overline{P}_c(\omega) =& \frac{1}{4}\left[F_{fi}(i\omega)U_i^*(i\omega) + U_i(i\omega)F_{fi}^*(i\omega)\right] - \frac{1}{2}U_i(i\omega)b_i(\omega)U_i^*(i\omega)\\ &- \frac{1}{2}\mu\left[U_i(i\omega)U_i^*(i\omega) - \beta^2\right]\end{aligned} \tag{6.85}$$

where μ is a Lagrange multiplier to be determined along with the constrained optimum $U_{ic}(i\omega)$. Realizing that absorbed power under velocity constraint in Equation (6.85) for a chosen frequency ω has an explicit dependence on U_i, U_i^*, and μ, a small variation $\delta\overline{P}_c$ can be expressed as

$$\delta\overline{P}_c = \frac{\partial\overline{P}_c}{\partial U_i}\delta U_i + \frac{\partial\overline{P}_c}{\partial U_i^*}\delta U_i^* + \frac{\partial\overline{P}_c}{\partial\mu}\delta\mu \tag{6.86}$$

where the parentheses containing ω and $i\omega$ have been dropped to keep the notation from getting cumbersome. When $\overline{P}_c$ is maximum, that is, at a "peak," an infinitesimal move in any "direction" about the peak causes no change in $\overline{P}_c$, or $\delta\overline{P}_c = 0$. For this to be true for any direction,

$$\frac{\partial\overline{P}_c}{\partial U_i} = \frac{\partial\overline{P}_c}{\partial U_i^*} = \frac{\partial\overline{P}_c}{\partial\mu} = 0. \tag{6.87}$$

The first two relations lead to thc conditions

$$\frac{1}{2}F_{fi}^*(i\omega) - b(\omega)U_i^*(i\omega) - \mu U_i^*(i\omega) = 0,$$
$$\frac{1}{2}F_{fi}(i\omega) - U_i(i\omega)b(\omega) - U_i(i\omega)\mu = 0. \tag{6.88}$$

The second of the two equations in (6.88) provides an expression for the optimum velocity $U_{ic}(i\omega)$ under the given constraint:

$$U_{ic}(i\omega) = \frac{F_{fi}(i\omega)}{2[b(\omega) + \mu]}. \tag{6.89}$$

The Lagrange multiplier μ is thus found to have the significance of additional damping μ that the PTO (or another mechanism) must apply, so that $D = b(\omega) + \mu$. To find the correct value of μ, it is argued that since at the "constrained optimum," $U_{ic}(i\omega) = \beta$ (both β and so μ may be given a frequency dependence), using Equation (6.89) and some algebra it is found that

$$\mu = \frac{|F_{fi}(i\omega)|}{2\beta} - b(\omega). \tag{6.90}$$

It is relatively easy for a single-mode device then to determine the extra damping needed ahead of time to satisfy the constraint β. The absorbed power under this constraint can be found by direct substitution of the constrained optimum velocity expression into Equation (6.85) to find

$$\overline{P}_c = \frac{F_{fi}F_{fi}^*}{4[b(\omega) + \mu]} - \frac{1}{2}U_{ic}[b(\omega) + \mu]U_{ic}^* + \frac{1}{2}\mu\beta^2. \tag{6.91}$$

Further algebra shows that

$$\overline{P}_c = \frac{|F_{fi}|^2}{8[b(\omega) + \mu]} + \frac{1}{2}\mu\beta^2. \tag{6.92}$$

We note that the extra damping load if applied by the PTO in order to restrict device excursion could be large if radiation damping is small. This would imply large actuator forces. There is thus a trade-off to be made among at least three design attributes, namely, the size and geometry, device/actuator excursion limits, and actuator forces that can be allowed. We note that for most devices the oscillations cannot in any case be so large that the entire hull emerges from water or becomes fully submerged. There is thus a natural "volume" constraint on oscillation, which can be used to define the quantity α. Actuator stroke length limits may often be smaller and so override the volume constraint. For heaving-point absorber type devices, the volume constraints define the so-called Budal limit on power capture [33, 83]. Additionally, as mentioned, independent of oscillation limits, PTO mechanisms may be limited in the maximum force or torque they can apply and support. Thus, any reactive forces required in control may also be limited by actuator capacity. A framework for addressing both oscillation and force constraints is provided in Chapter 13.

6.9 Oscillating Water Column Devices

Oscillating water column devices need a special treatment because of their operating principle. It is the rise and fall in the water level inside a chamber contained within a structure that causes air flow and pressure oscillations within the chamber. The containment structure may be fixed relative to the sea floor. If the containment structure is floating, the combined dynamics of the structure and the water surface within the chamber drive the subsequent conversion stages. Early work on the oscillating water column in the 1970s treated the entire water inside the chamber as a virtual "solid" body or defined an equivalent piston with a thickness depending on the depth of immersion of the front lip. A more satisfying treatment was proposed by Evans in the early 1980s [86], which more directly modeled the dynamics by extending the solid-body potential theory framework to an oscillating chamber pressure and the air-flow flux in and out of the chamber. Thus, the "equation of motion" related the total air-flow flux to the chamber pressure and its time rate of change. The total flow flux was a linear superposition of the flux with the chamber under zero load and zero (relative to atmospheric) chamber pressure (i.e., just the effect of the containment structure interacting with the wave field), the flow resulting just from a nonzero chamber pressure, and a further flow that resulted once the load was activated. The theory was extended to oscillating water columns within floating chambers, where in addition to the three effects above, there was the flow flux caused by the floating body oscillation modes, and essentially by Newton's third law, the effect of the flow flux on the floating body oscillations [88].

We again assume linearity here as before. We first consider a single oscillating water column within a single fixed chamber (Fig. 6.6). The total velocity potential $\phi \equiv \phi(\mathbf{r}; t)$ can be written as before as [86]

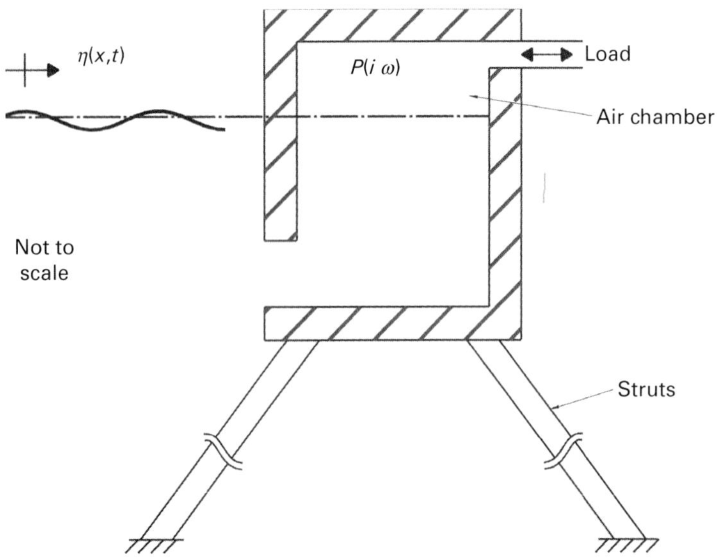

Figure 6.6 An oscillating water column device fixed to the sea floor containing a single chamber.

$$\phi = \phi_D + \phi_r \tag{6.93}$$

where ϕ_D represents the diffraction potential (incident plus just the disturbance due to the body), and ϕ_r the radiation potential representing the outgoing waves created by an oscillating pressure within the chamber. We assume harmonic excitation here, with $\phi = \varphi e^{i\omega t}$, and so on.

The total volume flow flux across the chamber inner surface S can be found as

$$\iint_S \frac{\partial \varphi}{\partial z} dS = \iint_S \frac{\partial \varphi_D}{\partial z} dS + \iint_S \frac{\partial \varphi_r}{\partial z} dS. \tag{6.94}$$

Defining the volume flow flux on the left side as q_t for total, and the two on the right as q_d and q_r for diffraction flow flux and radiation flow flux respectively,

$$q_t = q_d + q_r. \tag{6.95}$$

The diffraction flow flux q_d occurs solely due to the diffracted (incident plus disturbed) wave field and exists even with the chamber fully open to atmosphere at the top. Once the chamber is partially closed to atmosphere, however, an oscillating pressure develops, which will add its influence to the flow flux and force a wave field around the structure. The resulting radiated wave field is a function of the chamber pressure and the flow flux q_r caused by just the pressure. For harmonic or regular waves, these fluxes can be expressed as $q_t = \Re[Q_t(i\omega)e^{i\omega t}]$, $q_d = \Re[Q_d(i\omega)e^{i\omega t}]$, and $q_r = \Re[Q_r(i\omega)e^{i\omega t}]$. The radiation flow flux is directly related to the chamber pressure and can be expressed as

$$Q_r(i\omega) = -i\omega A(\omega)P(i\omega) - B(\omega)P(i\omega) \tag{6.96}$$

where $p(t) = \Re[P(i\omega)e^{i\omega t}]$ denotes chamber pressure. Note thus that $A(\omega)$ and $B(\omega)$ here are somewhat analogous to the added mass and radiation damping variations for solid bodies, except that here $A(i\omega)$ denotes the radiation susceptance and $B(\omega)$ denotes the radiation conductance in the frequency domain [88]. One also notices an electrical analogy between voltage and pressure, and flow-flux and current. A Kramers–Kronig relation also exists between $A(\omega)$ and $B(\omega)$ due to the causal nature of the radiation effect of the chamber pressure. Introducing a radiation admittance $Y(i\omega) = B(\omega) + i\omega A(\omega)$,

$$Q_t(i\omega) = Q_d(i\omega) - Y(i\omega)P(i\omega). \tag{6.97}$$

When there is a PTO load connected to the oscillating water column, it offers an impedance to the air flow or an admittance to the chamber pressure, expressed as

$$Q_t(i\omega) = \Gamma(i\omega)P(i\omega). \tag{6.98}$$

Substitution of Equation (6.98) into Equation (6.97) leads to

$$[Y(i\omega) + \Gamma(i\omega)]\, P(i\omega) = Q_d(i\omega). \tag{6.99}$$

We see thus that, just as the exciting force due to the diffraction potential drives velocity dynamics for floating bodies, the diffraction flow flux drives the chamber pressure dynamics for oscillating water columns. The power absorbed by the oscillating water column can be expressed as

$$P_c(\omega) = P_e(\omega) - P_r(\omega) \tag{6.100}$$

where $P_e(\omega)$ is the power supplied by the diffraction flow flux, while $P_r(\omega)$ is the power radiated away in the form of outgoing waves generated by the oscillating chamber pressure:

$$P_e(\omega) = \frac{1}{4}\left[P^*(i\omega)Q_d(i\omega) + Q_d^*(i\omega)P(i\omega)\right]. \tag{6.101}$$

Further,

$$P_r(\omega) = \frac{1}{4}\left[P^*(i\omega)Q_r(i\omega) + Q_r^*(i\omega)P(i\omega)\right]. \tag{6.102}$$

Using Equation (6.96) and realizing that $Y(i\omega) + Y^*(i\omega) = 2B(\omega)$,

$$P_r(\omega) = \frac{1}{2}P^*(i\omega)B(\omega)P(i\omega). \tag{6.103}$$

To find the condition for maximum absorbed power $P_c(\omega)$, Equation (6.100) is differentiated with respect to $P(i\omega)$ and $P^*(i\omega)$ to obtain

$$\frac{\partial P_c(\omega)}{\partial P(i\omega)} = 0, \quad \frac{\partial P_c(\omega)}{\partial P^*(i\omega)} = 0. \tag{6.104}$$

The second of the two conditions above leads to

$$Q_d(i\omega) = 2B(\omega)P(i\omega) \tag{6.105}$$

or

$$P(i\omega) = \frac{Q_d(i\omega)}{2B(\omega)} \tag{6.106}$$

for a single oscillating water column chamber. This relation again is analogous to the previously encountered relation for optimum velocity for single-mode floating body devices. Substitution of Equation (6.106) into Equation (6.100) via Equations (6.103) and (6.101) leads to

$$P_c(\omega) = \frac{Q_d^*(i\omega)Q(i\omega)}{8B(\omega)}. \tag{6.107}$$

It has been shown by Evans [86] that, for oscillating water columns contained within three-dimensional fixed structures (e.g., immersed cylinders open at the bottom),

$$B(\omega) = \frac{1}{8\lambda P_w}\int_0^{2\pi} |Q_d(i\omega)|^2 d\theta, \tag{6.108}$$

which for an axisymmetric structure simply becomes

$$B(\omega) = \frac{k|Q_d(i\omega)|^2}{8P_w}. \tag{6.109}$$

Thus, there is again a symmetry or reciprocity between the radiation and the diffraction problems, here expressed in terms of the radiation conductance and the diffraction flow flux. Use of Equation (6.109) with Equation (6.107) leads to the expression for maximum power absorption by oscillating water columns within fixed axisymmetric structures,

$$\eta_c = \frac{P_c(\omega)}{P_w} = \frac{1}{k} = \frac{\lambda}{2\pi}, \tag{6.110}$$

where η_c represents the maximum power capture length.

For two-dimensional (crest-spanning) single oscillating water columns within fixed structures,

$$B(\omega) = \frac{1}{2\rho\omega}\left[|\mathcal{K}_{UW}(i\omega)|^2 + |\mathcal{K}_{DW}(i\omega)|^2\right] \tag{6.111}$$

where $\mathcal{K}_{UW}(i\omega)$ and $\mathcal{K}_{DW}(i\omega)$, respectively, are the up-wave and down-wave Kochin functions for the pressure distribution, directly related to the far-field radiated wave amplitudes.

As pointed out earlier, this treatment was extended to oscillating water columns within floating structures by Falnes and McIver [88]. For ease of understanding, a structure in predominantly single-mode oscillation containing a single oscillating water column is considered below. Examples include heaving axisymmetric buoys enclosing oscillating water column tubes. Generalizations to six rigid body modes with multiple oscillating water column chambers (such as the Kaimei, Mighty Whale, and more recent applications of the backward bent oscillating water columns) can be derived using matrix algebra.

The dynamic modeling of floating oscillating water columns is complicated by the two-way interaction between the rigid-body motion of the structure and the oscillating pressure variation produced within the chamber. Chamber pressure in this case is influenced not only by the passing wave field, but also by the oscillation of the rigid body containing the oscillating water column. In turn, any forcing of the chamber pressure by the rigid body oscillation produces an equal and opposite reaction, so that the oscillating chamber pressure also influences the rigid body oscillation. It is recalled further from Equation (6.98) that the chamber pressure is also influenced by the PTO. Under linear theory, the total air-flow flux received by the PTO is thus a linear superposition of (1) the diffraction flow flux q_d occurring when the structure is held fixed in approaching waves and the chamber is fully open to atmosphere, (2) the radiation flow flux q_r caused by the chamber pressure, and (3) an additional contribution q_c forced by the structure containing the oscillating water column as it oscillates in the absence of approaching waves and with the chamber fully open to atmosphere. The third component depends on the rigid-body velocity of the structure and the geometric coupling between the structure and the chamber (as determined by the underwater shape of the structure, the placement of the chamber, the direction and depth of the underwater opening of the oscillating water column, etc.). The total flow flux can now be expressed as

$$q_t = q_d + q_r + q_c. \tag{6.112}$$

For harmonic excitation, we let $q_c(t) = \Re[Q_c(i\omega)e^{i\omega t}]$, and further represent $Q_c(i\omega) = S(i\omega)U_i(i\omega)$ where $U_i(i\omega)$ is the complex amplitude of the rigid-body velocity as before.

The frequency-dependent complex coefficient $S(i\omega)$ has units of area and is termed the "radiation coupling coefficient," as it represents the amount of coupling between the rigid-body oscillation of the structure and the dynamic response of the oscillating water column. Then recalling Equation (6.96),

$$Q_t(i\omega) = Q_d(i\omega) - Y(i\omega)P(i\omega) + S(i\omega)U_i(i\omega). \tag{6.113}$$

With the PTO opposing the flow via its own admittance $\Gamma(i\omega)$ so that $Q_t(i\omega) = \Gamma(i\omega)P(i\omega)$ as before,

$$[Y(i\omega) + \Gamma(i\omega)]\, P(i\omega) - S(i\omega)U_i(i\omega) = Q_d(i\omega). \tag{6.114}$$

Accompanying Equation (6.114) should be an equation of motion for the floating rigid body containing the oscillating water column. The force caused by the chamber pressure on the rigid body can be written in terms of complex amplitudes as $F_c(i\omega) = S(i\omega)P(i\omega)$. Recall that $S(i\omega)$ has units of area. Then, the equation of motion in harmonic excitation can be written as

$$\left[i\omega\,[m + \overline{a}(\infty) + a(\omega)] + \frac{k}{i\omega} + D + b(\omega)\right] U_i(i\omega) - S(i\omega)P(i\omega) = F_f(i\omega). \tag{6.115}$$

The radiation coupling coefficient $S(i\omega)$ can also be expressed as

$$S(i\omega) = G(\omega) + i\omega J(\omega). \tag{6.116}$$

Whereas the radiation conductance $B(\omega)$ is related to the diffraction flow flux $Q_d(i\omega)$ and the radiation damping $b(\omega)$ to the exciting force $F_f(i\omega)$, the imaginary part $J(\omega)$ of the coupling coefficient $S(\omega)$ is related to both diffraction quantities as [88] (but see also [82])

$$-\,iJ(\omega) = \frac{1}{8\lambda P_w}\int_0^{2\pi} Q_d(i\omega,\theta)F^*(i\omega,\theta)d\theta. \tag{6.117}$$

For axi symmetric oscillating water columns within axi symmetric structures, this reduces to

$$-\,iJ(\omega) = \frac{kQ_d(i\omega)F^*(i\omega)}{8P_w}. \tag{6.118}$$

The oscillating pressure distribution viewpoint thus enables a straightforward way to model oscillating water column/rigid body motion coupling. Since wave radiation due to internal pressure variation in an oscillating water column is causal, the radiation conductance $B(\omega)$ and $A(\omega)$ are related by Kramers–Kronig relations. This fact can be used to determine $B(\omega)$ and $A(\omega)$, and $b(\omega)$ and $a(\omega)$, if required, respectively, from the corresponding diffraction quantities $Q_d(i\omega)$ and $F_f(i\omega)$. Similarly, the impulse response function corresponding to $S(i\omega)$ is also causal. Hence, once $J(\omega)$ is determined $G(\omega)$ can be computed via the Kramers–Kronig relations between $J(\omega)$ and $G(\omega)$. The reciprocal relations also serve as a validation for numerical approaches to determine the hydrodynamic parameters.

6.10 Device Arrays

Use of a large number of device units arrayed together in some manner (e.g., arrayed end to end for terminators, side to side for attenuators, and in a matrix fashion for axi symmetric point absorbers) was first considered in the 1970s, with the goal of maximizing the overall energy capture from an approaching wave front. In addition, the interactions between individual devices within arrays could be used to increase the energy conversion by individual units, thereby minimizing the number of units required in an array. Point-absorber arrays have been considered particularly attractive, where a large number of small devices combine to provide the requisite amount of power. There is a further cost benefit from arrays due to the economy resulting from mass production of the structure, PTO, electronics, and so on. Furthermore, there is also the possibility of integrating control strategies with array configurations to maximize the economic benefits. This is of particular interest to this text and is considered in detail in Chapters 11 and 14. Underlying results primarily based on hydrodynamic theories of the early years of wave energy are summarized here, with the goal of understanding inter unit interactions. Fig. 6.7 shows an example of an array comprised of small three-dimensional devices.

Each device unit in the array experiences a force due to the wave radiated by any of the other units. The radiation damping and added mass thus become matrices with diagonal and non diagonal terms. Due to reciprocity, the force exerted on a unit n by the wave radiated by another unit m equals the force on the unit m due to the wave radiated by unit n. The radiation damping and added mass matrices are thus symmetric. It has been shown further [53] that the radiation damping $\leftrightarrow$ exciting force relation can be further generalized to relate each element in the radiation damping matrix to particular elements within the exciting force vector as

$$b_{nm} = \frac{1}{8\lambda P_w} \int_0^{2\pi} F_{fn}(\theta) F^*_{fm}(\theta) d\theta. \tag{6.119}$$

F_{fn} and F_{fm} are the complex amplitudes of the exciting forces on bodies n and m, respectively, in the presence of others. If we assume next that the device units are small enough or placed far enough apart so that the presence of one unit does not distort the far field created by any of the other units, then as shown by Evans [53], the maximum capture length of a point absorber buoy in such an array can be expressed as

$$l_{max} = \frac{P_{max}}{P_w} = \frac{\lambda}{2\pi} N q(\mu, \theta_I) \tag{6.120}$$

where N is the number of buoys in the array, $\mu = kd$, with $k = 2\pi/\lambda$ and d the spacing between adjacent buoys. θ_I is the approach angle of the incident wave. Evans finds an expression for the interaction gain coefficient or the "q-factor" $q(\mu, \theta_I)$ as

$$q(\mu, \theta_I) = \frac{1}{N} \mathbf{L}^* \mathbf{J}^{-1} \mathbf{L} \tag{6.121}$$

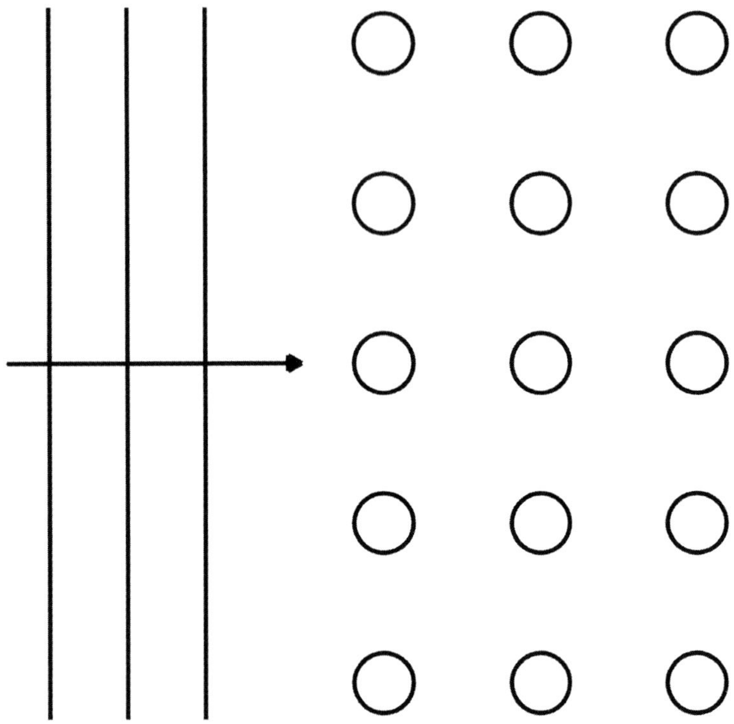

Figure 6.7 A large number of three-dimensional axisymmetric devices working in an array. Point absorber arrays are considered particularly attractive due to (1) the possibility of high overall efficiency and (2) the advantage of mass production.

where elements of the matrix **J** are

$$J_{nm} = \frac{1}{2\pi} \int_0^{2\pi} L_n(\theta) L_m^*(\theta) d\theta \tag{6.122}$$

where * has been used to denote complex-conjugate transpose when it is associated with the vector **L** and represents just a complex conjugate for the vector element L_m. Each element L_n is

$$L_n = \mathrm{e}^{i\mu n \sin\theta_I}. \tag{6.123}$$

Each element of **J** thus can be expressed in terms of the zeroth-order Bessel function as

$$J_{nm} = J_0 \left[\mu(n-m)\right]. \tag{6.124}$$

It was found, for just $N = 2$, that the interaction gain $q(\mu, 0)$ was about 1.6 for $d = 0.6\lambda$. Similar results have been obtained for thin attenuators (with appropriate modifications to the parameter definitions) [89].

As we consider Equation (6.120) further, we may realize that the array interaction could be favorable, $q(\mu, \theta_I) > 1$, or unfavorable, $q(\mu, \theta_I) < 1$, depending on μ and θ_I. The effect of wave direction on array interactions was discussed in [89]. In particular, assuming that the radiation damping matrix is not singular (e.g., for heaving point absorbers separated according to spacing μ it is not singular), the maximum power that can be converted is [53]

$$P_{max} = \frac{1}{8}\mathbf{F_f}^*(i\omega, \theta_I)\mathbf{b}(\omega)^{-1}\mathbf{F_f}(i\omega, \theta_I). \tag{6.125}$$

Then using Equations (6.125), (6.119), and (6.120), and some algebra, it can be seen that [89]

$$q(\mu, \theta_I) = \frac{1}{N}\sum_{i=1}^{N}\sum_{j=1}^{N} F^*_{fi}(i\omega, \theta_I)\left(\frac{1}{2\pi}\int_0^{2\pi} F_{fi}(i\omega, \theta)F^*_{fj}(i\omega, \theta)d\theta\right)F_{fj}(i\omega, \theta_I). \tag{6.126}$$

The forces $F_{fi}(i\omega, \theta_I)$ can be related to the far-field radiation amplitudes due to each body (see Equation (6.66)). As discussed in [89], for a row of heaving point absorbers comprised of $N = 5$ units, q is a function of the parameter ka where a denotes the radius of each unit and k the wave number, but is even more strongly affected by the separation parameter $\mu = kd$. In general, for waves approaching such that the wave crests are parallel to the line passing through the device centroids, namely, $\theta_I = 0$, the specific values of the parameter μ can be identified where the interaction is most favorable; for instance, for $ka = 0.4$, $\mu = 5.1$ leads to a q value of 2.25. For head-sea conditions, that is, with the crest lines perpendicular to the line through the device centroids ($\theta_I = \pi/2$), the interaction is generally unfavorable: $q \leq 1$ for most of the kd values in the test range. For intermediate directions, q shows an intermediate range of values. It is seen further that the bodies oscillate in phase in beam seas when $\mu \geq 2.1$ for the case studied, while the phase variations are greater in head seas.

The effect of oscillation constraints is also studied in [89]. The maximum capture length l_{max} falls appreciably if oscillations are constrained to three or two times the incident wave amplitude. In beam seas, favorable interaction ($q > 1$) occurs only near $kd = 5.1$, with the phase dependence being similar to that without constraints. Phases are particularly important in head seas under constraints, as there is a considerable drop in the available l_{max} compared with the optimum oscillation case.

As most practical applications require oscillation constraints, and wave directions approaching an array can vary, typical array layouts may be two-dimensional and may include rows and columns of elements or even elements arranged along sides of triangles. Further, control can be used to influence both amplitude and phase of individual oscillations, which in turn will influence the interaction parameter q and the overall capture length l. Therefore, not only can control influence the performance of particular array layouts, as shown in Chapter 11, but the gains from control and the ease of control application (in terms of necessary forces) could prompt certain choices of array layouts. Chapter 14 pursues this point further and explores ways in which the control strategy and array layout can be designed to maximize overall performance.

As a final note on the present discussion of arrays, an interesting comparison has been reviewed in [90] between a compact two-dimensional array comprised of a large number of point-absorber buoys and a single large buoy of the same effective volume. Based on an approximate theory, it has been found that the array can provide greater absorption over a greater bandwidth than the single buoy, when both the radiation and diffraction effects of the small buoys are small enough to be ignored. The comparison also ignored viscous friction for both cases. It is interesting to note that certain spacings that satisfy $\mu = kd = n\pi$ for uniformly spaced two-dimensional arrays lead to the Bragg scattering effect [91], where most of the incident energy is reflected and the overall conversion efficiency drops (note, for instance, also the head-sea behavior observed in [89]).

6.11 Conclusion

We tried to probe a little deeper into the heart of a wave energy device in this chapter. Our understanding was shaped by a somewhat limited view that the assumption of linearity provides. Although this view is still far-reaching and the best available if design is to be based on solid insight, we should recognize that the real behavior of wave energy devices at sea could be a little more complicated. The discussion of this chapter is offered in the faith that good engineering relies on the best available understanding, not so much of what an existing system is known to do as what a new system likely will, and another value of mathematical arguments in wave energy lies in their ability to encompass large classes of devices.

In addition, our attempt here was to provide an overview for an engineer, and many of the critical details were left out. For a comprehensive look at the linear hydrodynamics of wave energy devices, reference may be made to early works of Bessho, Evans, Newman, Mei, and Falnes (many of which are listed in the References at the end of the book) and to the text by Falnes [66].

The purpose of this chapter was to emphasize the crucial role hydrodynamics plays in the energy absorption performance of a device, and no amount of control, however sophisticated, can help overcome poor hydrodynamic design without excessive, unaffordable, and impractical penalties elsewhere. We saw that, for floating devices designed to perform mainly in the beam-sea configuration, a shape maximizing wave radiation into the direction from which most waves arrive will perform most efficiently (note that the pitching Salter duck performed efficiently by radiating mostly in front, so the radiated wave canceled the reflected wave). Alternatively, the device could be allowed to oscillate in two modes (one symmetric and one anti symmetric, such as heave and surge) so that the radiated waves combined to produce a radiated wave field in the preferred direction. We considered the Bristol submerged cylinder in beam-sea configuration with its centroidal axis in circular orbital motion, which could function at 100% efficiency by generating a radiated wave that canceled the transmitted wave, such a cylinder having been shown not to produce a reflected wave. For omni directional devices with axial symmetry, the hydrodynamics are slightly different. They can capture energy from a crest width larger than their own diameters. However, an axisymmetric heaving device produces a symmetric radiation pattern, while an axisymmetric surging device produces a pattern with a directionality. The surge mode therefore allows even greater (twice as much) energy capture, if unconstrained. These are possible maxima for single devices, which can be achieved when impedance matching conditions prevail.

For single-mode devices with no viscous damping when oscillating without constraints, impedance matching occurs at resonance if, in addition, the rate of energy absorption (damping) equals the rate of energy radiation. Control can extend the range of wave conditions under which impedance matching can be achieved. It cannot stretch the capture maxima beyond what is mathematically possible (e.g., for beam-sea devices with equal radiation in up-wave and down-wave directions, the best result control can provide is 50% absorption). Finally, it is important to recall that good energy conversion is possible with good energy radiation. High radiation damping is to be preferred in design, as then greater power can be captured without driving oscillation amplitudes past limits.

Given that a number of energy conversion concepts have used oscillating water columns for primary conversion, a summary was included of the approach that generalizes our hydrodynamic understanding of oscillating bodies to oscillating pressures. Further, given the recent resurgence of interest in arrays of WECs, a short overview of the main features was also included, and it was seen how individual devices within arrays can interact hydrodynamically to increase the overall capture such that total absorption exceeds the simple sum of the maxima achievable with individual units.

Part IV

Velocity Control Using a Hydrodynamic Model

In this part, we examine the control problem based on the fundamentals of the hydrodynamic model. In particular, Chapter 7 uses Cummins' equation as a starting point and develops the classic reactive control strategy leading to a velocity response that approaches the hydrodynamic velocity optimum. Chapter 7 also addresses the issue of the requirement for future wave elevation information and discusses two possible ways to utilize up-wave measurements for this purpose. One uses instantaneous up-wave measurements at a distance related to the prediction time ahead and assumes nondispersive propagation. The second accounts for dispersive propagation and uses up-wave measurements made over a duration of time and at a distance, both of which are directly related to the practical range of group velocities commonly found in wave spectra and the prediction time ahead. Chapter 8 uses a simplified hydrodynamic model, motivated by the idea that, while offering some compromise in performance, less detail in the model may offer improved robustness properties. The method proposed in Chapter 8 addresses the issue of future information by modeling the wave spectrum as a narrow-banded process, where an extended Kalman filter is used to make instantaneous measurements of wave period and amplitude. The control method can cater to device amplitude constraints, and a possible lower-level power take-off controller is presented.

7 Reactive Control in Time Domain

As we have seen repeatedly in the previous chapters, a differential equation description of a wave energy device holds only in regular waves, due to the nature of the radiation force. A frequency domain thought process leading to the cancellation of the reactive terms in the dynamics and matching of the radiation damping (ignoring viscosity) is accurate only on a frequency-by-frequency basis. As a real wave input at any given time in general contains the combined effect of numerous frequency components, reactive cancellation and resistive matching requires a time domain approach that works on a wave-by-wave basis. Several control approaches have emerged in recent years that are based on the integro-differential equation description (Cummins' equation) of a device [63], seeking to optimize energy absorption based on certain performance criteria. In light of our discussion in Chapters 4 and 6, any approach leading to the hydrodynamic optimum for the real-time velocity in irregular waves must seek reactive cancellation and resistive matching in real time. The degree of approximation and realism employed in attempting to reach the hydrodynamic optimum and the amount of information required as to the incoming waves have varied over the years. The goal of this chapter is to describe the closest practical approaches to the hydrodynamic velocity optimum in the time domain.

7.1 Approaching the Hydrodynamic Optimum

Reactive control was first attempted in the mid-1970s by Salter [92] and Budal [93]. Both were able to increase the bandwidth, that is, the range of frequencies where good conversion occurred. Salter first implemented it in the "narrow tank," using analog instrumentation to measure the instantaneous position and velocity of the duck and using both to determine the torque that opposed the duck oscillation. The torque contained a part proportional to displacement, a part proportional to velocity, and a part proportional to acceleration. The adjustable constants of proportionality constituted, respectively, the load spring, load damping, and load inertia. Negative spring and positive inertias were used to extend the bandwidth into the longer-period range. Effectively, this amounted to controlling both the amplitude and the phase of the force applied by the Power take-off (PTO). The approach was later extended to multiple-mode oscillation and power absorption in the "wide tank," and thus represented multiple-mode complex-conjugate control [94, 95]. Single-mode reactive control was later also implemented in tank tests on an oscillating water column device [47]. Devices could be tuned to the peak frequencies in particular wave spectra by controlling the reactive component in the load, which would allow a device to adjust to daily and/or seasonal changes in the approaching wave spectra. Such an approach was demonstrated on the Wave Star device recently, showing two- to three-fold increases in the annual energy conversion [45].

We have seen in the previous chapters why implementation of a complex-conjugate impedance match for optimum velocity requires information from the future. Specifically, if the amplitude, phase, and frequency of every one of the component waves combining to form an irregular wave input were known, the future wave input would be known, much as in the case of a regular wave of known amplitude, phase, and frequency. This information, however, is not available for an irregular wave input. How do we proceed to estimate/predict the information from the future? The following treatment formalizes some of this discussion and suggests some approaches based on basic hydrodynamic modeling that could provide estimates for the wave input and device response from the future in some circumstances. These approaches are here tested through simulation.

We consider a single mode device again in order to keep the focus on the fundamentals. The particular mode of energy absorption is not critical for this discussion as long as the parameter and motion units (mass/moment of inertia, linear/angular displacement, etc.) are accounted for in a consistent manner. The integro-differential equation appropriate for irregular waves may be written in terms of velocity v as

$$[m + \overline{a}(\infty)]\,\dot{v} + c_d\dot{v} + \int_0^{\infty} h_r(\tau)v(t-\tau)d\tau + k\int_{-\infty}^{t} v(\tau)d\tau = F_f + F_r \qquad (7.1)$$

with $\int_{-\infty}^{t} v(\tau)d\tau = x(t)$, the displacement. Small-amplitude waves and small oscillations are assumed in this section. Note further that the actuator is thought to be linear, ideal (i.e., without any power losses), and capable of providing the required force magnitudes. Any friction, hysteresis, and stiction effects in the PTO system are also ignored. Viscous losses due to body oscillation in water are assumed to be linear. The system is thus assumed to be fully linear. As discussed in Chapter 4, note that the radiation force is represented in part by means of the convolution integral with the kernel h_r operating on the velocity v, and in part by the inertial force term $\overline{a}(\infty)\dot{v}$ due to the infinite-frequency added mass. Due to the linearity of the radiation problem, the kernel h_r represents the impulse response function for the body radiating waves while oscillating in the absence of incident and diffracted waves. As discussed in Chapter 4, since both the radiation force and the velocity are real valued, $h_r(t)$ is real, and since the radiation force depends only on the present and past oscillations, $h_r(t)$ is causal, or $h_r(t) = 0,\ t < 0$. Causality of $h_r(t)$ also implies that its Fourier transform is analytic in the upper half of the complex frequency plane. The real and imaginary parts of the Fourier transform are therefore related and not independent of each other (see Chapter 4). The relations allowing one to be expressed in terms of the other are the Kramers–Kronig relations introduced in Chapter 4. Thus,

$$\int_{-\infty}^{\infty} h_r(t)\mathrm{e}^{-i\omega t}dt = H_r(i\omega) = i\omega a(\omega) + b(\omega). \tag{7.2}$$

As before the frequency-dependent added mass $\bar{a}(\omega) = \bar{a}(\infty) + a(\omega)$, so that $a(\omega)$ represents just the frequency-variable part. As we saw in Chapter 4, because $h_r(t)$ is real valued, $H_r(i\omega) = -H_r^*(i\omega)$. Hence, $b(-\omega) = b(\omega)$ and $-i\omega a(-\omega) = -i\omega a(\omega)$, implying that $b(\omega)$ is an even function, and $\omega a(\omega)$ is an odd function. Therefore, their inverse Fourier transforms $h_b(t)$ and $h_a(t)$ must also be real-valued even and odd functions, respectively, where

$$h_b(t) = \frac{1}{2\pi}\int_{-\infty}^{\infty} b(\omega)\mathrm{e}^{i\omega t}d\omega,$$

$$h_a(t) = \frac{1}{2\pi}\int_{-\infty}^{\infty} i\omega a(\omega)\mathrm{e}^{i\omega t}d\omega, \tag{7.3}$$

with

$$i\omega a(\omega) = \int_{-\infty}^{\infty} h_a(t)\mathrm{e}^{-i\omega t}dt,$$

$$b(\omega) = \int_{-\infty}^{\infty} h_b(t)\mathrm{e}^{-\omega t}dt. \tag{7.4}$$

Now, $h_r(t) = h_b(t) + h_a(t)$. Because $h_r(t) = 0,\ t < 0,\ h_b(t) + h_a(t) = 0,\ t < 0$. Therefore, $h_b(t) = h_a(t) = h_r(t)/2,\ t > 0$, and $h_b(t) = -h_a(t),\ t < 0$. Consequently,

$$h_r(t) = \frac{2}{\pi}\int_0^{\infty} b(\omega)\cos\omega t d\omega = -\frac{2}{\pi}\int_0^{\infty} \omega a(\omega)\sin\omega t d\omega. \tag{7.5}$$

The two relations above can be inverted to write

$$a(\omega) = -\frac{1}{\omega}\int_0^{\infty} h_r(t)\sin\omega t dt,$$

$$b(\omega) = \int_0^{\infty} h_r(t)\cos\omega t dt. \tag{7.6}$$

As we saw earlier in Chapter 4, optimum velocity requires that the reactive forces be canceled out while the rate of energy absorption is matched by the PTO load. $\omega a(\omega)$ here drives just the frequency-dependent reactive part of the radiation force due to the generation of waves that do not propagate into the far field. On the other hand, $b(\omega)$ represents the resistive part due to the generation of propagating waves that take energy away from the body into the far field. For a PTO load force to cancel the reactive force but match the resistive force, the phase relationship between the reactive and resistive forces must be a "conjugate" of the phase relationship between the two forces as imparted on the body by the fluid. In the time domain, this means that the impulse response function for the load force must be a mirror image ("time-reversed" image), about the $t = 0$ axis, of the radiation impulse response function. The mirror-image impulse response function is thus fully noncausal, or anticausal, and operates on velocity information of the future only. Information is required from as far into the future as the memory effect of $h_r(t)$ continues responding to the velocity of the past (recall that such knowledge is implicitly available for a regular wave of known phase and amplitude but is not readily available for irregular waves). Often, this duration could be on the order of a few tens of seconds. However, the forces involved are important to the annual productivity of a wave energy converter, so considerable effort has been directed at synthesizing versions of the correct load force to varying degrees of approximation.

As mentioned in Chapter 4, there is another aspect of the hydrodynamics that complicates control in irregular waves. In the frequency domain, it is customary to express the exciting force F_f per unit incident wave amplitude at the device centroid. Thus, in the frequency domain,

$$F_f(i\omega) = H_f(i\omega)\eta(x_B; i\omega) \tag{7.7}$$

where $H_f(i\omega)$ is the complex frequency response function for the exciting force and is just the exciting force (amplitude and phase) at frequency ω per unit incident wave amplitude at the body centroid x_B (see Chapter 6). $\eta(x_B; i\omega)$ is the actual incident wave surface elevation contribution at x_B of a particular frequency component at the correct phase. Not directly reflected in this representation are the facts that (1) the force is a result of fluid pressure acting over the entire wetted surface of the body and that (2) because wave action over the free surface is a continuum phenomenon, the body experiences a force before a wave elevation actually arrives at its wetted surface. Hence, if a time domain expression for the exciting force F_f is obtained via inverse Fourier transformation of Equation (7.7) (as is commonly done), then $F_f(t)$ is a noncausal force, beginning to be affected by the incident wave profile considerably before the wave reaches the centroid x_B or another designated point on the body. Thus,

$$F_f(t) = \int_{-\infty}^{\infty} h_f(\tau)\eta(x_B; t - \tau)d\tau \tag{7.8}$$

where

$$h_f(t) = \frac{1}{2\pi}\int_{-\infty}^{\infty} H_f(i\omega)e^{i\omega t}d\omega \tag{7.9}$$

and

$$\eta(x_B; t) = \frac{1}{2\pi}\int_{-\infty}^{\infty} \eta(x_B; i\omega)e^{i\omega t}d\omega. \tag{7.10}$$

As just discussed, $h_f(t) \neq 0, t < 0$; that is, h_f is noncausal. This is another challenge to real-time control in irregular waves (though not in regular waves), as it implies that information on the future wave surface elevation is needed if the exciting force is to be known at the present instant. As discussed in Chapter 9, the exciting force needs to be known ahead of time in order to compute the on-off switching action used in approaches such as latching and declutching. The question of incident-wave prediction is addressed in Sections 7.2.2–7.3.2.

It should be noted that in all of the work in this chapter, we assume that $\eta(\cdot; t) \to 0, t \to \pm\infty$, so that the Fourier transform $\eta(\cdot; i\omega)$ exists [51]. There is some practical justification for allowing such an assumption, as the control as discussed here will likely be applied only during periods of wave activity near the device occurring in between periods of negligible wave activity. In addition, for most realistic spectra, it is justifiable to assume that $\eta(\cdot; i\omega) \to 0$ as $\omega \to \infty$.

Returning to the equation of motion (7.1), we see that the first term $[m + \overline{a}(\infty)]\dot{v}$ is driven by the current acceleration only, as the fourth term $k \int_{-\infty}^{t} v(\tau)d\tau$ is by the current displacement only. These two terms are both reactive and demand no net energy except what might be dissipated in the actuator. The second term $c_d\dot{v}$ depends on the current velocity only and is resistive. As seen in Chapter 4, information from only the present would be required to synthesize the appropriate control forces if the response of the body was not complicated by the nature of the radiation and exciting forces. To cancel the effect of the inertial term, one may sense the acceleration of the body using accelerometers (if appropriate noise filtering techniques are used) and synthesize a force,

$$F_a(t) = [m + \overline{a}(\infty)]\,\dot{v}(t). \tag{7.11}$$

Similarly, a force may be synthesized using the instantaneous displacement of the body, which could be measured using transducers based on underwater acoustics, or simply via accelerometers if the zero-displacement reference is appropriately set. Thus,

$$F_s(t) = kx. \tag{7.12}$$

We assume here that m, $\overline{a}(\infty)$, and k are known exactly. As perfect knowledge is rare in practice, additional measures such as on-line parameter estimation may be employed so as to use the best parameter estimates at any time. This would constitute adaptive control and is not considered here. For the present, the overall control force F_r can then be expressed as

$$F_r = F_a + F_s + F_c. \tag{7.13}$$

F_a and F_s are causal and reactive. With F_r applied, the equation of motion now reduces in order, such that

$$c_d v(t) + \int_0^{\infty} h_r(\tau)v(t-\tau)d\tau = F_f(t) + F_c(t). \tag{7.14}$$

Decomposing $h_r(t)$ into its even and odd component parts $h_b(t)$ and $h_a(t)$, as discussed above,

$$c_d v(t) + \int_{-\infty}^{\infty} h_a(\tau)v(t-\tau)d\tau + \int_{-\infty}^{\infty} h_b(\tau)v(t-\tau)d\tau = F_f(t) + F_c(t). \tag{7.15}$$

It is convenient to include the term containing c_d within the force term F_c, where

$$F_c(t) = -c_d v(t) + \int_{-\infty}^{\infty} h_a(\tau)v(t-\tau)d\tau - \int_{-\infty}^{\infty} h_b(\tau)v(t-\tau)d\tau. \tag{7.16}$$

Note that the first integral is intended to cancel the contribution of the odd function h_a (related to $\omega a(\omega)$), while the second matches the contribution of the even function h_b

related to $b(\omega)$. Together, they make up the sign-reversed version of the anticausal force required. With this substitution for $F_c(t)$ in Equation (7.15),

$$2\left(c_d v(t) + \int_{-\infty}^{\infty} h_b(\tau)v(t-\tau)d\tau\right) = F_f(t), \tag{7.17}$$

which can be further expressed as

$$2\int_{-\infty}^{\infty} \left[c_d\delta(\tau) + h_b(\tau)\right] v(t-\tau)d\tau = F_f(t) \tag{7.18}$$

where $\delta(t)$ is the Dirac delta function. Recall that

$$F_f(t) = \int_{-\infty}^{\infty} h_f(\tau)\eta(x_B; t-\tau)d\tau. \tag{7.19}$$

Equation (7.17) represents the response of the device under full real-time control, and the solution $v(t) = v_o(t)$ to Equation (7.17) is the hydrodynamically optimum velocity expressed in the time domain. A time domain solution for $v_o(t)$ can be obtained indirectly via the following procedure:

$$v_o(t) = \frac{1}{2\pi}\int_{-\infty}^{\infty} v_o(i\omega)e^{i\omega t}d\omega \tag{7.20}$$

where

$$v_o(i\omega) = \frac{H_f(i\omega)\eta(x_B; i\omega)}{2\left[c_d + b(\omega)\right]}. \tag{7.21}$$

Then, an impulse response function $h_o(t)$ is defined as

$$h_o(t) = \frac{1}{2\pi}\int_{-\infty}^{\infty} \frac{H_f(i\omega)}{2[c_d + b(\omega)]}e^{i\omega t}d\omega. \tag{7.22}$$

An inverse Fourier transform of $v_o(i\omega)$ thus becomes a convolution integral leading to

$$v_o(t) = \int_{-\infty}^{\infty} h_o(\tau)\eta(x_B; t-\tau)d\tau. \tag{7.23}$$

Because $h_f(t)$ is not a causal function with respect to the incident wave surface elevation $\eta(x_B; t)$, neither is $h_o(t)$. Therefore, even the synthesis of the desired optimum velocity $v_o(t)$ to drive a feedback loop with constant gains around the device requires knowledge or prediction into the future of the wave surface elevation at the device.

As we noted earlier, the impulse response function $h_f(t) \to 0$, that is, becomes vanishingly small at a long enough time $t_{f1} > 0$. Similarly, in the $t < 0$ range, $h_{f2} \to 0$ when $|t| > t_{f2}$. The finite length of h_f for $t > 0$ implies that surface elevation information only up to t_{f1} from the past is required. The finite length of $h_f(t)$ in the $t < 0$ range means that only $\eta(x_B; t)$ up to t_{f2} into the future need be predicted. In Equation (7.19), the integral on the right may be truncated as

$$F_f(t) = \int_{-t_f}^{t} h_f(\tau)\eta(x_B; t-\tau)d\tau \tag{7.24}$$

where t_f is defined to be the larger of t_{f1} and t_{f2}, or $t_f = \max[t_{f1}, t_{f2}]$. The force $F_f(t)$ at the present instant can be expressed to practical accuracies if wave surface elevation information is available that goes back a duration t_f and looks forward all the way until t. For a device whose dynamic response is modeled by means of Equation (7.1), moreover, the velocity $v(t)$ at the present instant is approximately dependent on the wave elevation $\eta(x_B; t)$ until $t = t$ into the past and until $t = t_f$ into the future.

The control force $F_c(t)$ to be synthesized at any time t also requires knowledge or prediction of the velocity $v(t)$ from the future. However, it is known that in realistic terms, the radiation impulse response function $h_r(t)$ has a "finite" memory going back up to $t = t_c$, because $h_r(t) \to 0$, $t > t_c$. Consequently, since $h_a(t) + h_b(t) = h_r(t)$, both $h_a(t)$ and $h_b(t) \to 0$, $t > t_c$. Further, since $h_a(t)$ and $h_b(t)$ are odd and even functions, respectively, in the $t < 0$ range, $h_a(t)$ and $h_b(t) \to 0$, $t < t_c$. Both $h_a(t)$ and $h_b(t)$ can thus be truncated to zero past finite times $-t_c$. Thus, in a practical application, the force $F_c(t)$ could be expressed using a truncated form of Equation (7.16) as

$$F_c(t) = -c_d v(t) + \int_{-t_c}^{t} h_a(\tau)v(t-\tau)d\tau - \int_{-t_c}^{t} h_b(\tau)v(t-\tau)d\tau. \tag{7.25}$$

We see thus that synthesis and application of a force $F_c(t)$ with a PTO will require knowledge of the device velocity going back a duration t and looking ahead a duration t_c. However, the need to know $v(t)$ up to t_c ahead implies that $F_f(t)$ must therefore be known up to $t_c + t_f$ ahead into the future.

An approximate approach to synthesizing $F_c(t)$ based on first principles is outlined in the following section.

7.2 Control Force Synthesis

Fig. 7.1 illustrates the synthesis of the impulse response function that would truly represent the noncausal control force necessary. Approximations can really be obtained in practice, as outlined in the subsections below and in Section 7.3.

7.2.1 Right-Shifting of Impulse Response Functions

In a practical calculation or synthesis, the impulse response functions h_f, h_a, and h_b need to be expressed as causal functions. The fact that these functions can be truncated at finite lengths of time makes this process easier. For instance, with h_f truncated at $\pm t_f$, the entire function can be shifted to the right (this is equivalent to shifting the origin to the left), to define

$$h_{fd}(t) = h_f(t - t_f). \tag{7.26}$$

In the frequency domain, this procedure reflects as a frequency-response function $H_{fd}(i\omega)$, where

$$H_{fd}(i\omega) = H_f(i\omega)\mathrm{e}^{-i\omega t_f}. \tag{7.27}$$

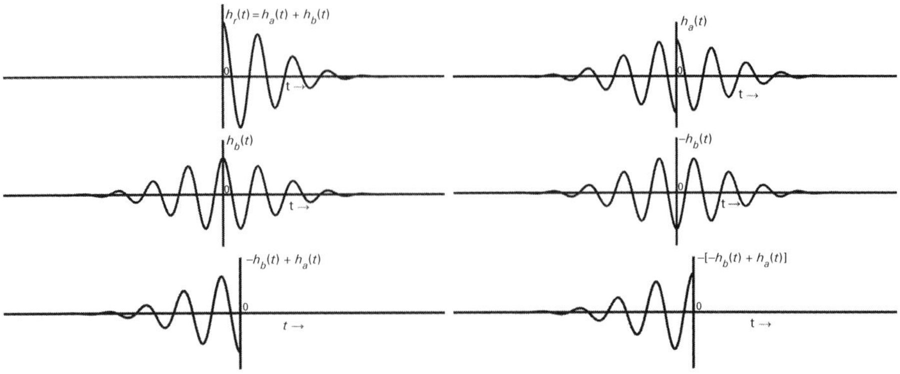

Figure 7.1 The final anticausal (or completely noncausal) impulse response function is needed for true optimal control in small-amplitude irregular waves. Approximations based on up-wave measurement and time-series prediction are needed if a control force based on the last impulse response function is to be synthesized in practice.

In a similar manner, right-shifted impulse response functions can be formed using the truncated h_a and h_b, so that

$$h_{ac}(t) = h_a(t - t_c),$$
$$h_{bc}(t) = h_b(t - t_c). \tag{7.28}$$

The corresponding frequency domain quantities $a_c(i\omega)$ and $b_c(i\omega)$ are

$$i\omega a_c(i\omega) = i\omega a(\omega)\mathrm{e}^{-i\omega t_c},$$
$$b_c(i\omega) = b(\omega)\mathrm{e}^{-i\omega t_c}. \tag{7.29}$$

For the convolution terms defining F_c in terms of the truncated impulse response functions as in Equation (7.25) to have the same values, however, the shift must be reflected in the velocity $v(t)$ on which the right-shifted impulse response functions operate. This can be seen easily in the frequency domain, starting with a Fourier-transformed version of Equation (7.25) as

$$F_c(i\omega) = -c_d v(i\omega) - b(\omega)v(i\omega) + i\omega a(\omega)v(i\omega). \tag{7.30}$$

It follows from Equations (7.29) that

$$i\omega a(\omega) = i\omega a_c(i\omega)\mathrm{e}^{i\omega t_c},$$
$$b(\omega) = b_c(i\omega)\mathrm{e}^{i\omega t_c}, \tag{7.31}$$

implying that

$$F_c(i\omega) = -c_d v(i\omega) - b_c(i\omega)\mathrm{e}^{i\omega t_c}v(i\omega) + i\omega a_c(i\omega)\mathrm{e}^{i\omega t_c}v(i\omega). \tag{7.32}$$

It is easy to see that a velocity of the future written as $v_c(i\omega)$ in the frequency domain can now be defined, such that

$$v(i\omega) = v_c(i\omega)\mathrm{e}^{-i\omega t_c}. \tag{7.33}$$

Then, in the time domain,

$$v(t) = v_c(t - t_c). \tag{7.34}$$

Combining Equations (7.35) and (7.33),

$$F_c(i\omega) = -c_d v(i\omega) - b_c(i\omega)v_c(i\omega) + i\omega a_c(i\omega)v_c(i\omega). \tag{7.35}$$

In the time domain, this can be written in the convolution form as

$$F_c(t) = -c_d v(t) - \int_0^{t+t_c} h_{bc}(\tau)v_c(t-\tau)d\tau + \int_0^{t+t_c} h_{ac}(\tau)v_c(t-\tau)d\tau. \tag{7.36}$$

Equation (7.36) highlights the need to know the velocity v_c from the future, as far ahead as t_c beyond the current instant. Estimates based on time-series analysis can be used to supply approximate values to the convolutions. Alternatively, as discussed a little later, approximate estimates can also be derived using a first-principles approach based on linearity of all elements taking part in the process.

A procedure similar to that followed above can be used to evaluate the exciting force $F_f(t)$ at the present instant. To gain an understanding of the convolution in Equation (7.24) under the right-shifted impulse response function h_{fd}, we apply the frequency domain treatment again. For the impulse response function h_{fd} as defined earlier

$$H_{fd}(i\omega) = H_f(i\omega)\mathrm{e}^{-i\omega t_f}. \tag{7.37}$$

Then, with

$$F_f(i\omega) = H_f(i\omega)\eta(x_B; i\omega) \tag{7.38}$$

we have

$$F_f(i\omega) = H_{fd}(i\omega)\mathrm{e}^{i\omega t_f}\eta(x_B; i\omega). \tag{7.39}$$

If we now define a surface elevation of the future at x_B using

$$\eta(x_B; i\omega) = \eta_A(x_B; i\omega)\mathrm{e}^{-i\omega t_f} \tag{7.40}$$

so that in the time domain

$$\eta(x_B; t) = \eta_A(x_B; t - t_f), \tag{7.41}$$

then

$$F_f(i\omega) = H_{fd}(i\omega)\eta_A(x_B; i\omega). \tag{7.42}$$

This has the time domain interpretation

$$F_f(t) = \int_0^{t+t_f} h_{fd}(\tau)\eta_A(x_B; t-\tau)d\tau. \tag{7.43}$$

The surface elevation of the future may again be obtained via time-series estimation based on past information. We examine two alternative approaches below. The first makes a long-wave approximation to simplify our treatment of the linear wave-propagation process. The second utilizes a more complete propagation model and considers wavelength and propagation-velocity ranges that cover most realistic wave spectra.

7.2.2 Wave Propagation and Future Information

An overarching assumption we make below is that over the distances of interest, wave propagation can be modeled deterministically, and any highly local effects of wind and other disturbances cause only small or negligible errors. A large portion of the development below is intended for applicability in swell-dominated wave spectra, for which the consequences of unmodeled wind forcing are not expected to be serious.

As we saw in Chapter 5, waves carry surface elevations occurring at one location to another location farther down in the propagation direction. Realistically, this transfer takes a finite length of time. Thus, surface elevations at a certain location at a given time will be "felt" at a down-wave location some time in the future. Conversely, one could say that the surface elevations occurring "now" at a certain location were felt some time ago at an up-wave location. A highly simplified analogy to this comes from wave propagation on a stretched string, where a pulse created by a disturbance at one end of the string travels unchanged down the length of the string all the way to the other end.

For surface waves in the ocean, the situation is a bit more complicated, not least because the pulse will change its shape considerably as it travels and, for far-enough locations, will experience some spreading before it arrives. Recall that this is because ocean waves are dispersive for the most part, and travel in groups of propagating waves in deep water, wherein the longer waves always overtake the shorter waves. A pulse is frequently understood as a superposition of many component waves of single frequency and single amplitudes. Since each travels at a different speed, the components quickly disperse out of the pulse as it propagates, and the pulse loses its original shape. Contrast this with electromagnetic waves or light through space, where all frequencies (and wavelengths) propagate at the same speed (namely, the speed of light), and packets of light remain as packets of light through the process. It is recalled that for ocean waves, generally speaking,

$$\omega^2 = gk \tanh kh \tag{7.44}$$

where ω is the angular frequency, g is the acceleration of gravity, $k = 2\pi/\lambda$ is the wave number with λ defining the wavelength, and h is the local water depth.

The consequences of the points above for unidirectional, long-crested waves are: (1) that the surface elevation η at point x_B and time t can be related to the up-wave elevations at a previous time, (2) η at x_B that will occur at a future time $t + t_f$ will be a result of elevations occurring up-wave currently at time t, and (3) there is a range of up-wave locations from which $\eta(x_B; t)$ came or from which $\eta(x_B; t + t_f)$ will come. Theoretically, this range spans from 0 to ∞ due to the mathematical presence of waves of zero frequency (infinite period) and infinite frequency (zero period). However, in practice, ocean waves are "band limited" in that there is an identifiable low frequency where the spectral density $\rightarrow$ 0, just as there is an identifiable high frequency beyond which the spectral density is negligible (this could be the realm of short ripples where surface tension effects rather than gravity drive wave propagation). Thus, there is a range of up-wave locations from which the surface elevation at the current time "derives" or "will derive" at future times.

While the first two points above offer a way to obtain information on future surface elevations (and thereby on the future exciting force and future velocities), the third complicates any approach that would be used to provide this information. Simplifying assumptions can be made, however, to enable easier progress toward an approximate, suboptimal implementation of control that would allow an approach to near-optimum velocities in conditions where the assumptions are satisfied.

Extending the discussion from Chapter 5, the surface elevation at any point along a long-crested irregular wave field propagating from left to right in the positive x direction can be represented as [96]

$$\eta(x;t) = \frac{1}{2\pi}\int_{-\infty}^{\infty} A(i\omega)\mathrm{e}^{-i[k(\omega)x-\omega t]}d\omega. \tag{7.45}$$

This is equivalent to a frequency domain representation for the surface elevation,

$$\eta(x;i\omega) = A(i\omega)\mathrm{e}^{-ik(\omega)x}. \tag{7.46}$$

For two points x_A and x_B along the x axis, thus

$$\eta(x_A;i\omega) = A(i\omega)\mathrm{e}^{-ik(\omega)x_A}, \quad \eta(x_B;i\omega) = A(i\omega)\mathrm{e}^{-ik(\omega)x_B}. \tag{7.47}$$

If x_A is up-wave of x_B, that is, closer to $x=0$, and $x_A = x_B - d$, then

$$\eta(x_B;i\omega) = A(i\omega)\mathrm{e}^{-ik(\omega)(x_A+d)} = \mathrm{e}^{-ik(\omega)d}\eta(x_A;i\omega). \tag{7.48}$$

Below, we choose d to be a specific distance d_f defined as the distance over which the longest wave in a group travels over a time t_f. Then with ω_l denoting the lowest frequency in an approaching wave spectrum (below which the spectral density has a negligible value), and k_l the corresponding wave number as defined by Equation (7.74), with the phase velocity v_l expressed as ω_l/k_l,

$$d_f = v_l t_f = \frac{\omega_l}{k_l}t_f \Rightarrow k_l d_f = \omega_l t_f. \tag{7.49}$$

If we assume that the spectrum is dominated by small wave numbers, or long wavelengths (i.e., swells), such that close to the energy period T_e of the spectrum, the wavelength is several times longer than the longest device dimension, then it may be justifiable to suppose that wave groups are dominated by waves that would be nondispersive at water depths at which neither the small-amplitude wave nor the small-amplitude device motion has a noticeable effect at distances exceeding about half the wavelength below the free surface. For instance, the wavelength corresponding to an energy period $T_e = 13$ s is around 250 m, and at a depth approaching half this wavelength, the water particle velocities will approach magnitudes about 4% of the values near the free surface. Broadly speaking, if we were to evaluate the hydrodynamic coefficients and exciting forces for the device in waters deeper than this, then only small differences would be observed when compared with the corresponding results obtained assuming infinite depth. Denoting such a depth (i.e., the depth beyond which the frequency variations of the hydrodynamic coefficients and exciting force are indistinguishable from those obtained for infinite depth) as h,

$$v_l \approx v_{gmx} = \sqrt{gh} \tag{7.50}$$

where v_{gmx} is the maximum group velocity at the defined depth h that is independent of frequency and equals the phase velocity, and where propagation is nondispersive. When the dominant portion of the spectrum as signified by the energy period T_e satisfies these conditions, for the surface elevation measured at a location x_A d_f up-wave of x_B,

$$kd_f \approx \omega t_f \tag{7.51}$$

and

$$\eta(x_B; i\omega) = \mathrm{e}^{-ik(\omega)d_f}\eta(x_A; i\omega) = \mathrm{e}^{-i\omega t_f}\eta(x_A; i\omega). \tag{7.52}$$

In other words, an up-wave measurement d_f up-wave of the device is here being used to approximate the surface elevation, as would occur at the device time t_f into the future. Thus,

$$F_f(i\omega) = H_f(i\omega)\eta(x_B; i\omega) = H_{fd}(i\omega)\eta(x_A; i\omega) \tag{7.53}$$

in the frequency domain, and

$$F_f(t) = \int_0^{t+t_f} h_{fd}(\tau)\eta(x_A; t-\tau)d\tau. \tag{7.54}$$

It is interesting to note that under the conditions assumed, $F_f(t)$ at the present instant can be found directly using the current and past (through $t - 2t_f$) surface elevation measurements at an up-wave location x_A, and no explicit prediction or time-series estimation is required. We note that this single-group velocity propagation approximation is used (but not not explicitly mentioned) in [97].

A similar approach could be used to compute the desired $v_o(t)$ to the approximation implied above. Returning to Equation (7.21) and recalling the arguments leading up to Equation (7.53),

$$v_o(i\omega) = \frac{H_{fd}(i\omega)\eta(x_A; i\omega)}{2[c_d + b(\omega)]} \tag{7.55}$$

in the time domain

$$v_o(t) = \int_0^{t+t_f} h_o(\tau)\eta(x_A; t-\tau)d\tau \tag{7.56}$$

where

$$h_o(t) = \frac{1}{2\pi}\int_{-\infty}^{\infty} \frac{H_{fd}(i\omega)}{2[c_d + b(\omega)]}\mathrm{e}^{i\omega t}d\omega. \tag{7.57}$$

Equation (7.56) enables calculation of the approximate desired optimum velocity for comparison with the actual velocity in a feedback loop, for instance. Such a feedback loop could complement the feedforward force F_r in Equation (7.1) above and provide corrections for external disturbances and measurement errors. The control force component related to noncausal effects, $F_c(t)$ in Equation (7.36), requires foreknowledge or prediction of $v_c(t)$ up to a time t_c into the future. The discussion above argues that in swell-dominated spectra, the wave surface elevation at an up-wave location x_A in long-crested waves at the present time t may be an adequate approximation for the wave elevation at the device location x_B at a time t_f into the future. Therefore, for a device responding as a linear system, the velocity of an "imagined" device sufficiently far up-wave should also be a reasonable approximation for the device velocity at the desired time into the future. It is possible thus to define a distance d_c such that $d_c = v_{gmx}t_c$ and $k_l d_c = \omega_l t_c$. As before, for swell-dominated spectra, extending this relationship to other wave numbers and frequencies, $k(\omega)d_c \approx \omega t_c$.

As a result, $v_c(t)$ (from the present instant through time $t + t_f$) in the past a distance d_c up-wave of the device location x_B may be used in the convolution integrals in Equation (7.36). However, in order to obtain $v_c(t)$ of the imagined device located d_c up-wave at the current instant, the exciting force $F_f(t)$ at the imagined device is required. This would in turn require a wave surface elevation at a distance d_f up-wave of the imagined device. Therefore, if a real-time wave surface elevation measurement at a point $d_R = d_c + d_f$ is available, $v_c(t)$ can be evaluated approximately for use in Equation (7.36).

7.2.3 Approximate Evaluation of Control Force

With the force components $F_s(t)$ and $F_a(t)$ determined using real-time displacement (direct or derived) and acceleration measurements, it remains to approximate the component $F_c(t)$. It is noted that all components of $F_r(t)$ are essentially feedforward forces, which can be applied regardless of whether or not a feedback loop exists around the device to attempt to minimize the difference between the desired and the actual velocities. Rather, with $F_s(t)$ and $F_a(t)$ in place, the goal has been to compute (to some approximation) the control force knowing the desired velocity (i.e., $v_o(t)$ in Equation (7.56)). However, the relation to be used to compute the $F_c(t)$ (Equation (7.36)) requires $v_c(t)$ to be computed using real-time measurement of the surface elevation d_R up-wave. One approach would be to use the desired velocity of the imagined device, that is, $v_{oC}(t)$, to drive the linear system in Equation (7.36). It is not difficult to see that if $v_o(t)$ at the device location can be computed using the impulse response function $h_o(t)$ and the surface elevation measurement history $\eta(x_A; t)$, then the same impulse response function can be used to compute the desired $v_c(t)$ for the imagined device located d_c up-wave. Thus, with the approximation

$$\eta(x_R; i\omega) = e^{ik(\omega)d_c}\eta(x_A; i\omega) \tag{7.58}$$

so that

$$v_{oC}(i\omega) = \frac{H_{fd}(i\omega)\eta(x_R; i\omega)}{2[c_d + b(\omega)]}, \tag{7.59}$$

in the time domain this has the form

$$v_{oC}(t) = \int_0^{t+t_R} h_o(\tau)\eta(x_R; t-\tau)d\tau. \tag{7.60}$$

An examination of Equation (7.35) shows that $F_c(i\omega)$ has, as expected, a resistive component and a reactive component. We let

$$F_l(i\omega) = -c_d v(i\omega) - b_c(i\omega)v_c(i\omega) \tag{7.61}$$

and

$$F_a(i\omega) = i\omega a_c(i\omega)v_c(i\omega). \tag{7.62}$$

Using the desired velocities v_o and v_{oC} to drive the two parts,

$$F_l(i\omega) = -\frac{c_d H_{fd}(i\omega)}{2[c_d + b(\omega)]}\eta(x_A; i\omega) - \frac{b_c(i\omega)H_{fd}(i\omega)}{2[c_d + b(\omega)]}\eta(x_R; i\omega) \tag{7.63}$$

and

$$F_a(i\omega) = \frac{i\omega a_c(i\omega)H_{fd}(i\omega)}{2[c_d + b(\omega)]}\eta(x_R; i\omega). \tag{7.64}$$

In the time domain, thus

$$F_l(t) = -\int_0^{t+t_f} h_{l1}(\tau)\eta(x_A; t-\tau)d\tau - \int_0^{2t_R} h_{l2}(\tau)\eta(x_R; t-\tau)d\tau \tag{7.65}$$

and

$$F_a(t) = \int_0^{t+t_R} h_a(\tau)\eta(x_R; t-\tau)d\tau \tag{7.66}$$

where the three impulse response functions can be found using the following relations:

$$h_{l1}(t) = \frac{1}{2\pi}\int_{-\infty}^{\infty} \frac{c_d H_{fd}(i\omega)}{2[c_d + b(\omega)]} e^{i\omega t} d\omega, \tag{7.67}$$

$$h_{l2}(t) = \frac{1}{2\pi}\int_{-\infty}^{\infty} \frac{b_c(i\omega)H_{fd}(i\omega)}{2[c_d + b(\omega)]} e^{i\omega t} d\omega, \tag{7.68}$$

$$h_a(t) = \frac{1}{2\pi}\int_{-\infty}^{\infty} \frac{i\omega a_c(i\omega)H_{fd}(i\omega)}{2[c_d + b(\omega)]} e^{i\omega t} d\omega. \tag{7.69}$$

A block diagram summarizing the overall approach above is shown in Fig. 7.3. Note that the control forces F_l and F_a are evaluated as feedforward inputs driving the actuator. These are based on real-time measurement of the wave surface elevation up-wave of the device. With the up-wave wave profile available, the desired velocity can be computed in real time for comparison with the actual velocity resulting from these forces. Any differences (i.e., due to unmodeled modes, measurement errors, actuator dynamics model errors, other external disturbances, etc.) would be used in a feedback control loop to generate forces seeking to narrow the difference. It should be noted, however, that the performance of even the closed-loop system would be subject to the assumptions made in the use of the instantaneous up-wave surface elevations following the approach above, as these assumptions are implicit also in the evaluation of the desired velocity.

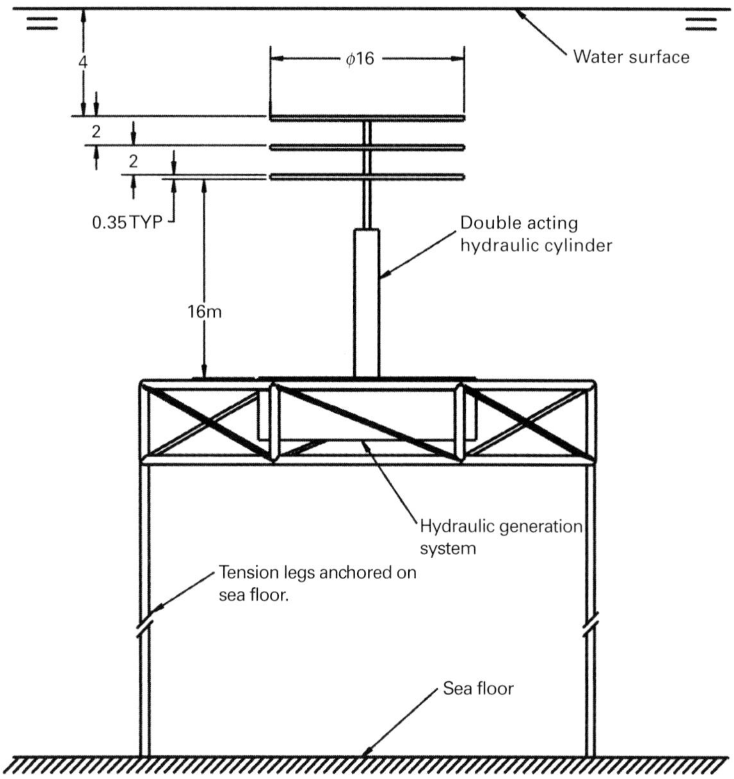

Figure 7.2 Side-view drawing of the device studied as an example case [97] (dimensions in meters) for the approximate control based on the long-wave assumption (single group velocity) and nondispersive propagation. Instantaneous wave measurements at one point up-wave are used. (Copyright: Elsevier Ltd.; reproduced by permission.)

The device velocity resulting from the present control strategy can be evaluated as in [97]. Denoting this velocity as $v_{act}(t)$, the power absorbed by the device can be computed as (ignoring actuator losses),

$$P_w = \frac{1}{T_f} \int_0^{T_f} F_l(t) v_{act}(t) dt. \tag{7.70}$$

The instantaneous absorbed power under the control above is shown in Fig. 7.4. Note the negative values, which are, in part, due to the approximations used through the process leading up to the impulse response functions and in part due to the use of instantaneous up-wave surface elevation measurements. Two-parameter Pierson–Moskowitz type spectra are used to generate the incident waves at the device location as well as up-wave.

The absorbed power under the present control can be compared with that found when no control is applied and damping is set to a constant value, and also when the device heave-mode response is selectively tuned to the peak frequency of the incoming spectra. Only continual spectral measurement is needed for such control at a point near

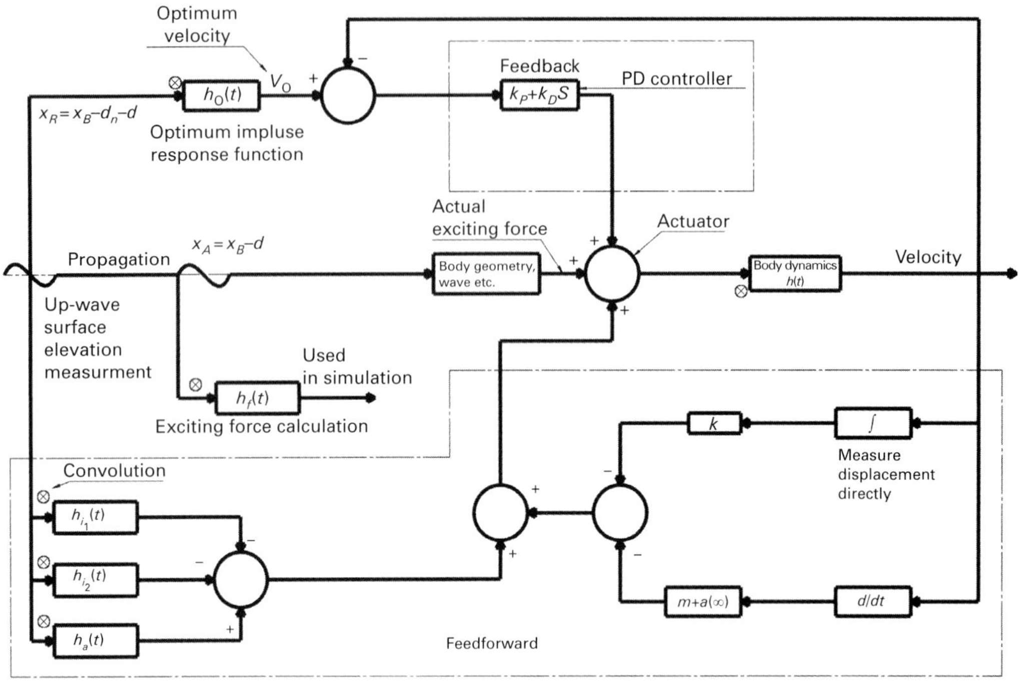

Figure 7.3 A block diagram showing an overall implementation based on up-wave surface elevation. Only the feedforward part is considered in this chapter.

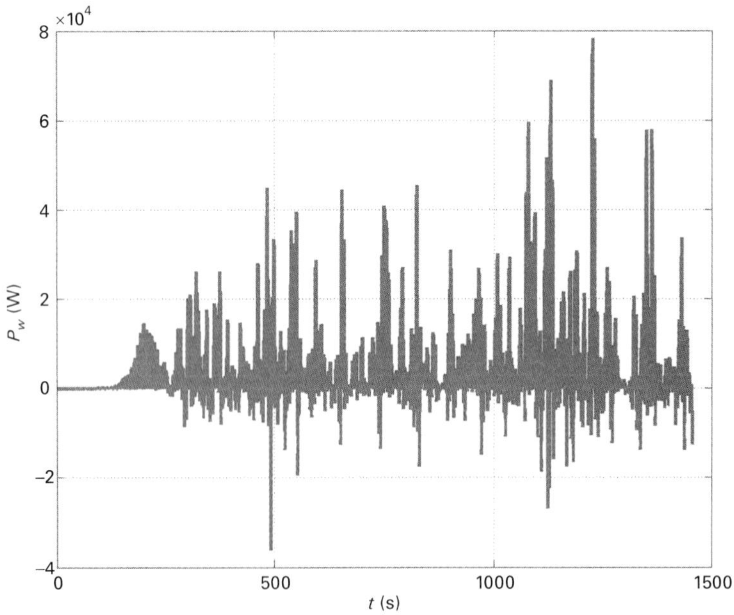

Figure 7.4 Instantaneous power absorbed under approximate control ($T_e = 11$ s, $H_s = 1$ m). Note that the control is clearly approximate because nondispersive propagation has been assumed [97]. (Copyright: Elsevier Ltd.; reproduced by permission.)

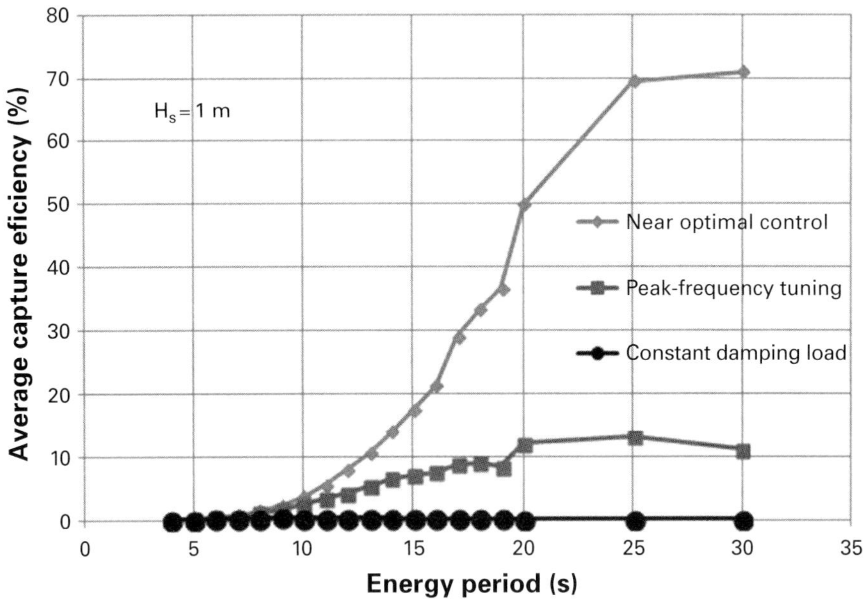

Figure 7.5 Average-power capture efficiency as a function of energy period ($H_s = 1$ m). Instantaneous up-wave measurements are used, and wave propagation is assumed to be nondispersive. [97]. (Copyright: Elsevier Ltd.; reproduced by permission.)

the device under such a "peak-frequency tuning" strategy. Tuning of the response here implies that a reactive force is applied such that the device now resonates at a frequency matching the spectral peak frequency, and the actuator-applied damping is set to a value equal to the radiation damping at the peak frequency plus the approximate viscous damping. These forces do not require a real-time calculation nor foreknowledge of the exciting force (see also [97]). There is thus no need for an up-wave surface elevation measurement (only wave spectra near the device need to be provided, and these can be up-dated every 10–20 min, as was done by the Wave Star group in their sea tests).

Fig. 7.5 compares the 10-min absorbed power averages in different two-parameter wave spectra, over a range of energy periods T_e. A significant wave height of $H_s = 1$ m is assumed. The damping value for the constant damping case is chosen to be of the same order of magnitude as the largest radiation damping value computed for the device in heave oscillation. We note that the present control strategy leads to significant improvements compared with the other two (more than 100% over peak-frequency tuning), particularly in the longer energy period spectra where the assumptions leading to it are more readily applicable.

7.3 Wave Prediction from Up-Wave Measurement Time History

Here we discuss a way to relax the nondispersive propagation assumption and to predict the wave elevation at the device location x_B for use in determining the optimum velocity and the control forces on a wave-by-wave basis. We again assume unidirectional propagation (i.e., long-crested waves) from left to right, and further suppose, as before, that linear superposition of regular wave components can be used to represent an irregular wave. Two approaches, one based on a measurement time history at a single up-wave location and one based on a "snapshot" of wave elevations covering an up-wave distance, are discussed in [98]. We consider an approach based on measurement time history at a single point here, given that such an approach may be potentially easier to implement from a hardware standpoint. Some of the steps from Section 7.2.2 are repeated below in the interest of clarity. The treatment below is adopted from [99]. We recall that our use of Fourier transformations for the wave elevations (as opposed to correlations as discussed in Chapter 5) requires an assumption that [51, 65].

$$\eta(x_m; t) \to 0, \ t \to \pm\infty, \tag{7.71}$$

where x_m is any point along the x axis. This assumption is realistic if the time horizon we choose for the Fourier transformation is long enough to span the length of time between periods of negligible wave activity. If this time horizon is shorter, we effectively impose a periodicity on η, which could affect the accuracy of predictions unless the period we introduce is long enough to be outside the band covered by realistic spectra (i.e., $> 2\pi/\omega_l$), where ω_l is the low-frequency cut-off representing realistic spectra. It should be noted that while our approach may be acceptable for situations where wave elevations only above a certain threshold are of practical interest, strictly speaking, the Fourier transform of a truly random signal with a nonzero mean square does not exist. Though the Fourier transform of the correlation function obtained from such an input does exist [51], recall that the Fourier transform of the autocorrelation function is the spectral density and that our interest here is in wave-by-wave control. Here we suppose that the time horizon we use for Fourier transformations is long enough to satisfy the assumption in Equation (7.71).

As discussed in [96], the contribution of a frequency component to the surface elevation at locations x_A and x_B along the x axis can be expressed as

$$\begin{aligned} \eta(x_A; i\omega) &= A(i\omega)\mathrm{e}^{-ik(\omega)x_A} \\ \eta(x_B; i\omega) &= A(i\omega)\mathrm{e}^{-ik(\omega)x_B} \end{aligned} \tag{7.72}$$

where point A is up-wave of B, that is, $x_A < x_B$,

$$x_A = x_B - d, \tag{7.73}$$

and the wave number $k(\omega)$ is related to the frequency ω through the dispersion relation

$$k \tanh kh = \frac{\omega^2}{g} \tag{7.74}$$

which in deep water takes the form

$$k = \frac{\omega^2}{g}, \quad \text{since } \tanh kh \to 1, \ h \to \infty. \tag{7.75}$$

The group velocity v_g in deep water is

$$v_g = \frac{d\omega}{dk} = \frac{1}{2}\frac{\omega}{k} = \frac{1}{2}\frac{g}{\omega}. \tag{7.76}$$

Impulsive disturbances to the free surface travel at the group velocity v_g [1, 4]. Now,

$$\eta(x_A; i\omega) = A(i\omega)\mathrm{e}^{-ik(\omega)(x_B-d)}, \tag{7.77}$$

so that

$$\eta(x_A; i\omega) = \mathrm{e}^{ik(\omega)d}\eta(x_B; i\omega),$$

$$\eta(x_B; i\omega) = \mathrm{e}^{-ik(\omega)d}\eta(x_A; i\omega). \tag{7.78}$$

7.3.1 Propagation Impulse Response Function

The treatment below follows some of the discussion in [96]. The surface elevation at location x_B as a function of time can be expressed as the inverse Fourier transform of $\eta(x_B; i\omega)$ as

$$\eta(x_B; t) = \frac{1}{2\pi}\int_{-\infty}^{\infty} \eta(x_B; i\omega)\mathrm{e}^{i\omega t}d\omega. \tag{7.79}$$

Substituting the second of Equations (7.78) into Equation (7.79),

$$\eta(x_B; t) = \frac{1}{2\pi}\int_{-\infty}^{\infty} \mathrm{e}^{-ik(\omega)d}\eta(x_A; i\omega)\mathrm{e}^{i\omega t}d\omega. \tag{7.80}$$

Alternatively,

$$\eta(x_B; t) = \frac{1}{2\pi}\int_{-\infty}^{\infty} H_l(i\omega)\eta(x_A; i\omega)\mathrm{e}^{i\omega t}d\omega \tag{7.81}$$

where $H_l(i\omega) = \exp[-ik(\omega)d]$. $H_l(i\omega)$ defines the linear propagation process that relates the wave profile at x_B to that at x_A in the frequency domain. $\eta(x_B; t)$ can then be expressed using Faltung's theorem as

$$\eta(x_B; t) = \int_{-\infty}^{\infty} h_l(\tau)\eta(x_A; t-\tau)d\tau \tag{7.82}$$

where $h_l(t)$ is the impulse response function that defines the linear propagation process in the time domain describing the surface profile created at x_B by an impulsive surface elevation at x_A. Note that this is a "line impulse" in that it spans the entire crest length (i.e., it is independent of the y coordinate). Thus, its propagation is unidirectional. A time-dependent wave profile $\eta(x_A; t)$ at x_A will thus result in a profile $\eta(x_B; t)$ according to the linear process described by Equation (7.82). The $h_l(t)$ defining this propagation process can be found as

$$h_l(t) = \frac{1}{2\pi}\int_{-\infty}^{\infty} H_l(i\omega)\mathrm{e}^{i\omega t}d\omega. \tag{7.83}$$

expected to be within $v_{gmn} \leq v_g \leq v_{gmx}$ as defined above. In summary, for a chosen "prediction interval" t_P ahead of the current time t,

$$\begin{aligned}
&\eta(x_B; t + t_p) = \int_0^T h_l(\tau)\eta(x_A; t - \tau)d\tau, \\
&x_B - x_A = d = t_P v_{gmx}, \\
&T = \frac{d}{v_{gmn}} - \frac{d}{v_{gmx}}, \\
&t > T.
\end{aligned} \tag{7.98}$$

Figs. 7.8 and 7.9 show the results using Equation (7.98) with h_l as given by Equation (7.92), except that $h_l(t) = 0$, $t < 0$, in Equation (7.98). Both for $t_P = t_A = 30$ s, as well as $t_P = t_R = 60$ s, good agreement is found between $\eta(x_B; t + t_P)$ as predicted using Equation (7.98) and the wave profile evaluated directly using the location x_B and time $t + t_P$ for the spectrum being considered.

With the wave elevation at the device location x_B thus predicted to a required time length into the future, the control forces can now be evaluated directly using the truncated impulse response functions of Equations (7.23) and (7.25). The desired optimum velocity $v_o(t)$ and the control force $F_c(t)$ can be computed once sufficient time has elapsed since the start of the up-wave measurements (i.e., sufficient to fill the entire arrays $\eta(x_B; t)$ from $[t - t_f, t + t_f]$ and $[t - t_R, t + t_R]$ in a discrete time implementation).

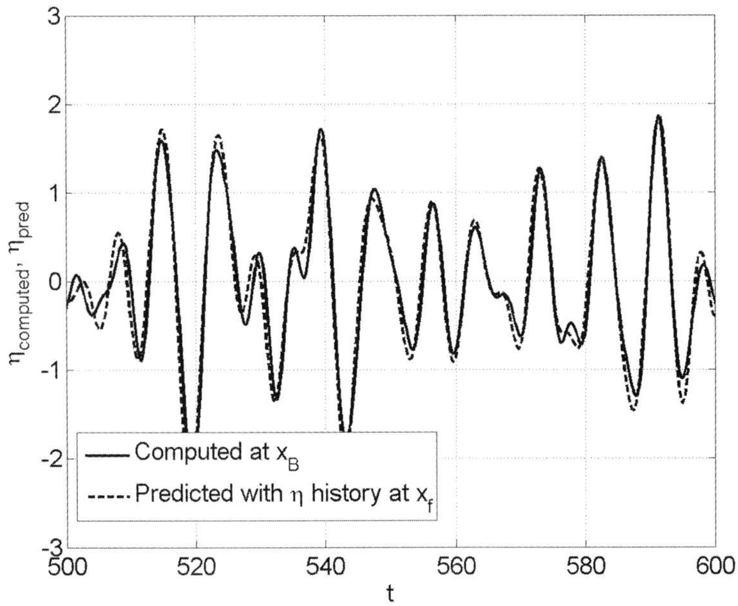

Figure 7.8 Surface elevation at x_B predicted using a measurement over T at an up-wave location x_A. Shown for comparison is the actual wave elevation at x_B for the spectral parameters used. The results are based on a simulation. A two-parameter spectrum with $H_s = 2.9$ m and $T_e = 11$ s with random phase distribution is used [99]. (Copyright: Elsevier Ltd.; reproduced by permission.)

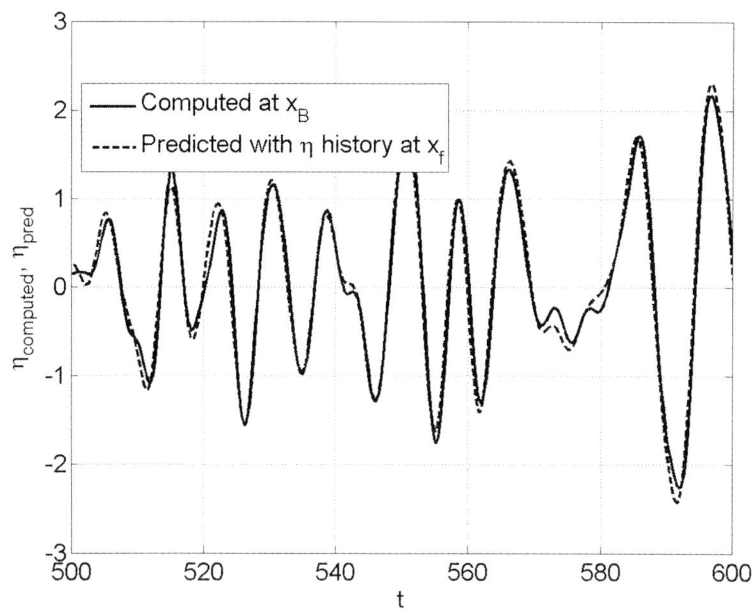

Figure 7.9 Surface elevation at x_B predicted using a measurement over T at an up-wave location x_A. Shown for comparison is the actual wave elevation at x_B for the spectral parameters used. The results are based on a simulation. A two-parameter spectrum with $H_s = 3.7$ m and $T_e = 13$ s with random phase distribution is used [99]. (Copyright: Elsevier Ltd.; reproduced by permission.)

As discussed in [104], the near-optimum velocity $v_{ro}(t)$ for a two-body axisymmetric device comprised of a heaving buoy reacting against a concentric submerged disc (see Fig. 7.10) can be found using

$$v_{ro}(t) = \int_{-\infty}^{\infty} h_o(\tau)\eta(x_B; t-\tau)d\tau \tag{7.99}$$

where

$$h_o(t) = \frac{1}{2\pi}\int_{-\infty}^{\infty} \frac{H_e(i\omega)}{2R_i(\omega)} e^{i\omega t} d\omega \tag{7.100}$$

and where $h_o(t)$ is non-causal. However, we realize that $\eta(x_B; t)$ is now available up to the required time into the future. For instance, h_o could be truncated at $-t_p$, where $t_p \sim 15-30$ s. The integration in Equation (7.100) could proceed from $-t_p$ to the present instant t. The control force part responsible for power absorption can be found as

$$F_l(t) = \int_{-\infty}^{\infty} h_{ri}(\tau)\eta(x_B; t-\tau)d\tau \tag{7.101}$$

where

$$h_{ri}(t) = \frac{1}{2\pi}\int_{-\infty}^{\infty} \frac{H_e(i\omega)}{2} e^{i\omega t} d\omega. \tag{7.102}$$

The reactive force to be applied on the relative oscillation is expressed as

$$F_a(t) = \int_{-\infty}^{\infty} h_{ci}(\tau)\eta(x_B; t-\tau)d\tau \tag{7.103}$$

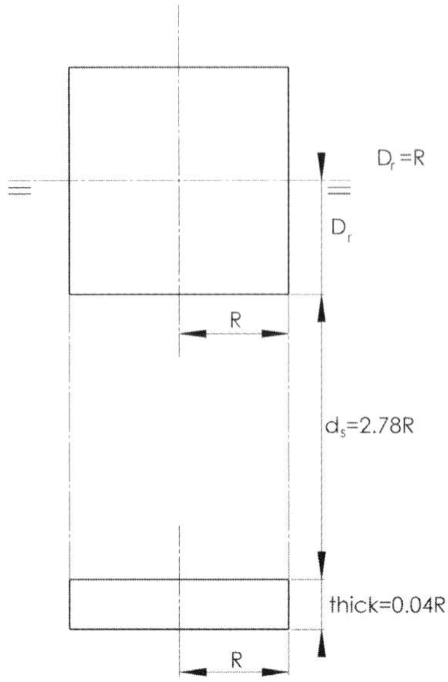

Figure 7.10 Two-body axisymmetric device in predominant heave oscillation; the relative heave is used for power conversion. Near-optimal control using the wave prediction approach of Section 7.3.2 was studied for this device [99]. (Copyright: Elsevier Ltd.; reproduced by permission.)

where

$$h_{ci}(t) = \frac{1}{2\pi}\int_{-\infty}^{\infty} \frac{iH_e(i\omega)C_i(\omega)}{2R_i(\omega)} e^{i\omega t} d\omega. \tag{7.104}$$

Note that h_{ri} and h_{ci} here are also noncausal. However, they can be truncated at $-t_P$, and the integration in Equations (7.102) and (7.104) can proceed from $-t_P$ to the current time t. The predicted wave elevation $\eta(x_B; t)$ up to t_P into the future is used in Equation (7.101). $H_e(i\omega)$, $R_i(\omega)$, and $C_i(\omega)$ represent the effective excitation-force frequency response function, the effective inherent resistive impedance, and the effective reactive impedance associated with the relative heave oscillation between the buoy and the disc. These functions are evaluated based on [105]. More details on the system used in the following figures are given in [99] and [104].

Considerable improvement in absorbed power can be observed, as illustrated in Figs. 7.14 and 7.15, where the former shows just the absorbed power time history using a prediction based on the method of the present section (i.e., accounting for the significant group velocity range). The latter compares the power conversion with the present method to that obtained using just the maximum group velocity as discussed in the previous section. As mentioned, these results are for a two-body heaving axisymmetric device in which the relative heave oscillation between a floating buoy and a submerged disc is used for energy conversion [99, 104]. Fig. 7.15 compares the energy capture lengths for the same device, for the control based on wave prediction of this section with that based on the single group velocity approximation. Relative oscillations are constrained to be less than 1/2 (or 1/4) of the draft of the floating

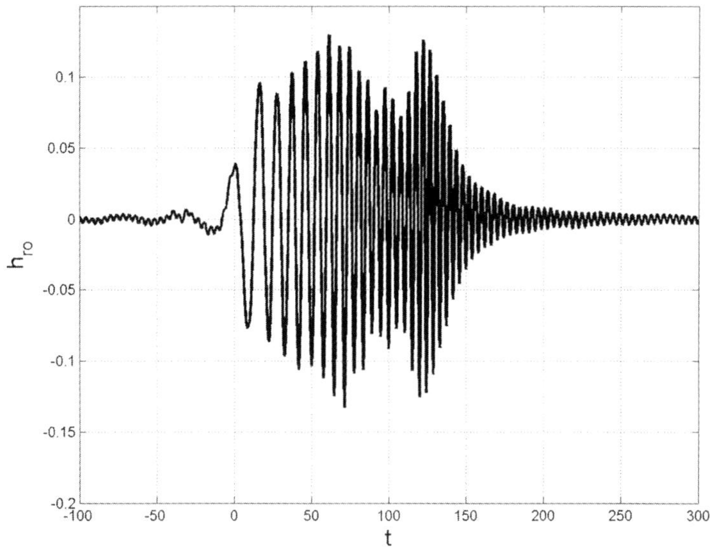

Figure 7.11 Impulse response function h_{ro} used in evaluating the near-optimum relative heave velocity [99]. (Copyright: Elsevier Ltd.; reproduced by permission.)

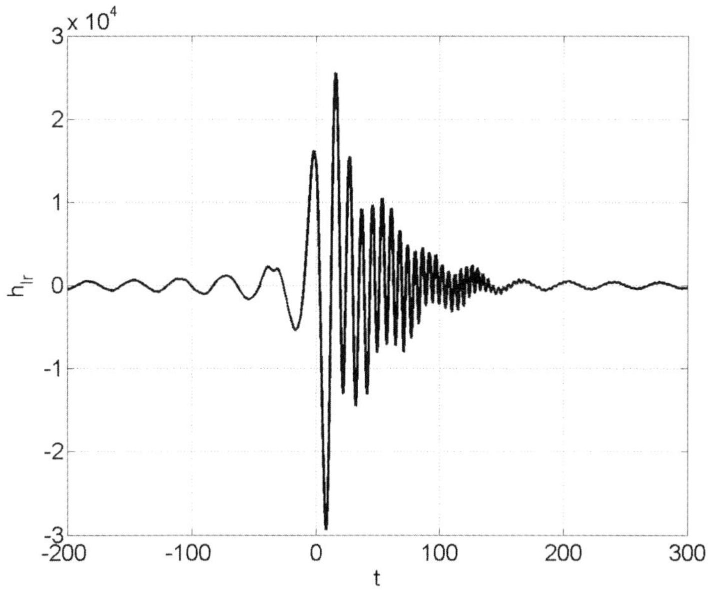

Figure 7.12 Impulse response function h_{ri} used in evaluating the resistive part of the control force applied on the relative heave oscillation [99]. (Copyright: Elsevier Ltd.; reproduced by permission.)

buoy. This plot shows that the more precise "velocity synchronization" enabled by the more precise control of this section enables better use of the available excursion in this case. Details regarding implementation of frequency domain constraints such as discussed in Chapter 6 are also included in [104]. Figs. 7.11–7.13 show the impulse

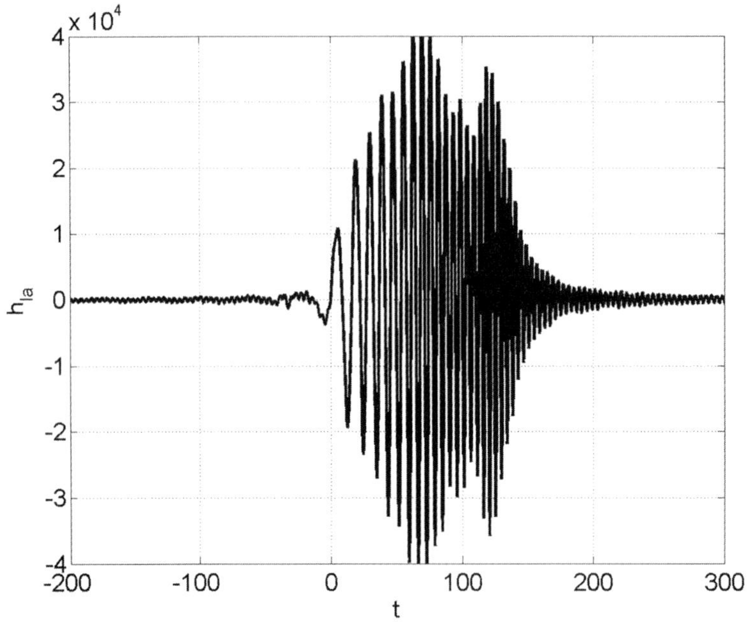

Figure 7.13 Impulse response function h_{ci} used in evaluating the reactive part of the control force applied on the relative heave oscillation [104]. (Copyright: Elsevier Ltd.; reproduced by permission.)

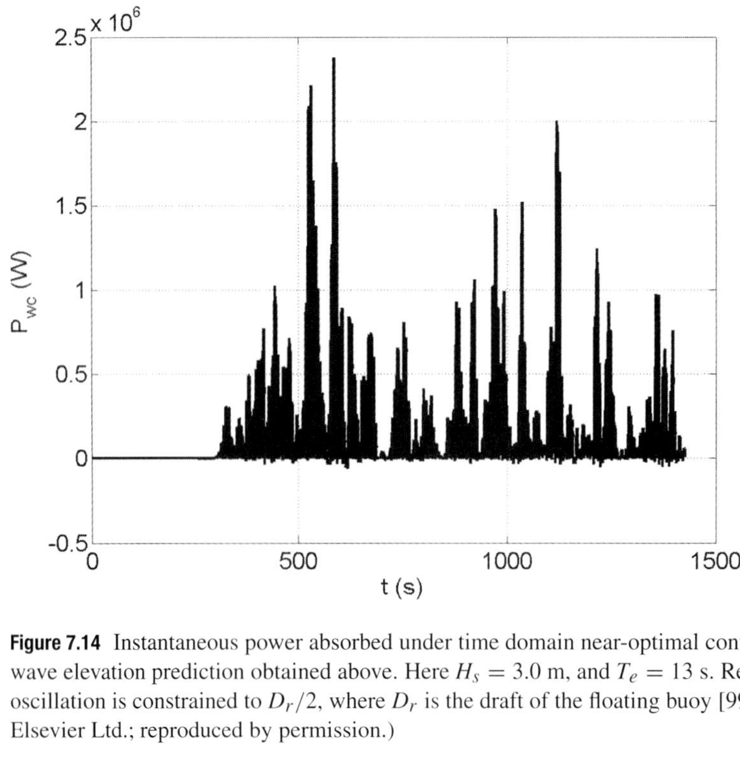

Figure 7.14 Instantaneous power absorbed under time domain near-optimal control based on the wave elevation prediction obtained above. Here $H_s = 3.0$ m, and $T_e = 13$ s. Relative heave oscillation is constrained to $D_r/2$, where D_r is the draft of the floating buoy [99]. (Copyright: Elsevier Ltd.; reproduced by permission.)

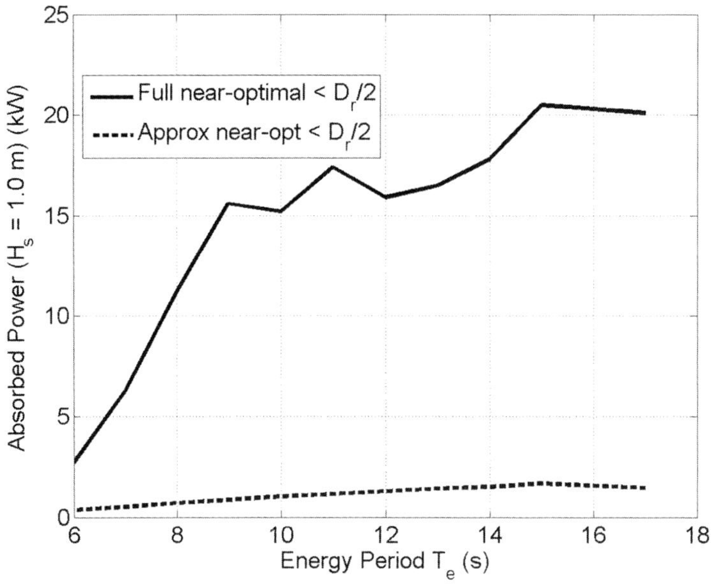

Figure 7.15 The 10-min averaged absorbed power for near-optimal time domain control based on the full-fledged wave elevation prediction above with the approximate near-optimal control based on up-wave wave measurement at a single point. Relative oscillations are constrained to a magnitude less than $D_r/2$, where D_r is the buoy draft [99]. (Copyright: Elsevier Ltd.; reproduced by permission.)

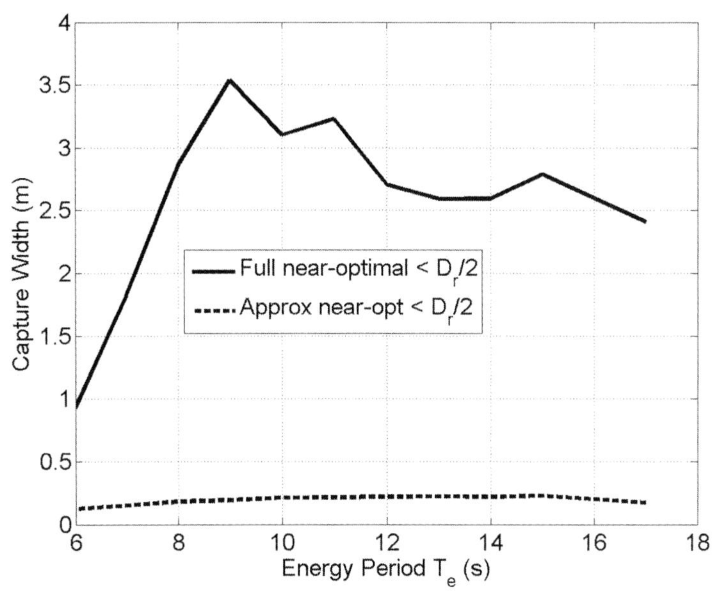

Figure 7.16 The 10-min averaged absorbed power capture length for near-optimal time domain control based on the full-fledged wave elevation prediction above with the approximate near-optimal control based on up-wave wave measurement at a single point. Allowable oscillation amplitude is constrained within $D_r/2$, with D_r denoting the buoy draft [99]. (Copyright: Elsevier Ltd.; reproduced by permission.)

response functions h_{ro}, h_{ri}, and h_{ci} for the relative heave oscillation for the device in Fig. 7.10. h_{ro} is used to compute the near-optimum relative heave velocity. h_{ri} and h_{ci} used in the evaluation of the control forces F_l and F_a to be applied on the relative heave oscillation. Note that the noncausality of these impulse response functions can now be handled effectively using the predicted wave elevation as obtained using Equation (7.98).

7.4 Conclusion

In this chapter our focus was on reactive control, as applied on a wave-by-wave basis, for devices where a single oscillation mode (or its equivalent) was used for power capture. It was assumed implicitly that sufficient stored energy was available so that a continuous exchange of reactive power could be sustained between the oscillating device and the PTO. It was also assumed that the PTO was able to apply the large forces required for reactive control. The PTO actuator was further assumed to be ideal, so that none of the reactive power being exchanged with the device was dissipated. It is easy to see that forcing a small body with a short resonant period to oscillate such that its velocity peaks and troughs are synchronous with the exciting force peaks and troughs in swells is no mean task and would require large reactive forces and large instantaneous reactive power; and even though the resulting gains in captured power are attractive, these need to be weighed against the design effort and the hardware requirements involved in the process. We were able to incorporate oscillation constraints in the approach here, though force constraints were not considered. More work is needed on reactive forces and power values required to provide control leading to near-optimal performance in irregular waves. Actuator designs that can provide large forces with minimum power loss pose an interesting engineering design challenge.

All of our discussion was based on linear systems reasoning. We discussed how the required fully noncausal or anticausal impulse response function could be constructed using odd and even impulse response functions, where effectively, the even part and the negative of the odd part were used in the synthesis of the control force to be applied by the actuator. Our treatment also recognized that the exciting force kernel is noncausal with respect to the wave surface elevation at the device centroid. We also assumed that the approaching waves were long-crested and unidirectional. In one of the approaches described, we used instantaneous wave surface elevation at up-wave stations with right-shifted versions of the impulse response functions required for the instantaneous control force evaluation. Implied in this approach was the assumption that the approaching waves were mostly long swells, long enough that most of the energy propagated at a single group velocity, namely, the shallow-water limit where propagation is nondispersive. The second approach accounted for a range of group velocities and employed a wave elevation prediction long enough ahead into the future using wave measurements made over a duration long enough at an up-wave location far enough. The required prediction time into the future was determined by the length of the noncausal impulse response functions along the $t < 0$ axis and, in particular, the $t < 0$

value where they could reasonably be truncated. Assuming realistic frequency limits for the low and high ends of practically encountered wave spectra, we evaluated the practically important range of propagation group velocities. The fastest group velocity was used along with the prediction time length to determine the up-wave distance at which measurements were required, and the slowest group velocity was used to set the length of time over which the measurement was needed. Device location (i.e., its centroid) was assumed to be at the x location where all group velocity lines intersected on the space–time diagram, the t coordinate being the time value obtained by adding the prediction time to the time up to which measurements are available (measurements are thought here to be available up to the present time). An impulse response function describing the propagation from the point of measurement to the device location was convolved with the measurements made over the required duration above to obtain the predicted wave elevation. The predicted wave elevation was used in deriving the instantaneous control forces, and captured-power results showed considerably better absorption than with the first approach in a range of wave spectra (T_e = 6–17 s). Of particular interest was the absorption in the presence of oscillation constraints, illustrating the value of correct synchronization of the device oscillations with the exciting force as enabled by the second approach. Finally, note that the control based on deterministic wave prediction represents a way to approach, in the time domain, the "hydrodynamic velocity optimum" discussed in Chapter 6. We recall that, at the hydrodynamic velocity optimum, the interaction among the radiated, incident, and diffracted wave fields is such that the best energy transfer (between the incident waves and the device) occurs.

8 A Causal Real-Time Controller for Wave Energy Converters

8.1 Introduction

As documented in Chapter 7, exact real-time implementation of complex-conjugate control, due to its noncausality [66, 106], requires exact wave foreknowledge. Chapter 7 considers two approaches to using information on approaching waves: one assuming nondispersive propagation and the other using a deterministic wave prediction scheme based on a more complete propagation model. Nevertheless, the possibility of designing a control scheme that obviates the requirement for future information, may be attractive. In addition, it would be desirable if the limitations imposed by the physics of the system (e.g., amplitude of motion or velocity, applicable forces) are explicitly considered.

As considered in Chapter 9, latching control [107]–[109] presents one alternative, where the oscillation in the system is delayed so to be in phase with the excitation from the waves. While, in theory, latching allows the achievement of high energy capture, its applicability has been questioned [110] due to the excessive loads on the latching mechanism, the disappointing performance in the case of self-reacting point absorbers [111], and the long (some minutes) prediction horizon required for the real-time optimization in irregular waves to converge [107, 109]. In addition, it may be difficult to observe physical system constraints under a latching philosophy. More recent causal control schemes include that reported in [112], though physical constraints are not explicitly addressed.

Controllers that adjust the damping and stiffness coefficients of the power take-off (PTO) mechanism, in real time, have also been proposed. In [113], a fuzzy-logic controller tunes the damping based also on predictions of the incident wave elevation. However, results are not compared with latching or model predictive control (MPC) types of controllers, and the assumption of linear damping can be unrealistic with most PTOs (hydraulic in particular).

Solutions based on MPC or numerical optimization (see Chapter 10), on the other hand, have shown a lot of benefits, including the ability to handle constraints in a straightforward way [114, 134]. Such techniques, however, need to solve a constrained optimization problem at each time step, which can be time-consuming if a complex model of the (WEC) is utilized. In addition, the performance of the controller depends on a variety of variables (cost function, optimization algorithm, prediction horizon, specification of the constraints, etc.) that are not always related to the physical domain of the problem and can be tuned only by extensive simulation. MPC-based controllers also require prediction of the excitation force, which, although possible for one to two wave periods [115], introduces additional uncertainty in the solution, as documented in Chapter 7.

In this chapter, we propose an alternative controller that, while being suboptimal, is simple and effective at the same time, which makes it well suitable for real-time applications. In particular, the proposed controller has a hierarchical structure, where a reference velocity is calculated from the current value of the wave force through a proportional constant, which is tunable in real time for performance optimization and constraint handling. The reference velocity is then imposed on the system by a low-level control loop based on classical control theory. Based on a nonstationary, harmonic model of the excitation force, the reference generation does not require an optimization, like MPC, and does not rely heavily on the model, which makes it inherently more robust to model uncertainties, through an appropriate design of the low-level controller.

The controller is of the reactive type and requires a PTO capable of implementing a bidirectional energy flow. Typical hydraulic PTOs and linear electrical generators that have been studied for applications in wave energy have the ability of inverting the energy flow [116, 117]. Non-ideal PTO efficiency, however, needs to be taken into account for an appropriate evaluation of the performance of the device [118, 119]. While the current study assumes ideal efficiency, since the focus is on the control concept, a quantitative discussion about the requirement of active power is proposed in Section 13.5.

8.1.1 Model Definition

A single-body floating system, as in Fig. 8.1, is considered where energy is extracted from the relative motion with the sea bottom, through a generic PTO mechanism. The system is constrained to move in the heave direction only. The external forces acting on the WEC are the excitation from the waves and the control force produced by the PTO. Under the assumption of linearity, neglecting viscosity and other losses, the force–velocity model can be expressed, in the frequency domain, as follows [66]:

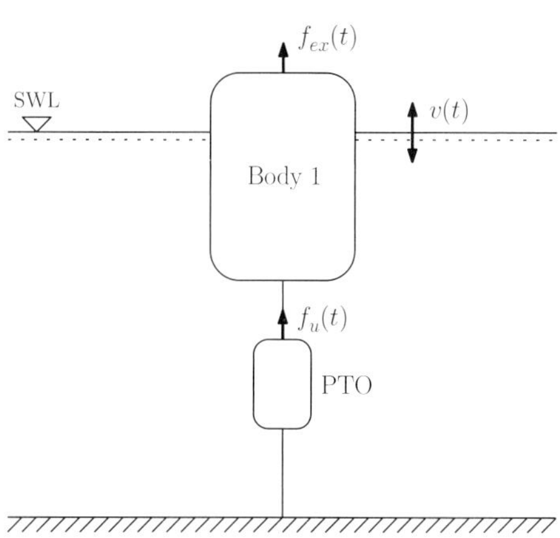

Figure 8.1 One-degree-of-freedom (heave) floating system for wave energy conversion.

$$\jmath\omega MV(\omega) + Z_r(\omega)V(\omega) + \frac{K_s}{\jmath\omega} = F_{ex}(\omega) + F_u(\omega), \tag{8.1}$$

following the development in Chapter 4.

In (8.1), $V(\omega)$ is the heaving velocity, while $F_{ex}(\omega)$ and $F_u(\omega)$ are the wave excitation and PTO forces, respectively. The imbalance between gravity and the hydrostatic restoring force is modeled by the stiffness coefficient K_s. The radiation force, due to waves radiated by the body's motion, is expressed through the radiation impedance $Z_r(\omega)$. Note that $Z_r(\omega)$ can be considered as a Fourier transform only in a generalized sense, as its imaginary part is unlimited for $\omega \to \infty$, so it is often decomposed as

$$Z_r(\omega) = B(\omega) + \jmath\omega[M_a(\omega) + M_\infty] = H_r(\omega) + \jmath\omega M_\infty \tag{8.2}$$

where $B(\omega)$ is the radiation resistance (real and even) and $M_a(\omega)$ is the added mass after the singularity at infinite frequency, M_∞, is removed. The function $H_r(\omega)$ is a valid Fourier transform.

The model in Equation (8.1) can be expressed in the compact form

$$V(\omega) = \frac{1}{Z_i(\omega)}\left[F_{ex}(\omega) + F_u(\omega)\right] \tag{8.3}$$

where the intrinsic impedance, $Z_i(\omega)$, is defined as

$$Z_i(\omega) = B(\omega) + \jmath\omega\left[M + M_a(\omega) + M_\infty - \frac{K_s}{\omega^2}\right]. \tag{8.4}$$

The excitation force is the effect of the incident wave on the floating system:

$$F_{ex}(\omega) = H_{ex}(\omega)\Xi(\omega), \tag{8.5}$$

where $\Xi(\omega)$ is the Fourier transform of the incident wave elevation, $\eta(t)$. The transfer function $H_{ex}(\omega)$ is a property of the WEC and has low-pass filter characteristics. Note, however, that $H_{ex}(\omega)$ is noncausal, since the incident wave at the device centroid is not the direct cause of the force (the pressure distribution is) [66] and cannot be implemented in real-time simulations.

The example WEC system considered in this study consists of a heaving cylinder with radius $R = 3$ m, height $H = 5$ m, draft $h = 4$ m, and mass $M = 3.2 \times 10^5$ kg. The radiation and excitation transfer functions, $H_r(\omega)$ and $H_{ex}(\omega)$, are identified numerically through the hydrodynamic software WAMIT [120] and are shown in Fig. 8.2.

8.1.2 Model Identification

The linear model in Equation (8.3) is known numerically, but only for some frequencies. This is due to the fact that the radiation impedance, $Z_r(\omega)$ or $H_r(\omega)$, can be derived only numerically from hydrodynamic simulation software, or from experiments.

From a finite-order approximation of the radiation impedance, $\hat{H}_r(s) \approx H_r(\jmath\omega)$, the floating system in Equation (8.3) can be approximately modeled as the following single input, single output system:

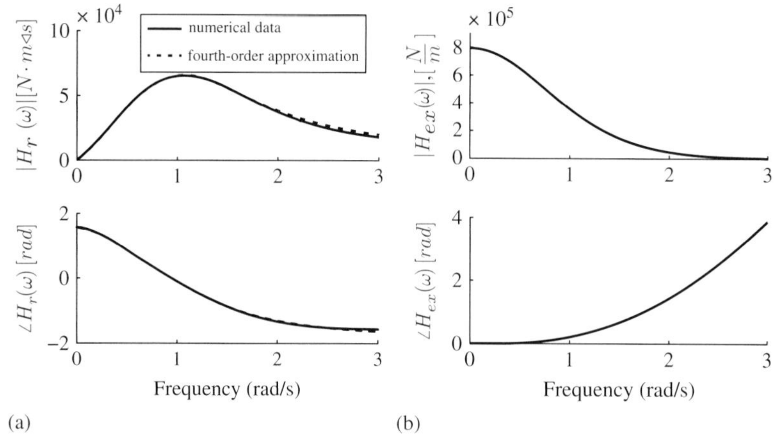

Figure 8.2 Radiation (a) and excitation (b) frequency responses for a floating cylinder with radius $R = 3$ m, height $H = 5$ m, draft $h = 4$ m, and mass $M = 3.2 \times 10^5$ kg, calculated using the hydrodynamic software WAMIT [120].

$$\frac{V(s)}{F_{ex}(s) + F_u(s)} = \frac{s}{s^2(m + M_\infty) + s\hat{H}_r(s) + K_s} \triangleq \frac{1}{Z_i(s)}, \tag{8.6}$$

where s is the complex frequency in the Laplace domain. The input is the external force (from the waves or from the PTO), $F_{ex}(s) + F_u(s)$, while the output is the system's velocity, $V(s)$.

For the geometry specified in Fig. 8.2, a fourth-order approximation of the radiation is calculated using frequency domain identification, following the approach in [121]. The result of the identification, shown in Fig. 8.2(a), gives a very accurate estimation of the radiation response of the system. A sixth-order force–velocity system is therefore obtained from Equation (8.6), and is specified by the following transfer function (coefficients are rounded to one decimal place for brevity):

$$\frac{1}{Z_i(s)} = \frac{1.8 \times 10^{-6} \cdot s(s^2 + 1.1s + 0.4)(s^2 + 1.6s + 1.8)}{(s^2 + 1.2s + 0.4)(s^2 + 1.4s + 1.6)(s^2 + 0.1s + 1.5)}. \tag{8.7}$$

8.2 Real-Time Controller

This section outlines the proposed strategy for real-time control of a WEC of the type modeled in Section 8.1.1. As shown in Fig. 8.3, the approach consists of the generation of a reference velocity, $v_{ref}(t)$, from knowledge of the current wave excitation force acting on the system, and then the imposition of such a velocity profile, through the PTO force, via a lower-level feedback controller, $K(s)$. After an analysis of the conditions for maximum wave energy absorption, in Section 8.2.1, the reference-generation strategy

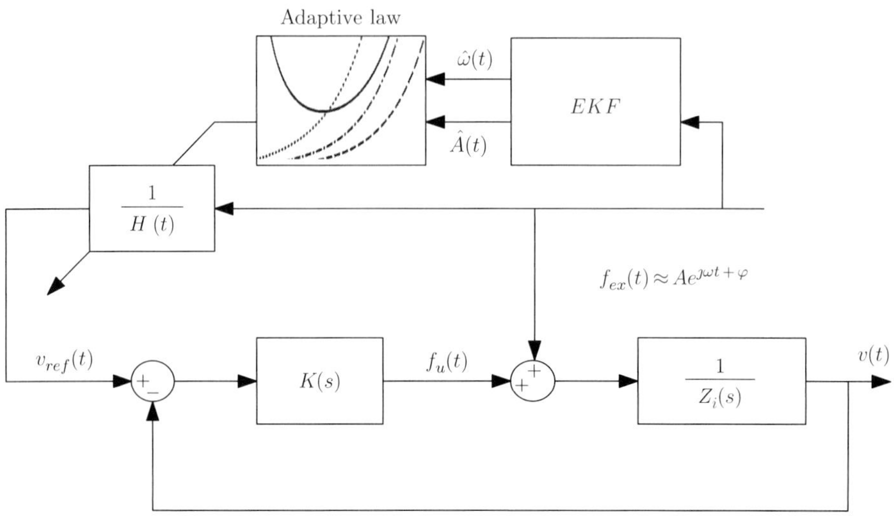

Figure 8.3 Proposed control architecture.

is derived in Section 8.2.2. Section 8.2.3 shows the possibility of handling constraints, and Section 8.2.4 describes the design of $K(s)$.

8.2.1 Maximum Wave Energy Extraction

As extensively studied in [66], maximum energy transfer from the waves to the system PTO is obtained when

$$F_u(\omega) = -Z_i^*(\omega)V(\omega), \tag{8.8}$$

where $Z_i^*(\omega)$ is the complex conjugate of the intrinsic impedance of the system. The condition in Equation (8.8), however, is anticausal and cannot be implemented in real-time (without prediction), as documented in Chapter 7.

The control law in Equation (8.8) can be identified as *complex-conjugate* control (note the electrical analogy from impedance matching) or *reactive* control, from the requirement of a bidirectional energy flow, which means that negative energy absorption would be experienced for parts of the cycle, as highlighted in Chapter 6.

An alternative, and equivalent, condition for maximum energy extraction [66] is a phase and amplitude relation between oscillation velocity and wave excitation force (see Equation (6.46)):

$$V(\omega) = \frac{1}{Z_i(\omega) + Z_i^*(\omega)} F_{ex}(\omega) = \frac{1}{2B(\omega)} F_{ex}(\omega). \tag{8.9}$$

The radiation resistance, $B(\omega)$, is a real and even function in the frequency domain. As a consequence, the phase and amplitude condition in Equation (8.9) is noncausal, and, in particular, the corresponding time domain impulse response function is also real and even [66].

Under conditions (8.8) or (8.9) the average absorbed power, $P_u(\omega)$, is maximum and it is half the average excitation power, $P_{ex}(\omega)$, the other half being lost in radiation, $P_r(\omega)$ [66].

It has been shown [115, 106] (and see Chapter 12) that the noncausality can be addressed with simple regressive predictors of excitation force, and some controllers use such information, for example, the controller formulation in Chapter 10. However, there are two more characteristics of reactive control that make its practical implementation difficult (if not impossible) in some situations:

- The optimal velocity, position, and PTO force can have very large amplitudes, well beyond physical constraints, particularly in large waves and for small systems.
- Since the energy flow in the PTO is bidirectional, power needs to be injected back into the system for part of the cycle, in order to achieve the optimum. Inefficiencies in the PTO, when reversing the energy flow, may therefore undermine the overall efficiency of a reactive conrol solution.

In Sections 8.2.2 and 8.2.3, a nonoptimal approximation of reactive control is proposed, where the noncausality and the constraints are overcome in a simple, but effective, way. Note that, in the following, we assume that the WEC, modeled in Section 8.1.1, has a PTO machinery capable of implementing a reversible energy flow with ideal efficiency.

8.2.2 A Simple and Effective Realization of Reactive Control

Based on Equation (8.9), the velocity should always be in phase with the excitation force and should have an amplitude that is modulated in the frequency domain by the inverse of double the radiation resistance, $1/2B(\omega)$, which is noncausal. While predictions of the excitation force can be utilized to resolve the noncausality [106, 115] (and see Chapter 12), a suboptimal simpler solution is proposed here, which does not require prediction.

The relation in Equation (8.9) is a frequency-dependent gain, shown in Fig. 8.4, to be applied to the excitation force, to obtain the optimal velocity profile. The noncausality

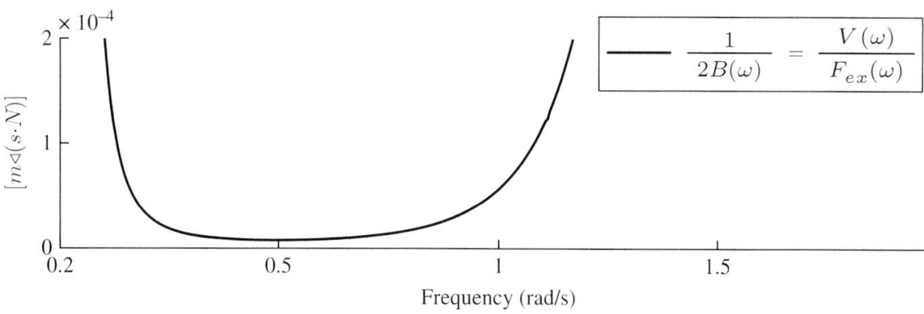

Figure 8.4 Optimal relation between excitation force and oscillation velocity. The phase is zero.

derives from the fact that, in theory, all the frequencies may be present at the same time. In practice, the excitation force is contained within a restricted band of frequencies, mostly within the flat part of $1/2B(\omega)$. A constant approximation of $B(\omega)$, based on a second-order model of the system, was proposed in [122]. If a real-time estimate of the peak frequency of the excitation force is available, however, we can adapt the constant from the curve $1/2B(\omega)$.

Assume that the excitation, $f_{ex}(t)$, is a narrow-band harmonic process, defined by time-varying amplitude $A(t)$, frequency $\omega(t)$, and phase $\varphi(t)$:

$$f_{ex}(t) = A(t)\cos(\omega(t)t + \varphi(t)). \tag{8.10}$$

Then, the reference velocity can be generated from the following adaptive law:

$$v_{ref}(t) = \frac{1}{H(t)} f_{ex}(t), \qquad \frac{1}{H(t)} = \frac{1}{2B(\hat{\omega})}, \tag{8.11}$$

where the value of the constant $H(t)$ is calculated from the curve $1/2B(\omega)$, based on a real-time estimate of the peak frequency of the wave excitation force.

An on-line estimate of the frequency, $\hat{\omega}$, and amplitude, $\hat{A}$, is obtained with an extended Kalman filter (EKF) [123], proposed in [115, 124] for modeling the wave elevation. Although suboptimal, the estimation obtained with the EKF is satisfactory in our case, as shown in Fig. 8.5. Alternative methods that give a solution closer to optimal have also been proposed, as reviewed in [125, 126].

Note that the sinusoidal assumption in Equation (8.10) was considered by a number of researchers [127, 128], in order to remove the noncausality of complex-conjugate control. However, simulations were always carried out in regular waves, to satisfy the assumption. Here the controller is tested with irregular waves (simulated and real data) that do not necessarily satisfy the assumption, as discussed in Section 13.5.

8.2.3 Constraint Handling

The control solution proposed in Section 8.2.2, where the velocity is simply calculated from the excitation force through a variable gain, allows the handling of physical

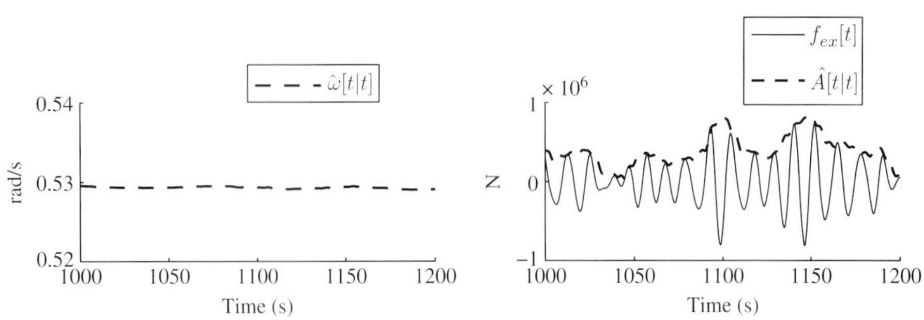

Figure 8.5 Real-time estimation of the amplitude and frequency of a stochastic excitation-force signal, generated from a power spectral distribution centred around 5.3 rad/s.

constraints in a quite intuitive and effective way. Based on the narrow-band assumption of Equation (8.10), the excitation force can be expressed, in complex notation, as

$$f_{ex}(t) = \Re\left\{Ae^{\jmath\varphi}e^{\jmath\omega t}\right\}, \qquad \hat{F}_{ex} \triangleq Ae^{\jmath\varphi}, \tag{8.12}$$

where $\hat{F}_{ex}$ is termed the complex amplitude of $f_{ex}(t)$.

As a consequence of the proportional reference-generation law, in (8.11), the complex amplitude of the velocity, $\hat{V}$, and position, $\hat{X}$, can be expressed as

$$\hat{V} = \frac{A}{H}e^{\jmath\varphi}, \tag{8.13}$$

$$\hat{X} = \frac{\hat{V}}{\jmath\omega} = \frac{A}{\jmath\omega H}e^{\jmath\varphi}. \tag{8.14}$$

Suppose that the vertical excursion of the WEC is limited to $\pm X_{lim}$ m from equilibrium. From Equation (8.14), the position constraint can be written as an equivalent velocity constraint:

$$\hat{X} = \frac{\hat{V}}{\jmath\omega} \leq X_{lim} \quad \Leftrightarrow \quad |\hat{V}| \leq \omega X_{lim}, \tag{8.15}$$

and the following upper bound for the variable gain, $1/H$, involving the amplitude and frequency of the excitation, can be derived from Equation (8.13):

$$\frac{1}{H} \leq \frac{\omega X_{lim}}{A}. \tag{8.16}$$

The reference generation strategy, based on Equations (8.9), (8.11), and (8.16), can therefore be modulated in order to keep the amplitude of the velocity within the bound specified in (8.15). A real-time estimate of the frequency, $\hat{\omega}$, and amplitude, $\hat{A}$, of the excitation is obtained through the EKF [124, 115], as previously discussed. The complete reference generation strategy is:

$$\frac{1}{H(t)} = \begin{cases} \dfrac{1}{2B(\hat{\omega})}, & \text{if } \dfrac{\hat{\omega}X_{lim}}{\hat{A}} > \dfrac{1}{2B(\hat{\omega})} \\ \dfrac{\omega X_{lim}}{\hat{A}}, & \text{otherwise.} \end{cases} \tag{8.17}$$

According to Equation (8.17), when in the unconstrained region, the velocity is tuned to the optimal amplitude given by complex-conjugate control, as in Equation (8.9). Otherwise, the maximum allowed velocity (lower than the optimal) is imposed, while keeping the velocity in phase with the excitation force. Note that a velocity calculated from Equation (8.17) is optimal within the proposed parametrization, which means that it gives the maximum energy extraction among all the velocities proportional to the excitation force and smaller than the constraint.

Given the geometry specified in Section 8.1.1, and a constraint $X_{lim} = 0.8$ m, Fig. 8.6(a) shows the profile of the reference velocity, for two frequencies, as a function of the amplitude of the incident wave elevation, which is related to the amplitude of the excitation force, $|\hat{\Xi}| = |AH_{ex}(\omega)|$, from Equation (8.5). For small waves, the velocity is calculated based on the optimal condition given by reactive control while, above waves of a certain size, it is kept at a constant maximum amplitude (smaller than optimal),

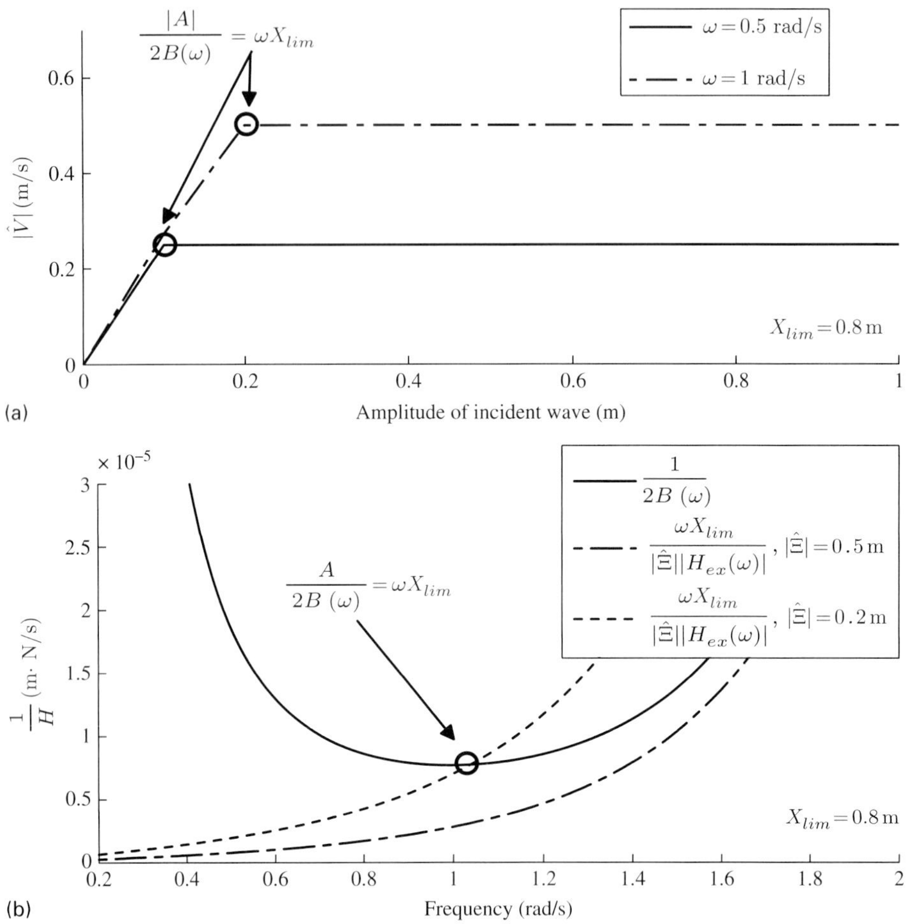

Figure 8.6 Graphical representation of adaptive law (8.17). Note that magnitude of the excitation force, $|\hat{A}|$, is translated in amplitude of the incident wave elevation, $\hat{\Xi} = |H_{ex}(\omega)A|$, from Equation (8.5). (a) Magnitude of velocity with respect to amplitude of incident wave elevation. (b) Ideal optimal value and upper limit, for a constraint $X_{lim} = 0.8$ m, of the adaptive constant $1/H$, as from Equation (8.17).

so that the constraints are satisfied. The limit of the unconstrained region depends on the frequency, as seen in Fig. 8.6(b). In the case of 0.2 m waves, the controller would work at a maximum allowed speed for $\omega < 1$ rad/s, where $1/2B(\omega) = \omega X_{lim}/A$, while it would follow the optimal speed for higher frequencies. When the system is excited by 0.5 m waves, the velocity is always tuned at the maximum allowed by the constraint (lower than the optimum).

Note that:

- Ultimately, the controller is regulated only through a single parameter, $1/H$, that is the constant of proportionality between the excitation force and the device velocity. Such a parameter can be tuned for best performance, as in

Equation (8.17), but could also be based on other requirements. It could, for example, be tuned for the regulation of the power at a desired level (absorbed power is proportional to the velocity squared).

- Unlike optimization-based solutions, the proposed controller relies on the model of the system only in the unconstrained region. In the constrained region, the low-level control loop depends on the model, while the reference generation does not, and a robust design would benefit from this separation. Further discussion on improving the robustness is mentioned in Section 8.2.4.
- Prediction of the excitation force is not necessary.
- A practical implementation would only require estimation of the excitation force, which is not a directly measurable quantity. However, an observer can be built from measurements of the motion (velocity, acceleration) of the system [114].

8.2.4 Velocity-Tracking Control Loop

The desired velocity, generated from Equation (8.11), is imposed on the system by the low-level controller that acts on the PTO force, $f_u(t)$. The controller, namely $K(s)$, is based on the concept of internal model control (IMC) [129], which was also proposed for the control of the Archimedes Wave Swing WEC [130].

Fig. 8.7 shows the equivalent structure of the IMC. The compensator, $K(s)$, is given by

$$K(s) = \frac{Q(s)}{1 - Q(s)/Z_i(s)} \tag{8.18}$$

where $1/Z_i(s)$ is the model of the system, given in Equation (8.7), and $Q(s)$ is a filter that can be appropriately chosen. In the case of a perfect model and no disturbances, it can be shown [129] that the controller works in open loop and the feedforward compensator simplifies to the filter $Q(s)$. Feedback appears only in the presence of model uncertainties or disturbances.

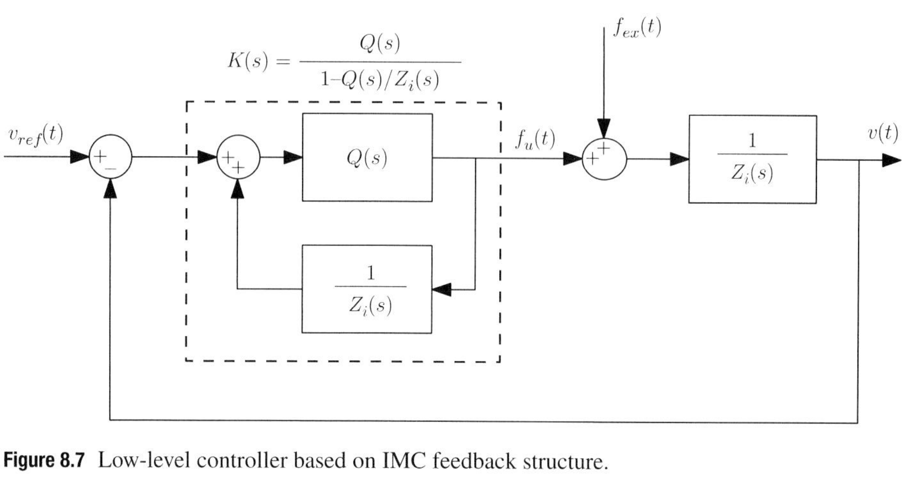

Figure 8.7 Low-level controller based on IMC feedback structure.

$$\eta(kT_s) = \sum_{j=0}^{N/2} a_j \sin(\omega_j k T_s + \phi_j), \tag{8.22}$$

where the amplitudes a_j are specified by the spectral shape $S_{\eta\eta}(\omega)$, and the phases ϕ_j are randomly chosen from a uniform distribution in the interval $[0, 2\pi]$. The number of frequencies N, equally spaced in the interval $[0, \pi f_s]$, is chosen such that the statistical properties of random waves are satisfied and that the signal $\eta(k)$ does not repeat itself within the simulation [133]. In particular, the signal $\eta(k)$ generated from Equation (8.22) is periodic with period $N \cdot T_s$ s, so N was chosen as $N = 3600 \cdot 2.56 = 9216$.

8.3.2 Performance in the Unconstrained Case

The performance of the controller proposed in Section 8.2.2 is initially validated for the unconstrained case. Complex-conjugate control, outlined in Section 8.2.1, is utilized as a comparison. The controllers are implemented, as shown in Fig. 8.3, as a reference-velocity generator followed by a velocity-tracking loop, as designed in Section 8.2.4. In the case of complex-conjugate control, the reference velocity is generated from the optimal noncausal law (8.9). The noncausality is resolved as proposed in [106], by assuming *exact* knowledge of the future excitation force, for the ideal case, and by using real-time predictions with an autoregressive (AR) model [115], labeled as the noncausal controller in Fig. 8.10(a).

The controller performance is measured in terms of the relative capture width (RCW):

$$\text{RCW} = \frac{\overline{P}_u}{2RP_w}, \tag{8.23}$$

which is the ratio between average absorbed power, $\overline{P}_u$, and average wave power, P_w, over a front as wide as the cylinder ($2R$). Note that deep water is assumed, so that the wave power, per meter of wave front, is calculated from the wave spectral distribution, $S_{\eta\eta}(\omega)$, as [66]:

$$P_w = \frac{\rho g^2}{2} \int_0^\infty \frac{S_{\eta\eta}(\omega)}{\omega} d\omega \qquad \left[\frac{W}{m}\right], \tag{8.24}$$

where ρ is the water density and g is the acceleration due to gravity.

Fig. 8.10(a) shows the RCW obtained with the nonoptimal (and causal) controller compared with the ideal optimum, for wide-banded ($\lambda = 0.5$) and narrow-banded ($\lambda = 5$) wave spectra, centered around different periods, $2\pi/\omega_0$. The proposed controller performs very close to the optimum, particularly for narrow-banded waves. For wide-banded sea states, the performance is very close to the ideal maximum only for short to medium waves, while it significantly drops for very long waves. Note that only a minor improvement in performance is obtained with a noncausal approximation of complex-conjugate control, using wave predictions.

Note that the RCW is quite large in most situations, and this is due to the fact that physical constraints are not taken into account, which is one of the limitations of ideal reactive control. Fig. 8.10(b) shows the distribution of the heaving position and

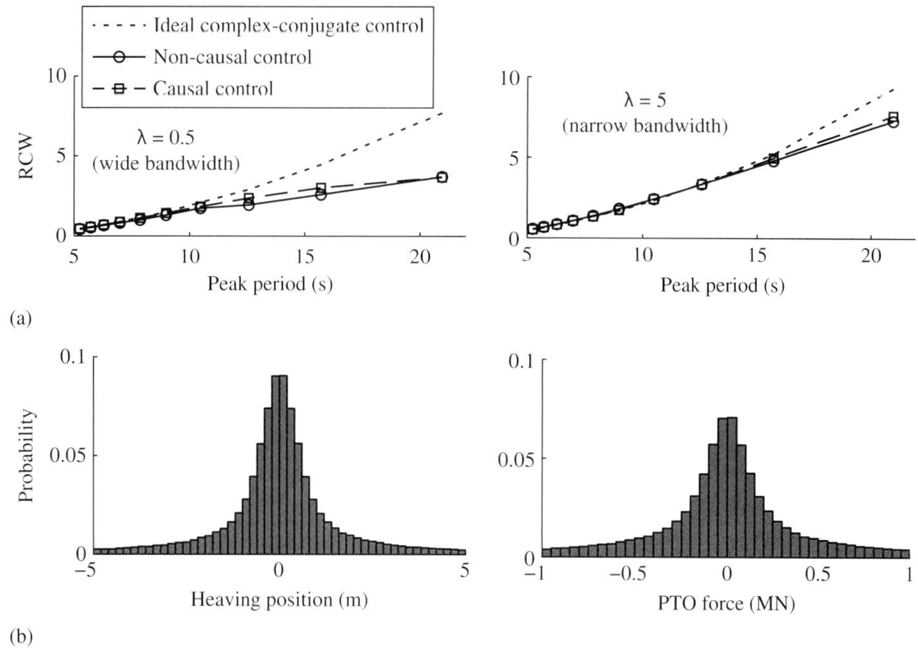

Figure 8.10 Performance of proposed nonoptimal causal control. (a) Relative capture width (RCW). (b) Distribution of heaving excursion and PTO force.

PTO force required to obtain the results of Fig. 8.10(a). The position typically ranges from −5 to 5 m (with bigger excursions in some situations!), which is quite impractical for a floating cylinder 5 m high and with a 4-m draft. Also, PTO forces of more than 1×10^6 N are experienced in some occasions, which can require large and expensive PTO machinery.

The performance of the proposed controller in double-peaked seas is also assessed. Fig. 8.11 shows the behavior of the RCW, relative to the ideal optimum from reactive control. Clearly, the performance is negatively affected (below 50% of optimum) when the two wave systems have a similar energy content and when they are widely spaced.

A final analysis is carried out in order to determine the amount of energy that needs to be injected from the PTO back into the system, during part of the cycle, known to be an issue with complex-conjugate control. Fig. 8.12 shows the ratio of the average reverse power, that is, the average absorbed power of negative sign, to the overall average absorbed power (which is always positive if the controller is effective). Most of the time a higher reversed energy flow is required than that actually absorbed (between 1 and 5 times!). The ratio appears to increase with the average power absorption of the system, but, in general, it looks quite unreasonable, since it would require an excessive over-specification of the PTO.

High levels of reverse power, together with the excessive motion required, make unconstrained complex-conjugate control impractical most of the time, apart from the case of very small waves (how small depends on the geometry and rating of the specific WEC).

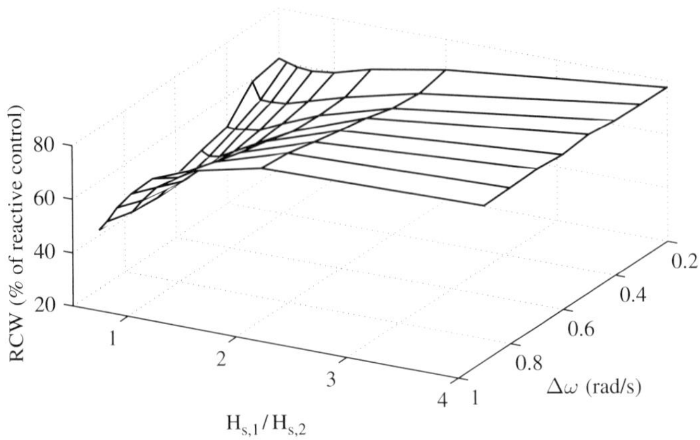

Figure 8.11 Performance of causal control, as a percentage of ideal reactive control, with double peak spectra: $H_{s,1}/H_{s,2}$ is the relative energy between the two wave systems; $\Delta\omega$ is the frequency spacing.

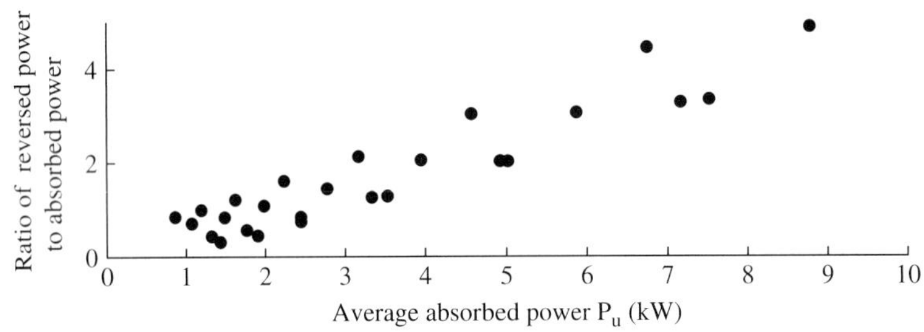

Figure 8.12 Ratio of average reversed power (energy from PTO to system) to absorbed power (energy from system to PTO).

8.3.3 Introduction of Constraints

The proposed causal, suboptimal controller is also validated when physical constraints are imposed, as described in Section 8.2.3. MPC, which solves a constrained optimization problem at each time step, is utilized as a comparison, following the approach proposed in [134]. In particular, the future horizon for the optimization is set to 30 steps ($\approx$ 11.77 s), which is about double the resonance period of the floating cylinder ($\approx$ 5.2 s). For a fair comparison, predictions are not assumed to be ideal and a forecasting algorithm, based on AR models, is implemented, as proposed in [115].

Fig. 8.13(a) shows the RCW obtained from the two controllers, in wide- ($\lambda = 0.5$) and narrow-banded ($\lambda = 5$) sea states for different peak periods, $2\pi/\omega_0$. Two significant wave heights, $H_s = 1$ and $H_s = 3$ m, are also utilized. Due to the constraints, the performance is similar to that obtained in the unconstrained case only for short waves (small period), while it shows a peak for medium waves and decreases again for longer waves, as theorized by Budal [135]. The suboptimal controller performs quite closely to MPC for short and long waves. It is less efficient, however, around the peak. Note that,

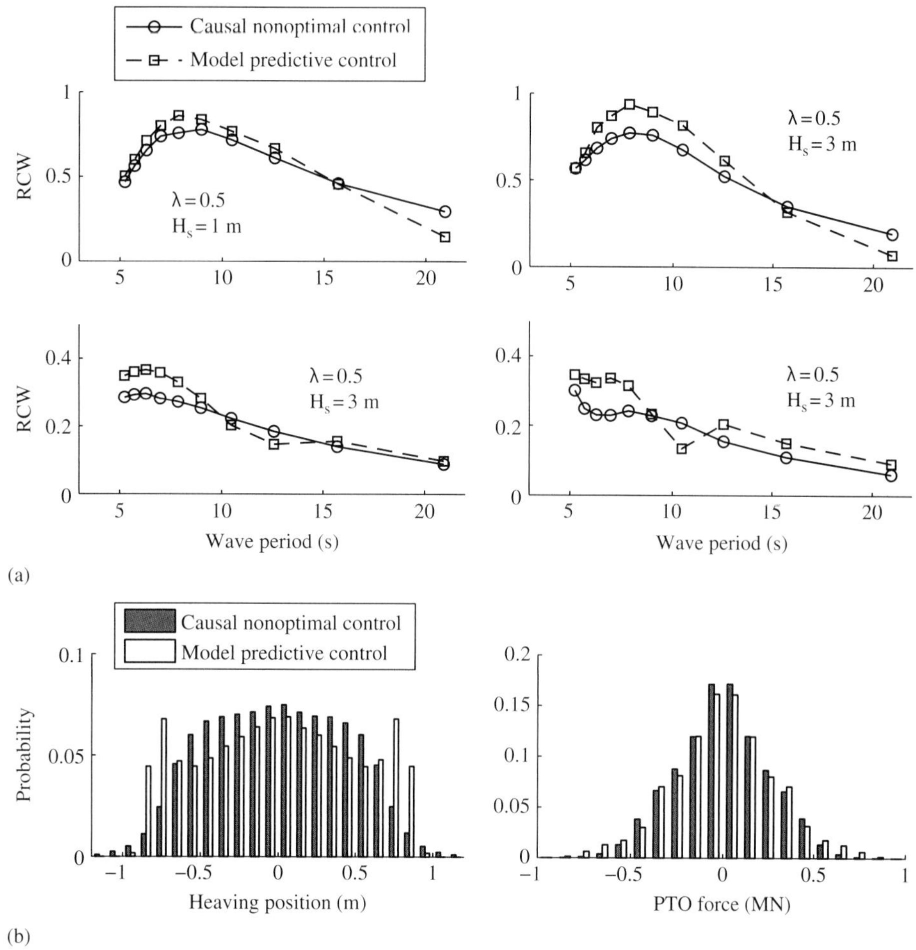

Figure 8.13 Performance of proposed nonoptimal causal control, compared with optimal MPC. (a) RCW. (b) Distribution of heaving excursion and PTO force.

unlike the unconstrained case, the bandwidth does not seem to affect the performance for the constrained case, although it affects the relative efficiency compared with MPC, which offers better results for narrow-banded waves, $\lambda = 5$, most likely due to better prediction of the excitation force [115] for this condition. Note that, for bigger waves, $H_s = 3$ m, the trend of the power capture variation with wave period is similar, but the peak moves toward smaller-period waves.

An important point that emerges from Fig. 8.13(a) is that, for very long waves, the suboptimal controller seems to outperform MPC. The reason behind this unexpected behavior may be found in the way MPC is specified, which depends on many parameters (optimization algorithm, future horizon, specification of functional, etc.) whose choice is not always straightforward, and our choice is probably not optimal enough for those waves. Such a behavior highlights one of the main benefits of the proposed controller, that although suboptimal, requires tuning of only a single parameter that has intuitive physical meaning ($1/H$), representing the modulation of the magnitude of the velocity with respect to the excitation force.

A further advantage of the proposed controller is the linearity achieved between excitation force, velocity, position, and PTO force. Motion and force have the same harmonic behavior as the incident wave, with no abrupt variations that may adversely affect the survivability of the system components. The probability distribution of position, Fig. 8.13(b), shows how MPC keeps the system close to the constraints, with higher velocities in moving from one extreme to the other. The proposed controller, on the other hand, imposes a smooth and harmonic oscillation in the system, so that the position is almost normally distributed around zero. Such a behavior can also be observed, though to a much lesser extent, in the force. Note that, for both controllers, the position goes slightly beyond the imposed constraint of 0.8 m, due to model uncertainties. This unwanted behavior occurs slightly more frequently with the suboptimal controller, obviously, but can be overcome by adopting a more conservative reference-generation strategy than in Equation (8.17). A sample time-series simulation that clarifies such considerations is shown in Fig. 8.14, where the evolution of the position and the required control force are plotted for the two controllers.

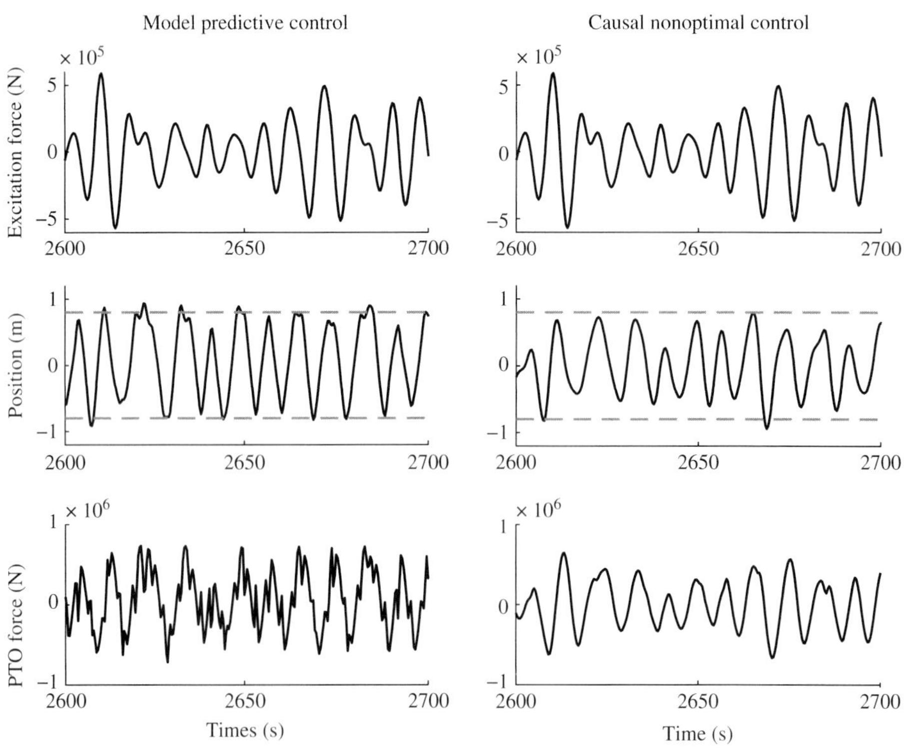

Figure 8.14 Time series of a specific simulation. Input wave comes from an Ochi spectrum with $H_s = 2$ m, $\omega_0 = 0.7$ rad/s, and $\lambda = 1$.

The performance of the suboptimal controller, relative to MPC, with double-peaked wave systems, is shown in Fig. 8.15. When the two peaks have a comparable size, the suboptimal controller outperforms MPC, particularly when $\Delta\omega$ is small. Such a behavior could be due to higher inaccuracies in the prediction of the excitation force, which is more challenging for multipeaked sea states [115].

The amount of reverse power required, in the presence of constraints, is shown in Fig. 8.16, where the ratio of reverse power to absorbed power is compared, for both the proposed controller and MPC. Much less energy needs to be injected back from the PTO to the WEC, between 10 and 40% of the absorbed energy for the constrained case, than in the unconstrained case (100–500%). Hydraulic or electrical generators can be designed to accomplish such levels of reversible energy flow, and the proposed controller can be optimized by taking into account inefficiencies in the two modes of operations, by acting on the constant $1/H$.

8.3.4 Performance with Real Wave Data

The suboptimal controller is also validated with real wave data, provided by the Irish Marine Institute and coming from a data buoy at the Belmullet wave energy test site,

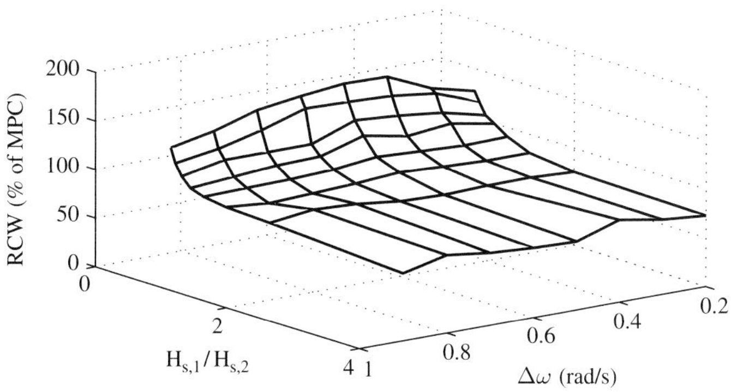

Figure 8.15 Performance of causal control, as a percentage of MPC performance, with double-peak spectra: $H_{s,1}/H_{s,2}$ is the relative energy between the two wave systems; $\Delta\omega$ is the frequency spacing.

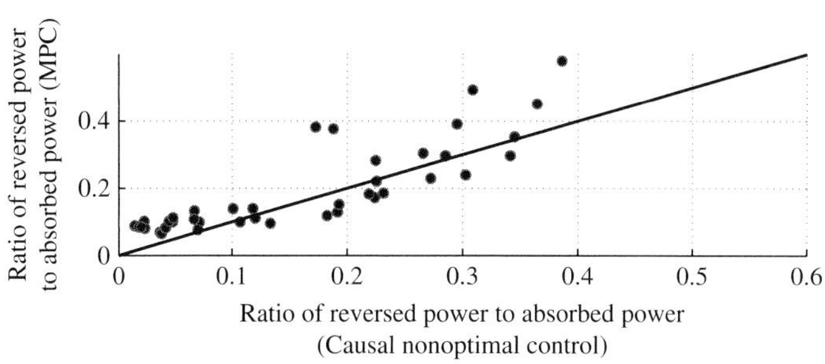

Figure 8.16 Ratio of average reversed power (energy from PTO to system) to absorbed power (energy from system to PTO).

Table 8.1 Performance with real wave data of Fig. 8.17.

Set	P_w [kW/m]	Causal nonoptimal control		MPC	
		$\overline{P}_u$ (kW)	RCW	$\overline{P}_u$ (kW)	RCW
S_1	7.58	36.5	0.48	43.6	0.58
S_2	17.0	57.2	0.34	62.7	0.37
S_3	44.3	64.7	0.146	65.4	0.148

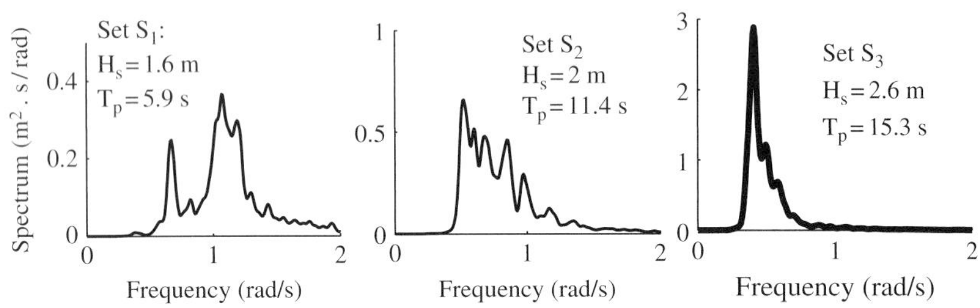

Figure 8.17 Wave spectra of selected real wave data.

off the west coast of Ireland. Three wave-elevation records of 30 min, sampled at 1.28 Hz, are selected. As shown in Fig. 8.17, a variety of sea conditions is covered, with the peak period going from 6 to 15 s and with different spectral bandwidths.

The performance of both the proposed controller and MPC is evaluated in the same conditions utilized for the simulated waves in Section 8.3.3. Table 8.1 summarizes the results, by showing the average power absorption and the RCW with each of the wave records selected. In line with what might be expected, based on the general results of Fig. 8.13(a), the suboptimal controller performs almost as well as MPC.

The energy absorption over a 3-hour simulation in real waves is also shown, in Fig. 8.18. Although real waves are nonstationary and the spectral distribution changes with time, the suboptimal controller maintains its performance, consistent with the MPC controller.

8.3.5 Sensitivity to Parameter Variations

In wave energy, the typical operation of a device involves large motions. Therefore, the typical assumptions of linearity, on which the model of Section 8.1.1 is based, may be invalid.

The model of Equation (8.6) is similar to a damped mechanical oscillator, with relatively small damping, and it makes sense to apply variations to one of the main terms that define the system, which is the hydrostatic stiffness, K_s. Relative variations ranging from $\Delta K = -50\%$ to $\Delta K = +50\%$ are applied, resulting in a change in the dynamics, as shown in Fig. 8.19. A large variation is verified around the resonance frequency of the system, in particular.

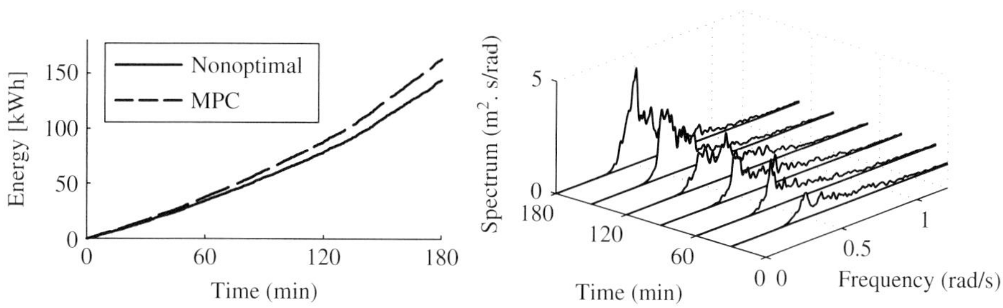

Figure 8.18 Energy capture over a 3-hour simulation with real wave data. The data were collected at Belmullet on October 1, 2010, 0:00 am.

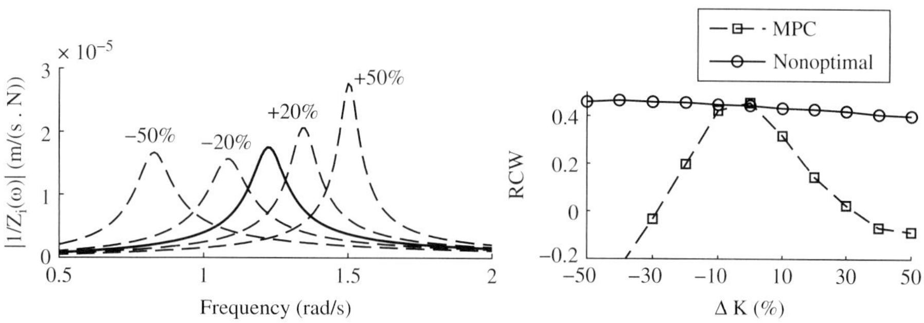

Figure 8.19 Performance of the proposed nonoptimal controller against MPC, in the case of variations of the buoyancy coefficient, ΔK.

A robustness analysis is carried out with the WEC being excited by irregular waves centered at a frequency of 1.25 rad/s (resonance frequency of the model). Fig. 8.19 shows how the RCW obtained with the proposed controller and with MPC is affected by parametric variations in such a wave condition. While MPC is quite robust for variations up to $\Delta K = \pm 10\%$, the robustness of the proposed controller is clearly superior. This is due to the fact that the reference velocity produced by the proposed controller, in the constrained region, does not depend on the model of the system, while MPC needs to solve an optimization based on the system model.

A more comprehensive robustness analysis, involving the synthesis of a robust lower-level controller to deal with nonlinear hydrostatic stiffness, is given in [131]. However, since the study in [131] essentially involves a control solution to a *nonlinear* hydrodynamic model, it is not covered in this volume, where the focus is on linear hydrodynamic models.

8.4 Conclusion

This chapter presents a simple strategy for the real-time control of oscillating WECs. The controller forces the velocity of the oscillator to be in phase with the excitation force and with an amplitude proportional to the wave force. The ratio between the magnitudes of velocity and excitation force can be tuned in real time for performance and constraint handling. While results were shown only for a bottom-referenced body, there is scope for generalizing the approach to self-referenced oscillating systems [136].

Based on a harmonic, nonstationary assumption about the wave excitation force, the proposed controller is suboptimal in real waves. However, the simple controller allows the achievement of levels of power capture close to MPC in most sea states, particularly in narrow-banded waves, without the need for wave prediction nor a real-time solution to an optimization problem. At the same time, the calculation of the reference velocity does not rely heavily on the model of the WEC, resulting in a superior robustness to model uncertainties. Indeed, this relative insensitivity to modeling errors may have played a significant part in the outperformance of MPC by the simple controller presented here in tank tests [137, 138].

In addition, unlike MPC, the resulting motion and forces are linear with respect to the excitation force, giving velocities and forces that follow the same harmonic behavior as waves, with benefits in terms of long-term survivability of the WEC equipment. The effectiveness and computational simplicity make the proposed controller ideal for real-time applications.

While ideal efficiency in the bidirectional energy flow of the PTO mechanism was assumed, it was highlighted how the nature of the controller allows the possibility of including the efficiency information in the adaptive strategy of the high-level control. Also, the structure of the low-level control loop allows the robustness to model uncertainties to be explicitly addressed, which is further developed in [131].

Part V

Control by Optimizing a Performance Index

In this part, optimal controllers, based on the maximization of an energy performance index, are presented. The optimization problem can be solved either algebraically or numerically. Chapter 9 examines the special case of switching control, where the control force is a braking force to be switched on and off, and has proved popular in the WEC control systems literature. The approach seeks an optimum switching sequence by maximizing a performance index (absorbed energy) under a basic variational formulation, and explicitly uses the Pontryagin principle to derive that sequence. The need for information from the future is discussed, even though the control applied is nonreactive. The optimum switching sequence here represents an optimum latching sequence, where velocity is alternately locked and released to maximize the performance index. The use of a numerical optimization framework allows system amplitude, force, and velocity constraints to be explicitly considered in the optimized design, by virtue of a constrained optimization solution. In order to achieve an efficient numerical solution (considering real-time control requirements), a pseudo spectral representation is chosen for the system variables, which parameterizes the variables as a set of (Fourier) basis functions. Chapter 10 presents the basic constrained optimal control formulation for a single device (which can be multibody), while Chapter 11 examines the opportunity for coordinated control of arrays of WECs, examining both independent and global control strategies. As in Part IV, future wave elevation is required, and Chapter 12 examines wave forecasting strategies based on both autoregressive (using only historical information) and up-wave philosophies.

9 Nonreactive Switching Control

Switching control has existed for many years in controls practice and has frequently been a manifestation of "bang-bang" control, where the state of the system is controlled by switching on the full value of the control input in between periods where no control input at all is applied. The origins of this approach in wave energy conversion can be traced back to early studies on small heaving point absorber buoys with short natural resonant periods. The technique consisted of locking the buoy displacement until the approach of a crest (or a trough) and releasing it so it achieved full velocity at the crest (or trough), and then relocking the displacement until the approach of a trough or crest. The presence or absence of a full-valued braking force was thus the only control required. An independent early application of this approach was recorded in the United States by Dedger Jones in the 1970s [139], while by the late 1970s it was already being vigorously developed by Budal and Falnes in Norway and by other workers elsewhere. A formal theoretical foundation was established in the mid-1980s by the work of Nichols and Hoskin [140]. In this chapter we examine the general theory of switching control for enhancing power absorption and, by means of an example of a body oscillating in air, understand why prediction of the future input is required in order to evaluate the optimal switching sequence. The theory is then applied to "latching" of wave energy converters (WECs).

9.1 Latching and Declutching

Because the control input can take only two values (on or off; 1 or 0, etc.) for any type of input forcing, latching is an example of nonlinear control. The other way in which it differs from the complex conjugate–type implementations in frequency or time domain is that no reactive control is possible when only braking forces are alternately switched on and off. Consequently, because optimum velocity as defined in Chapter 6 requires reactive loads off resonance, it generally cannot be achieved with latching. Latching control is thus suboptimal in the hydrodynamic control sense, as it does not generally lead to hydrodynamically optimum velocities. Latching sequences can be derived within a classical optimal control formulation, where a performance criterion (e.g., based on the absorbed energy over a duration) is maximized, or minimized, depending on the way it is defined. The control derived thus may be optimum within that formulation and for the chosen performance criterion, without being the hydrodynamic optimum control for the device under the given input conditions.

Further, since the formulation can proceed from start to finish in the time domain and no Fourier transformation is required or implied, this approach can account for device nonlinearities such as viscous friction, actuator saturation, and stiction, without resorting to linearized approximations. Furthermore, braking forces are generally easier to apply than reactive forces, and because the strategy does not involve exchange of large reactive forces and energy between the power take-off (PTO) and the primary converter, no on-board energy storage or continuous energy exchange with the grid are required. On the other hand, each time a braking force is applied to stop the motion of a large mass, every vibration mode within the entire power conversion chain can potentially be excited; some of these modes may require special attention, as they could cause fatigue and slow structural damage, growing misalignments in mechanical components, and so on, which could result in frequent service and maintenance visits to the device.

While latching control received considerable attention in the context of point absorber buoys championed by Budal and Falnes, and was also being adopted in other devices such as the oscillating water column, a reverse methodology was investigated by the Edinburgh Wave Power group [141], which involved alternately declutching and clutching the primary converter during its oscillations. Thus, the primary converter was allowed to move freely under no load (as opposed to being locked in place) during parts of the cycle, and moved under full load when clutched to the PTO. Because free motion would quickly drive the primary converter/PTO to any physical end stops before the clutches could be engaged, this approach could work only in the absence of end stops or when the no-load oscillation was known not to exceed the total available excursion between end stops. In general, therefore, while latching can be used, in a sense, to slow down the natural short-period response of a device so the correct phase relationship is approached over parts of long-period wave cycles, declutching allows speeding up the natural response of a device that resonates at periods longer than the expected energy periods. It was thus a preferred choice for the Mace device studied by the Edinburgh group and other devices utilizing surge oscillations (since no hydrostatic stiffness is present for this mode, and the PTO and/or the mooring cables must provide the needed stiffness, surging devices are expected to have long natural periods). Finally, however, sudden clutching can be as great a source of vibration as sudden braking, so declutching is not necessarily superior to latching from this standpoint.

Early implementations of latching-type switching control by Budal and Falnes [33] made use of a Kalman filter–type algorithm to derive estimates and predictions of an approaching wave crest/trough to time the release of the buoy. The estimator was based on the measurement of the wave surface elevation up-wave of the buoy and involved assumptions about the heave force on the buoy. Hoskin and Nichols [140] recognized that the latching sequence for a device could be evaluated to provide optimality to a performance index within a formal structure based on classical optimal control theory. Latching and declutching control were applied in sea tests on the Pico-Azores oscillating water column device [142, etc.]. A number of works studied latching on floating body–type devices [71, 143, 144, etc.]. Methods to acquire estimates of the future values of exciting force were examined in [145, 146, etc.], and geometry variations to minimize the exciting force prediction horizon for the buoy-type device have also been considered (for example, see Chapter 14).

9.2 Optimal Control Formulation and Need for Future Information

Our goal is to optimize a performance criterion/index over a prescribed time duration $[0, T]$. At any given time $t \in [0, T]$, we wish to apply a control input (here on or off) such that it helps us optimize the performance criterion over the entire prescribed duration $[0, T]$. As we might expect, then, in order to evaluate an optimal switching input at $t < T$, we need estimates or predictions of the exciting force from the future, whether the body is in air or on the free surface. Note that this need for "foreknowledge" is independent of the nature of the radiation phenomenon and also independent of the continuum/distributed nature of the wave–body interaction. It simply derives from our desire to optimize a performance index over a prescribed time interval (or time horizon). The control input at the present time cannot generally be set so it will contribute positively to the optimality over the time horizon unless we know what its effect into the future will be. Such an evaluation requires knowledge or prediction of the future forcing on the body. We review here a formal procedure from optimal control theory that is just right for our overall purpose.

We begin with a body supported in air by spring and damping elements and excited by a sinusoidal external force here (Fig. 9.1). Also acting on the body is an additional energy-absorbing force. Latching and unlatching are carried out by applying and releasing braking forces by means of the vise grip–type surfaces shown in the figure.

Our goal here is to maximize the power absorbed from the oscillation of m by the power absorbing force $f_a(t)$ over a time duration T. f_a is thought to be purely resistive and is expressed as

$$f_a(t) = C_a v(t). \tag{9.1}$$

The braking force is taken to be provided by a combination of a large damping force and a large stiffness force, such that

$$f_b(t) = C_\infty v(t) + k_\infty x(t) \tag{9.2}$$

with $C_\infty, k_\infty \to \infty$ to approximate a rigid vise grip. By defining a control input s such that the only admissible values of s are 1 and 0, and the two forces f_a and f_b can be combined into a single expression,

$$f_{ab}(t) = sC_a v(t) + (1 - s)\left[C_\infty v(t) + k_\infty x(t)\right]. \tag{9.3}$$

Thus, when $s = 1$, only the power absorption component is active, while for $s = 0$, large damping and stiffness forces oppose motion and no power is absorbed. The equation of motion for the mass m can be expressed as

$$m\ddot{x} + [c + sC_a + (1 - s)C_\infty]\dot{x} + [k + (1 - s)k_\infty]x = f(t). \tag{9.4}$$

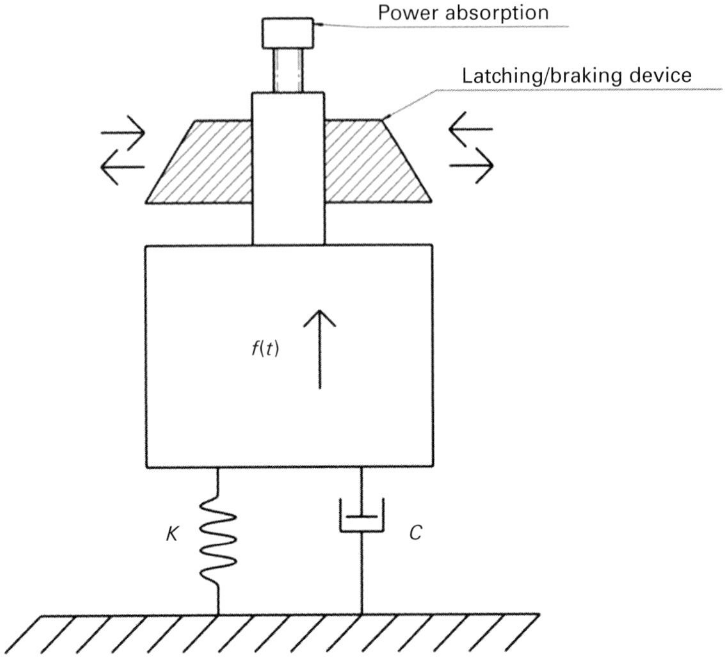

Figure 9.1 Latching-type switching control applied to a body oscillating in air.

It is helpful to rewrite this equation of motion in state-space form as

$$\dot{x} = v,$$
$$\dot{v} = -\frac{1}{m}\left[c + C_a + (1-s)C_\infty\right]v - \frac{1}{m}\left[k + (1-s)k_\infty\right] + f(t) \tag{9.5}$$

where x and v (respectively, displacement and velocity) represent the two state variables for the system and form the "generalized coordinates" for the work below. $t = 0$ is the time at which the clock starts. We set the initial conditions to zero, so that $x(0) = v(0) = 0$. The power absorbed by the actuator is

$$P_{abs} = sC_a v^2(t). \tag{9.6}$$

This function can be integrated to obtain the energy absorbed over a given time duration T as

$$E_a(T) = \int_0^T sC_a v^2(t)dt. \tag{9.7}$$

$E_a(T)$ defines the performance criterion for this problem, and P_{abs} forms the Lagrangian. To maximize our performance criterion (i.e., the absorbed energy $E_a(T)$ through T) while making sure the system states evolve according to Equations (9.5), an augmented performance criterion $\overline{E}_a(T)$ may be defined such that the state equations are included within it, as

$$
\begin{aligned}
\overline{E}_a(T) = & \int_0^T \left\{ sC_a v^2(t) + \lambda_1(\dot{x} - v) \right\} dt \\
& + \int_0^T \left\{ \lambda_2 \left[\dot{v} + \frac{1}{m} [[c + sC_a + (1-s)C_\infty] v \right.\right. \\
& + \left.\left. [k + (1-s)k_\infty] x - f(t)] \right] \right\} dt.
\end{aligned} \tag{9.8}
$$

This is a standard technique from the calculus of variations [147, 148], and the so-called co-state variables λ_1, and λ_2 are the "to-be-determined" Lagrange multipliers here. The integral in Equation (9.8) is a "functional" that is to be maximized in the present case. We note that a functional is in general an operation on a function, resulting in a numerical value. Here that value represents the total energy absorbed from $t = 0$ to $t = T$.

Our goal now is the following: of all possible functions $x(t)$ and $v(t)$ describing possible state-space trajectories (x, v) that transition our system from $x(0) = v(0) = 0$ to $x(T), v(T)$, to find the $[x(t), v(t)]$ trajectory that maximizes the functional above. The state Equations (9.5) restrict our set of choices to only those functions that satisfy Equations (9.5). They can thus be thought of as constraints representing the dynamics of our particular spring–mass system in Fig. 9.1. The augmented performance criterion in Equation (9.8) provides a rational way to apply these constraints at all t. The Lagrange multipliers λ_1 and λ_2, or the co-states, are also functions of time. In fact, we now need to determine not only functions $x(t)$, $v(t)$, but also $\lambda_1(t)$ and $\lambda_2(t)$ such that the augmented functional is maximized. The necessary condition for a maximum resembles that in ordinary calculus, in that a small "variation" in any of the "directions" defined in the admissible function space must be zero. This is frequently termed the "first variation." Note that the same condition, that is, that the first variation be zero, holds also for a minimum. Use of this condition thus can guarantee only an "extremum." Just as in ordinary calculus, it is the "second variation" that provides further information on whether this extremum is a maximum or a minimum, or neither (as with an inflection point where the second variation is also zero). Often, however, verification through direct evaluation of the functional for the trajectory as determined suffices and may be more convenient. Finally, if the state equations are nonlinear, the extremum may be only a "local maximum" or a "local minimum," often over confined neighborhoods, and the numerical algorithm used to determine the desired control variation then may need to be designed to approach a maximum ranging over some predefined large neighborhood.

In any case, the necessary condition for $\overline{E}_a$ to be maximum is that the first variation $\delta\overline{E}_a$ be zero for the "optimum trajectory" as provided by the "optimum control" to be determined. This condition is expressed as

$$\delta\overline{E}_a = 0. \tag{9.9}$$

Because $t = 0$ and $t = T$ are both prescribed time instants, it is the variation of the integrand that determines the variation of the integral. Defining the integrand as

$\mathcal{L} \equiv \mathcal{L}(x, \dot{x}, v, \dot{v}, \lambda_1, \lambda_2, s)$ (with the arguments in the parentheses being just those on which $\mathcal{L}$ has an explicit dependence),

$$\delta \overline{E}_a = \int_0^T \delta \mathcal{L} dt = \int_0^T \left[\frac{\partial \mathcal{L}}{\partial x} \delta x + \frac{\partial \mathcal{L}}{\partial \dot{x}} \delta \dot{x} + \frac{\partial \mathcal{L}}{\partial v} \delta v + \frac{\partial \mathcal{L}}{\partial \dot{v}} \delta \dot{v} \right] dt$$
$$+ \int_0^T \left[\frac{\partial \mathcal{L}}{\partial \lambda_1} \delta \lambda_1 + \frac{\partial \mathcal{L}}{\partial \lambda_2} \delta \lambda_2 + \frac{\partial \mathcal{L}}{\partial s} \delta s \right] dt. \tag{9.10}$$

The goal of this current exercise is to relate the first variation of the augmented performance criterion $\overline{E}_a$ to the first variations of the generalized coordinates and the control variable defining $\overline{E}_a$ (basically, all quantities in the parentheses attached to $\mathcal{L}$). To this end, integrating the above by parts and collecting like variations, we have

$$\delta \overline{E}_a = \left[\frac{\partial \mathcal{L}}{\partial \dot{x}} \delta \dot{x} \right]_0^T + \int_0^T \left[\frac{\partial \mathcal{L}}{\partial x} - \frac{d}{dt} \left(\frac{\partial \mathcal{L}}{\partial \dot{x}} \right) \right] \delta x dt$$
$$+ \left[\frac{\partial \mathcal{L}}{\partial \dot{v}} \delta \dot{v} \right]_0^T + \int_0^T \left[\frac{\partial \mathcal{L}}{\partial v} - \frac{d}{dt} \left(\frac{\partial \mathcal{L}}{\partial \dot{v}} \right) \right] \delta v dt$$
$$+ \int_0^T \frac{\partial \mathcal{L}}{\partial \lambda_1} \delta \lambda_1 dt + \int_0^T \frac{\partial \mathcal{L}}{\partial \lambda_2} \delta \lambda_2 dt + \int_0^T \frac{\partial \mathcal{L}}{\partial s} \delta s dt. \tag{9.11}$$

The only conditions on the variations δ are that they are small, independent, compatible with any kinematic constraints, and contemporaneous [147].[1] Note that the first and third terms depend only on the boundaries of the interval $[0,\ T]$. The integrals clearly depend on the entire trajectory $[x(t),\ v(t)]$, the corresponding co-states $[\lambda_1(t),\ \lambda_2(t)]$, and the pattern of s variation from $t = 0$ to $t = T$. For $\delta \overline{E}_a = 0$, thus, (1) the boundary-dependent terms must both equal zero so that arbitrary choices can be prescribed for the integration limits and (2) the integrands must identically be zero over the entire state and co-state trajectories and over the entire s variation. This argument thus states that for $\overline{E}_a$ to be maximum, the following conditions must be met:

$$\frac{d}{dt} \left(\frac{\partial \mathcal{L}}{\partial \dot{x}} \right) - \frac{\partial \mathcal{L}}{\partial x} = 0,$$
$$\frac{d}{dt} \left(\frac{\partial \mathcal{L}}{\partial \dot{v}} \right) - \frac{\partial \mathcal{L}}{\partial v} = 0,$$
$$\frac{\partial \mathcal{L}}{\partial \lambda_1} = \frac{\partial \mathcal{L}}{\partial \lambda_2} = 0,$$
$$\frac{\partial \mathcal{L}}{\partial s} = 0, \quad \left[\frac{\partial \mathcal{L}}{\partial \dot{x}} \delta x \right]_0^T = 0, \quad \left[\frac{\partial \mathcal{L}}{\partial \dot{v}} \delta v \right]_0^T = 0. \tag{9.12}$$

Since the boundary conditions at $t = 0$ are prescribed as $x(0) = v(0) = 0$, both δx and $\delta v = 0$ at the lower limit in the two boundary-dependent terms above. This leaves just

$$\left[\frac{\partial \mathcal{L}}{\partial \dot{x}} \right]_{t=T} = 0, \quad \left[\frac{\partial \mathcal{L}}{\partial \dot{v}} \right]_{t=T} = 0. \tag{9.13}$$

[1] This is reminiscent of, and a generalization of, the method of virtual work in dynamics.

Equation (9.13) states the "natural boundary conditions" for the present case, which here have an important significance. The first two of Equations (9.12) are written with a sign reversal and are the Euler–Lagrange equations for this variational problem. The equations resulting from the partial derivatives of $\mathcal{L}$ with respect to λ_1 and λ_2 are simply the state Equations (9.5). The Euler–Lagrange equations here lead to

$$\dot{\lambda}_1 = \frac{1}{m}\left[k\lambda_2 + (1-s)k_\infty\lambda_2\right],$$
$$\dot{\lambda}_2 = 2sC_a v - \lambda_1 + \frac{1}{m}\left[c + C_a + (1-s)C_\infty\right]\lambda_2. \tag{9.14}$$

Equations (9.14) represent the "adjoint equations" here. The natural boundary conditions at $t = T$ (9.13) lead to the "transversality conditions"

$$\lambda_1(T) = 0, \quad \lambda_2(T) = 0. \tag{9.15}$$

The transversality conditions form the boundary conditions for the co-state equations in λ_1 and λ_2. This point is discussed further near the end of this section. It is interesting to note that the condition arising from the partial derivative of $\mathcal{L}$ with respect to s does not contain s at all and would lead to a singular solution connecting the state variables and the co-state variables that is independent of s. We must therefore adopt an alternative approach. We start by defining the Hamiltonian $\mathcal{H}$ associated with the system above as [149]

$$\mathcal{H} = sC_a v^2 + \lambda_1 v - \lambda_2\frac{1}{m}\left[(c + sC_a + (1-s)C_\infty)v\right] - \lambda_2\frac{1}{m}\left[(k + (1-s)k_\infty)x - f(t)\right]. \tag{9.16}$$

The Pontryagin maximum principle then provides a condition for a maximum for the original $\overline{E}_a$ (see, e.g., [149]), which can be expressed as

$$\frac{\partial\mathcal{H}}{\partial s}\delta s \leq 0 \tag{9.17}$$

for all possible directions of δs. In other words, changing the control variable to any "nearby" value would reduce $\mathcal{H}$. For switching control, we recall that there are only two possible values of s in the available set of control inputs. Therefore, only two directions exist here, namely, $\delta s < 0$ and $\delta s > 0$. In fact, when $s = 1$, only $\delta s < 0$ is possible, and when $s = 0$, only $\delta s > 0$ is possible. Then we could ensure a maximum by selecting an s value such that when $\partial\mathcal{H}/\partial s$ is positive, δs would be negative, so that s must then be set equal to 1. Otherwise, when $\partial\mathcal{H}/\partial s < 0$, s should be zero so δs is positive. In case $\partial\mathcal{H}/\partial s = 0$, s can be zero or 1. Here,

$$\begin{array}{lll} \text{If } C_a v^2 - \frac{\lambda_2}{m}\left(C_a - C_\infty\right)v + \frac{\lambda_2}{m}x & > 0, & s = 1, \\ \text{else} & \leq 0, & s = 0. \end{array} \tag{9.18}$$

Some important observations based on Equations (9.18) are in order here. First, we note that the condition used to set an s value at any given instant needs the value of

λ_2 at that instant. It also needs values of the state variables x and v at that instant, and perhaps these could be obtained by direct measurement. However, λ_2 cannot be known unless Equations (9.14) are solved for λ_1 and λ_2 subject to the transversality conditions at $t = T$. Equations (9.14) need to be solved at each instant in backward time, that is, from $t = T$ to $t = 0$. Recall that the state Equations (9.5) are to be solved from $t = 0$ to $t = T$. Altogether, we thus have a "two-point boundary value problem" situation, where we know $x(0), v(0), \lambda_1(T)$, and $\lambda_2(T)$ and are to find $x(t), v(t), \lambda_1(t)$, and $\lambda_2(t)$ over $[0, T]$.

Because the state variable v appears in the adjoint Equations (9.14), however, at any time $t < T$, $v(t)$ up to $v(T - t)$ needs to be known, for us to be able to solve (9.14). In short, switching or bang-bang control, that is, latching-type control meeting an optimality criterion over a specified time duration $[0, T]$, requires, at any time $t \in [0, T]$, knowledge of the state variables over the entire time duration, including into the future. This is true regardless of whether the body is in water subject to waves or in air subject to some other excitation.

A basic-level approach to such control is: (1) specify a trial sequence for the control variable s from $t = 0$ to $t = T$, (2) solve the state equations for this sequence from $t = 0$ to $t = T$, (3) solve the adjoint equations backward from $t = T$ to $t = 0$, (4) apply the condition (9.18) at each instant to evaluate a "better" s sequence, (5) solve the state equations from $t = 0$ to $t = T$ for the revised s-sequence, (6) solve the adjoint equations again from $t - T$ to $t = 0$, (7) reapply the condition (9.18) to improve the s sequence, (8) re-solve the state equations, and so on. This procedure is repeated until it converges to a particular s sequence to apply on the system. The procedure as outlined is open loop, though state feedback to correct for any model errors in the state equations or disturbances can be implemented within, for instance, a Kalman filter–type strategy or another estimation scheme in a practical application.

9.3 Optimal Switching Control in Waves

The overall method of Section 9.2 translates without much difficulty to switching control of bodies in waves, as long as the wave forces are accounted for correctly. The need for future knowledge now implies that the wave-applied exciting forces need to be known or predicted ahead of time. The resulting control is suboptimal in the hydrodynamic sense, as discussed near the beginning of this chapter. We consider an example of a latching application here, where the heaving oscillations of a hemispherical buoy are used to drive an air flow over a Wells turbine. The oscillations are reacted against a fixed reference, which supports the pneumatic chamber and the turbine as a buoy-driven piston oscillates within the chamber to cause bidirectional air flow over the turbine. Fig. 9.2 shows a schematic for this system. A valve is driven by a small electric motor to open or close the air passage to the turbine. With the valve fully closed, the large pressure build-up inside the chamber provides a large force opposing the buoy heave. With the valve fully open, the buoy oscillation can drive the turbine via the air flow through the chamber.

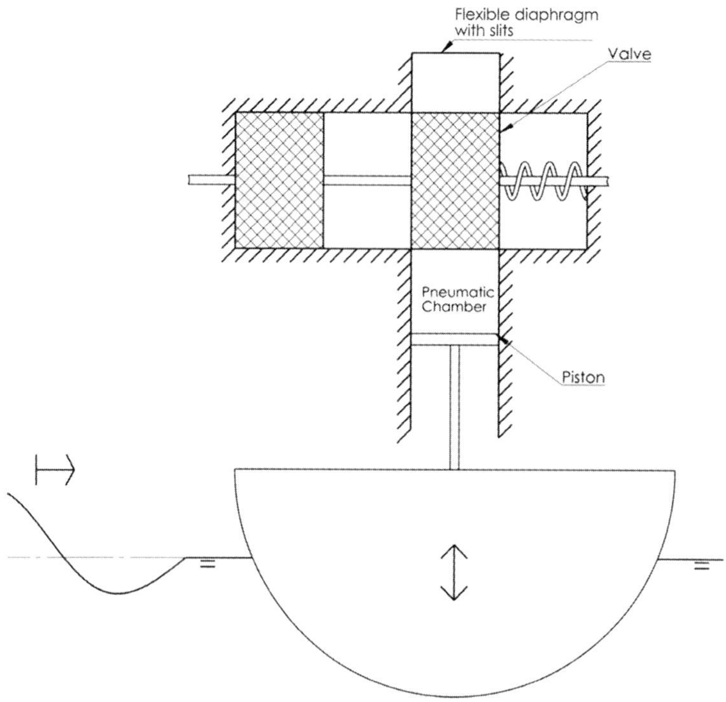

Figure 9.2 Latching-type switching control applied to a body oscillating in water. The air flow past a turbine is controlled by means of a valve.

With the instantaneous pressure inside the chamber denoted as $p(t)$, the heave response of the buoy can be described as

$$[(m + \overline{a}(\infty)]\,\dot{v} + \int_{-\infty}^{\infty} h_r(\tau)v(t-\tau)d\tau + kx + A_p p = F_f \tag{9.19}$$

where A_p represents the chamber cross-section area. Several points are important to note here. As observed earlier, an integro-differential equation with a convolution representation of the radiation force is essential here, as switching control is nonlinear and therefore precludes a frequency domain treatment. The second important point is more subtle and pertains to the integration limits for the convolution, which are particularly important to the present goal of determining a control strategy that is optimal (in the sense of Section 9.2) over a chosen time duration $[0, T]$. Finally, the pressure in the chamber is an important variable in the overall dynamics, and its dependence on the buoy oscillation and the control (i.e., whether $s = 0$ or $s = 1$; the former blocking the air flow and the latter allowing it) needs to be modeled appropriately. We discuss the second point below.

In the general representation of the convolution in Equation (9.19), the arguments of the impulse response functions $h(\cdot)$ and velocity $v(\cdot)$ are freely interchangeable, as convolution in this full form is a commutative operator. This can be verified by a variable change from τ to $\overline{\tau} = t - \tau$. Thus,

$$\overline{F}_R(t) \equiv \int_{-\infty}^{\infty} h_r(\tau)v(t-\tau)d\tau = \int_{-\infty}^{\infty} h_r(t-\tau)v(\tau)d\tau. \tag{9.20}$$

Recognizing now that $h_r(t)$ is a causal function, that is, $h_r(t) = 0$, $t < 0$, the form on the left really implies,

$$\overline{F}_R(t) = \int_{0}^{\infty} h_r(\tau)v(t-\tau)d\tau. \tag{9.21}$$

Next a further restriction is introduced that the body was at rest until $t = 0$. We use the initial conditions $[x(0) = 0, v(0) = 0]$. If $v(t) = 0$, $t < 0$, then the upper limit on the integral needs to go only as far as t, the current instant, so that

$$\overline{F}_R(t) = \int_{0}^{t} h_r(\tau)v(t-\tau)d\tau. \tag{9.22}$$

If the convolution representation on the right side of Equation (9.20) is to be used with the knowledge that $h_r(t)$ is causal, then $h_r(t-\tau) = 0$ for any value of $\tau > t$. Then,

$$\overline{F}_R(t) = \int_{-\infty}^{t} h_r(t-\tau)v(\tau)d\tau. \tag{9.23}$$

With the body assumed at rest until $t = 0$ so that $v(\tau) = 0$ for all $\tau < 0$, the lower limit on the integral in (9.23) can be replaced by 0, so that

$$\overline{F}_R(t) = \int_{0}^{t} h_r(t-\tau)v(\tau)d\tau. \tag{9.24}$$

The two forms thus have identical limits for the present problem, though such may not be the case in general. The particular convolution form and the integration limits used are still important, as they affect the integration limits in the convolution integral that occurs in the adjoint equations.

The dependence of the chamber pressure on buoy oscillation and vice versa are discussed next. This pressure is affected by the valve opening and closing, the displacement and velocity of the piston, and the PTO load when the valve is open. The piston velocities, even when the valve is open, are generally small compared with the speed of sound in air. Further, the temperature in the chamber is assumed to be unaffected by the motion, and the thermal effects of air compressibility are assumed to be small. However, with the valve closed, some compression of the air within the chamber does occur, allowing small movement of the piston. This helps provide a cushioning effect against the customary sudden starts and stops associated with latching, but could influence the overall dynamics and so needs to be modeled. The behavior of the air inside the chamber is here assumed to be isentropic and is modeled using [140, 143]

$$\left(\frac{p+p_0}{p_0}\right) = \left(\frac{\rho_a}{\rho_0}\right)^{\gamma}. \tag{9.25}$$

The net air-flow flux through the chamber and the pressure together drive the PTO when the valve is open. This action can be modeled using a conductance-like coefficient C such that [25, 140]

$$\dot{m} = \rho_a \frac{dV}{dt} + V\frac{d\rho_a}{dt} = -sCp. \tag{9.26}$$

The air-chamber volume V at time t is

$$V(t) = V_0 - A_p x(t). \tag{9.27}$$

From Equation (9.25),

$$\rho_a = \rho_0 \left(1 + \frac{p}{p_0}\right)^{1/\gamma}. \tag{9.28}$$

For the type of velocities and pressures expected in such a device, $p \ll p_0$ generally, and a linearized approximation for ρ_a may be derived:

$$\rho_a = \rho_0 \left(1 + \frac{1}{\gamma}\frac{p}{p_0}\right). \tag{9.29}$$

This leads to

$$-\rho_0 \left(1 + \frac{1}{\gamma}\frac{p}{p_0}\right) A_p v + \frac{V_0 \rho_0}{\gamma p_0}\dot{p} = -sCp. \tag{9.30}$$

Furthermore, consistent with the assumption of small velocities and pressures, $vp \approx 0$. Then Equation (9.26) can be linearized and rewritten in the state-space form as

$$\dot{p} = -\frac{s\gamma p_0 C}{V_0 p_0} p + \frac{\gamma A_p p_0}{V_0} v \tag{9.31}$$

where the coupling between the chamber pressure p and the buoy velocity v is linear. The dependence of $\dot{p}$ on the control variable s and the conductance-like constant C modeling the PTO is captured explicitly in the first term. Equation (9.19) can be broken down into two state equations as before, and the overall system can be described as

$$\begin{aligned}
\dot{x} &= v, \\
\dot{v} &= \frac{1}{m + \bar{a}(\infty)} \left[F_f(t) - \int_{-\infty}^{\infty} h_r(\tau) v(t-\tau) d\tau - kx - A_p p \right], \\
\dot{p} &= -\frac{s\gamma p_0 C}{V_0 p_0} p + \frac{\gamma A_p p_0}{V_0} v.
\end{aligned} \tag{9.32}$$

The initial conditions are

$$x(0) = v(0) = p(0) = 0. \tag{9.33}$$

The instantaneous absorbed power can be found by multiplying the instantaneous chamber pressure by the air-flow flux through the chamber. The latter quantity is just the right side of Equation (9.26) divided by the instantaneous air density ρ_a. By approximating this as ρ_0, the instantaneous absorbed power can be written as

$$P_{abs}(t) = \frac{sCp^2}{\rho_a} \approx \frac{sCp^2}{\rho_0}. \tag{9.34}$$

A performance index can now be defined as the energy absorbed over a specified time interval $[0, T]$. As before, this energy is denoted as

$$E_a(T) = \int_0^T \frac{sCp^2}{\rho_0} dt. \tag{9.35}$$

To maximize this subject to the dynamic constraints implied by the state equations, an augmented performance criterion $\overline{E}_a(T)$ can be defined:

$$\begin{aligned}\overline{E}_a(T) = &\int_0^T \left[\frac{sCp^2}{\rho_0} + \lambda_1 (\dot{x} - v)\right] dt \\ &+ \int_0^T \left[\lambda_2 \left(\dot{v} - \frac{1}{m + \overline{a}(\infty)} \left[F_f(t) - \int_{-\infty}^{\infty} h_r(\tau) v(t-\tau) d\tau - kx - A_p p\right]\right)\right] dt \\ &+ \int_0^T \left[\lambda_3 \left(\dot{p} - \left[-\frac{s\gamma p_0 C}{V_0 \rho_0} p + \frac{\gamma A_p p_0}{V_0} v\right]\right)\right] dt,\end{aligned} \tag{9.36}$$

using Lagrange multipliers λ_i, $i = 1, \ldots 3$, which form the co-state variables here. Note that the co-states are time dependent. To maximize the augmented performance criterion, a similar procedure as before is adopted, with the goal of seeking the stationary value of $\overline{E}_a(T)$ such that $\delta \overline{E}_a = 0$, where

$$\delta \overline{E}_a = \int_0^T \delta \mathcal{L}(x, \dot{x}, v, \dot{v}, p, \dot{p}, s; \lambda_1, \lambda_2, \lambda_3) dt = 0. \tag{9.37}$$

The resulting Euler–Lagrange equations have the form

$$\begin{aligned} &\frac{d}{dt}\left(\frac{\partial \mathcal{L}}{\partial \dot{x}}\right) - \frac{\partial \mathcal{L}}{\partial x} = 0, \\ &\frac{d}{dt}\left(\frac{\partial \mathcal{L}}{\partial \dot{x}}\right) - \frac{\partial \mathcal{L}}{\partial x} = 0, \\ &\frac{d}{dt}\left(\frac{\partial \mathcal{L}}{\partial \dot{p}}\right) - \frac{\partial \mathcal{L}}{\partial p} = 0, \\ &\frac{\partial \mathcal{L}}{\partial s} = 0; \quad \frac{\partial \mathcal{L}}{\partial \lambda_i} = 0, \text{ i} = 1, \ldots 3, \\ &\left[\frac{\partial \mathcal{L}}{\partial \dot{x}}\right]_T = \left[\frac{\partial \mathcal{L}}{\partial \dot{v}}\right]_T = \left[\frac{\partial \mathcal{L}}{\partial \dot{p}}\right]_T = 0. \end{aligned} \tag{9.38}$$

The top three of Equations (9.38) lead to the adjoint or co-state equations for the Lagrange multipliers/co-states λ_i. The first of the fourth equations leads to a singular condition as before, while the next three simply lead back to the state equations. The last line contains the boundary terms at $t = T$, which lead to the natural boundary conditions for the Lagrange multipliers λ_i.

The convolution term requires special handling because there are now integrals involved [151, 152]. We proceed by changing the order of integration in the term

$$\delta\left[\frac{1}{m+\bar{a}(\infty)}\int_0^T \lambda_2(t)\int_0^t h_r(\tau)v(t-\tau)d\tau dt\right]$$
$$=\frac{1}{m+a(\infty)}\int_0^T\left[\int_0^{T-t} h_r(\tau)\lambda_2(t+\tau)d\tau\right]\delta v dt. \tag{9.39}$$

Returning briefly to an earlier discussion (Equations (9.20)–(9.24)), if we had used the alternative form of the convolution for the radiation force, the order change above would have resulted in

$$\delta\left[\frac{1}{m+\bar{a}(\infty)}\int_0^T \lambda_2(t)\int_0^t h_r(t-\tau)v(\tau)d\tau\right]$$
$$=\frac{1}{m+\bar{a}(\infty)}\int_0^T\left[\int_t^T h_r(\tau-t)\lambda_2(\tau)d\tau\right]\delta v dt. \tag{9.40}$$

It can be verified that the two forms are equivalent. In sum, the adjoint equations for λ_i are

$$\dot{\lambda}_1=\frac{k}{m+\bar{a}(\infty)}\lambda_1,$$
$$\dot{\lambda}_2=-\lambda_1+\frac{1}{m+\bar{a}(\infty)}\left[\int_0^{T-t} h(\tau)\lambda_2(t+\tau)d\tau\right]-\rho_0 p\lambda_3,$$
$$\dot{\lambda}_3=\frac{2sC}{\rho_0}p+\frac{A_p}{m+\bar{a}(\infty)}\lambda_2+\frac{s\gamma p_0 C}{V_0\rho_0}\lambda_3. \tag{9.41}$$

The natural boundary conditions result in the transversality conditions for λ_i at $t=T$:

$$\lambda_1(T)=\lambda_2(T)=\lambda_3(T)=0. \tag{9.42}$$

Once more it is necessary to use the Hamiltonian $\mathcal{H}$ to make further progress, as the condition resulting from $\partial\mathcal{L}/\partial s=0$ does not lead to a viable way to determine the control variable s. Thus,

$$\mathcal{H}=\frac{sCp^2}{\rho_0}+\lambda_1 v+\lambda_2\frac{1}{m+\bar{a}(\infty)}\left[F_f(t)-\int_{-\infty}^{\infty} h_r(\tau)v(t-\tau)d\tau-kx-A_p p\right]$$
$$+\lambda_3\left[-\frac{s\gamma p_0 C}{V_0 p_0}p+\frac{\gamma A_p p_0}{V_0}v\right]. \tag{9.43}$$

By the Pontryagin maximum principle,

$$\frac{\partial\mathcal{H}}{\partial s}\delta s\leq 0 \tag{9.44}$$

for all possible directions of δs. If the only values of s to choose from are 0 and 1, then $\delta s<0$ when $s=1$, and $\delta s>0$ when $s=0$. Therefore, Equation (9.44) leads to the condition

10 Optimal Numerical Control of Wave Energy Converters

Not unlike the energy maximization problem for wind turbines, the optimal control problem for wave energy converters (WECs) involves the maximization of a cost functional of the form

$$J = -\int_0^T \dot{\mathbf{x}}(t)^\mathsf{T}\, \mathbf{f}_{pto}(t)\, dt \tag{10.1}$$

where $\dot{\mathbf{x}}(t)$ is the device velocity, $\mathbf{f}_{pto}(t)$ is the power take-off (PTO) force, and J represents the useful energy converted by the PTO. However, Equation (10.1) is a departure from the standard optimal (regulator) control problem, which is normally quadratic in the arguments of the integral and therefore represents a convex optimization problem.

Early work on the optimal constrained control of WECs in irregular waves focussed on the application of Pontryagin's maximum principle and the numerical solution, using Lagrange multipliers, to the resulting two-point boundary value problem [154]. The main difficulty with such indirect methods (which attempt to solve the necessary conditions for optimality) is that the set of necessary conditions for optimality, which must be derived analytically, are problem specific and the region of convergence for the optimal solution is small, requiring a good initial guess.

More recently, the popular model predictive control (MPC) philosophy has been adapted for use by the wave energy community, for example, in [110, 155, 156]. However, attempts to adopt discrete-time versions of performance functional of (10.1) have attracted various modifications, including the specification of a time offset between $\dot{\mathbf{x}}(t)$ and $\mathbf{f}_{pto}(t)$ (namely, $\dot{\mathbf{x}}(k+1)\mathbf{f}_{pto}(k)$, where k is the discrete time index [110]), or the addition of extra terms to make the cost-functional quadratic [156].

This chapter describes the application of a direct transcription method to the optimal control of WECs and develops a generalized discretization framework within which the WEC control problem can be solved. The transcription method used for the discretization of the control problem employs the mean weighted residuals as described in [157]. The first reported application of a direct transcription method to the wave energy control problem was in [158], which parameterizes device motion and PTO force using Fourier basis functions (and employs amplitude constraints only), while [159] uses a pseudospectral method to approximate motion and PTO force using polynomials, requiring the addition of a further constraint to ensure periodicity.

The chapter proceeds as follows: the discretization is carried out for a generic configuration of WECs and for generic sets of basis and test functions; detailed derivations of the quadratic program resulting from the direct transcription are also included here. As an illustrative example, Section 10.2 describes the application of the Galerkin method to a single-body WEC, where the basis functions are truncated Fourier series. Derivations of the matrices composing the quadratic program and a discussion about some of their properties are also included (e.g., convexity).

Simulation results are presented in Section 10.2.2 for both regular and irregular incident waves, including restrictions on both PTO force and oscillation amplitude. Section 10.3 provides a general discussion on the computational aspects associated with the algorithm, with Section 10.3.1 in particular providing a simplification of the convolution integral associated with the radiation force, when the velocity is approximated with a generic expansion. It is shown that, because of linearity, the computations involving the numerical integration of the convolution integral can be carried out off-line, thus significantly reducing the computational load when solving the nonlinear programming (NLP) problem.

10.1 General Case: Mean Weighted Residual Formulation of a System of WECs

The objective of this section is to present the steps for the direct transcription of the optimal control problem of a generic system of WECs. Following the initial presentation in Equation (4.27), the model is described by Cummins' equation [63], which, including the vector of PTO forces, $\tilde{\mathbf{f}}_p(t)$, is

$$\mathbf{M}\ddot{\mathbf{x}}(t) + \mathbf{B}\dot{\mathbf{x}}(t) + \int_0^t \mathbf{K}(t-\tau)\dot{\mathbf{x}}(\tau)d\tau + \mathbf{S}\mathbf{x}(t) = \tilde{\mathbf{f}}_p(t) + \mathbf{f}_e(t) \tag{10.2}$$

where $\mathbf{x}(t) \in \mathbb{R}^n$ is the position vector of the WEC and n is the number of degrees of freedom of the system. The matrix $\mathbf{M} \in \mathbb{R}^{n \times n}$ is the sum of the generalized mass matrix and the added mass at infinite frequency; $\mathbf{B}$ and $\mathbf{S}$ are positive constant $n \times n$ diagonal matrices describing the linear damping and the stiffness, respectively. The elements of the $n \times n$ matrix of radiation impulse responses $\mathbf{K}(t)$ are continuous functions in $[0, \alpha)$ and zero for $t < 0$, where $\alpha \leq \infty$.

The vector of excitation forces $\mathbf{f}_e(t)\colon [0, t] \to \mathbb{R}^n$ is assumed to be continuous, and the vector of the PTO forces is considered to be $\tilde{\mathbf{f}}_p(t) = \mathbf{F}_P \mathbf{f}_{pto}$, where $\mathbf{F}_P$ is an $n \times m$ constant matrix. The role of the matrix $\mathbf{F}_P$ is to allow a general combination of PTO components, particularly where the number of PTO forces (m) are different from the number of modes of oscillation of the system (n). In general, there are fewer PTO forces than modes of oscillations ($m \leq n$). A more detailed explanation regarding the matrix $\mathbf{F}_P$ will be provided in Section 10.1.1. The matrix $\mathbf{F}_P$ also takes part in the definition of the total absorbed energy over the time interval $[0, T]$, which is defined as the sum of the mechanical work done by all the PTO forces:

$$J = -\int_0^T \dot{\mathbf{x}}(t)^\mathsf{T}\, \mathbf{F}_P\, \mathbf{f}_{pto}(t)\, dt. \tag{10.3}$$

In practice, $\mathbf{F}_P$ specifies how the PTO forces and the velocities are combined to produce the usable power.

The optimal control problem is to find the PTO force vector $\mathbf{f}_{pto}$ that maximizes the total absorbed energy J, subject to the equation of motion (10.2) and the additional control and motion path constraints described as

$$\mathbf{h}(\mathbf{x}, \dot{\mathbf{x}}, \mathbf{f}_{pto}, t) = 0, \tag{10.4}$$

$$\mathbf{g}(\mathbf{x}, \dot{\mathbf{x}}, \mathbf{f}_{pto}, t) \leq 0. \tag{10.5}$$

Before proceeding with the discretization of the optimal control problem, the equation of motion (10.2) has to be rewritten as a system of first-order integro-differential equations. This is achieved, as usual, by introducing an additional state variable describing the velocity ($\mathbf{v}$), and the resulting system dynamic is

$$\dot{\mathbf{x}} = \mathbf{v}, \tag{10.6}$$

$$\mathbf{M}\,\dot{\mathbf{v}} = -\mathbf{B}\,\mathbf{v} - \int_{-\infty}^{t} \mathbf{K}(t-\tau)\mathbf{v}(\tau)\,d\tau - \mathbf{S}\,\mathbf{x} + \mathbf{F}_P\mathbf{f}_{pto} + \mathbf{f}_e. \tag{10.7}$$

The discretization of the control problem is performed by approximating the position ($\mathbf{x}$) and the velocity ($\mathbf{v}$) vectors with a linear combination of the basis functions $\phi_k(t)$, and the PTO force ($\mathbf{f}_{pto}$) vector with a linear combination of the functions $\phi_k^P(t)$; the ith components of these vectors are

$$x_i(t) \approx x_i^N(t) = \sum_{k=1}^{N} x_{ik}^x\,\phi_k(t) = \Phi(t)\hat{x}_i^x, \quad i = 1, \ldots, n, \tag{10.8}$$

$$v_i(t) \approx v_i^N(t) = \sum_{k=1}^{N} x_{ik}^v\,\phi_k(t) = \Phi(t)\hat{x}_i^v, \quad i = 1, \ldots, n, \tag{10.9}$$

$$f_{pto_i}(t) \approx f_{pto_i}^{N^P}(t) = \sum_{k=1}^{N^P} u_{ik}\,\phi_k^P(t) = \Phi^P(t)\hat{u}_i, \quad i = 1, \ldots, m, \tag{10.10}$$

where N and N^P are the orders of the expansions for the states (position and velocity) and the control input (PTO force), respectively,

$$\hat{x}_i^x = \left[\hat{x}_{i1}^x, \hat{x}_{i2}^x, \ldots, \hat{x}_{iN}^x\right]^{\mathsf{T}}, \quad \hat{x}_i^v = \left[\hat{x}_{i1}^v, \hat{x}_{i2}^v, \ldots, \hat{x}_{iN}^v\right]^{\mathsf{T}}, \quad \hat{u}_i = \left[\hat{u}_{i1}, \hat{u}_{i2}, \ldots, \hat{u}_{iN_P}\right]^{\mathsf{T}},$$

and

$$\Phi(t) = \left[\phi_1(t), \phi_2(t), \ldots, \phi_N(t)\right],$$
$$\Phi^P(t) = \left[\phi_1^P(t), \phi_2^P(t), \ldots, \phi_{N_P}^P(t)\right].$$

Using the approximated velocity and PTO force, from Equations (10.9) and (10.10), respectively, the total absorbed energy J^N is

$$J^N = -\int_0^T \Phi(t)\mathbf{X}^v\,\mathbf{F}_p\,\mathbf{U}^{\mathsf{T}}\Phi^{P^{\mathsf{T}}}(t)\,dt = \sum_{i=1}^{N^P}\sum_{j=1}^{N} w_{ij}\Gamma_{ij} \tag{10.11}$$

where w_{ij} are the elements of the matrix $\mathbf{W} = \mathbf{X}^v \mathbf{F}_p \mathbf{U}^\mathsf{T}$, which depends on the coefficients of the velocity ($\mathbf{X}^v$) and the PTO force ($\mathbf{U}$), while Γ_{ij} are the elements of the constant matrix Γ, which depends on the bases Φ and Φ^P as

$$\Gamma = \int_0^T \Phi^\mathsf{T}(t)\, \Phi^P(t)\, dt. \tag{10.12}$$

The matrices $\mathbf{X}^x$, $\mathbf{X}^v$ and $\mathbf{U}$ are defined as

$$\mathbf{X}^x = [\hat{x}_1^x, \ldots, \hat{x}_N^x], \qquad \mathbf{X}^v = [\hat{x}_1^v, \ldots, \hat{x}_N^v], \qquad \mathbf{U} = [\hat{u}_1, \ldots, \hat{u}_{N_P}].$$

The derivatives of the approximated state variables (positions and velocities) are

$$\dot{x}_i^N(t) = \sum_{k=1}^{N} x_{ik}^x\, \dot{\phi}_k(t) = \dot{\Phi}(t)\hat{x}_i^x, \qquad i = 1, \ldots, n, \tag{10.13}$$

$$\dot{v}_i^N(t) = \sum_{k=1}^{N} x_{ik}^v\, \dot{\phi}_k(t) = \dot{\Phi}(t)\hat{x}_i^v, \qquad i = 1, \ldots, n, \tag{10.14}$$

since the vectors $\hat{x}_i^x$ and $\hat{x}_i^v$ are independent of time. Substituting the approximated states (10.8), (10.9), their derivatives (10.13), (10.14), and the PTO forces (10.10) into the dynamic Equations (10.6) yields the residual form

$$r_i^x = \dot{x}_i^N - v_i^N, \tag{10.15}$$

$$\begin{aligned} r_i^v =& \sum_{j=1}^{n} m_{ij}\dot{v}_j^N + \mathrm{B}_{ii} v_i^N + \sum_{j=1}^{n} \int_0^t \mathrm{K}_{ij}(t-\tau) v_j^N(\tau)\, d\tau \\ &+ \mathrm{S}_{ii} x_i^N(t) + \sum_{j=1}^{m} f_{P_{ij}} f_{pto_j}^{NP} + f_{e_i} \end{aligned} \tag{10.16}$$

where m_{ij} are the elements of the matrix $\mathbf{M}$; B_{ii} and S_{ii} are the diagonal elements of the diagonal matrices $\mathbf{B}$ and $\mathbf{S}$, respectively; $f_{P_{ij}}$ are the elements of $\mathbf{F}_P$; and K_{ij} are the elements of $\mathbf{K}$.

In applying the mean weighted residual method, for any given values of the PTO force described by the coefficients $\mathbf{U}$, the coefficients of the velocity and position satisfying the equation of motion are calculated by solving the linear system of equations

$$\langle r_i^x, \psi_j \rangle = 0, \tag{10.17}$$

$$\langle r_i^v, \psi_j \rangle = 0, \qquad \text{for } i = 1, \ldots, n \text{ and } j = 1, \ldots, N, \tag{10.18}$$

where ψ_j are linearly independent test functions and $\langle\ \rangle$ denotes the inner product. In particular, two specific choices for the test functions lead to two well-known methods [157], which are the Galerkin or spectral method and the collocation or pseudospectral method [160]. The Galerkin method is discussed in detail in Section 10.2, while the pseudospectral method uses translated Dirac delta test functions, illustrated in [161, 162]. The system of Equations (10.17) and (10.18) is linear because the dynamic equation is linear, and they form a system of $2nN$ equations in $2nN$ variables $(\mathbf{X}^x, \mathbf{X}^v)$.

The result of the discretization is the finite-dimensional nonlinear program described by the quadratic cost function J^N in Equation (10.11), the linear equality constraints due to the dynamic equations in (10.17) and (10.18), and the additional equality and inequality path and control constraints in (10.4) and (10.5), which are now functions of the vectors $\mathbf{X}^x, \mathbf{X}^v, \mathbf{U}$, and time, as

$$\mathbf{h}(\mathbf{X}^x, \mathbf{X}^v, \mathbf{U}, t) = 0, \tag{10.19}$$

$$\mathbf{g}(\mathbf{X}^x, \mathbf{X}^v, \mathbf{U}, t) \leq 0. \tag{10.20}$$

10.1.1 Definition of the PTO Configuration Matrix

The role of the matrix $\mathbf{F}_P$ in Equations (10.3) and (10.7) is to combine the PTO forces and the velocities of the oscillating modes for which energy is absorbed, and this section illustrates how to build the matrix for some common configurations of WECs.

In the case of a single-body device referenced to the seabed, such as the heaving buoy in Fig. 10.1, the system has one degree of freedom and there is only one velocity and one PTO force. The absorbed energy is

$$J = \int_0^T \dot{x}(t) f_{pto}(t)\, dt,$$

with $\mathbf{F}_P = 1$. A self-reacting WEC composed of two heaving bodies, which are restricted to oscillate in heave only (Fig. 10.2), has two degrees of freedom. In this two-body configuration, there is only one PTO force, and it acts on each of the bodies

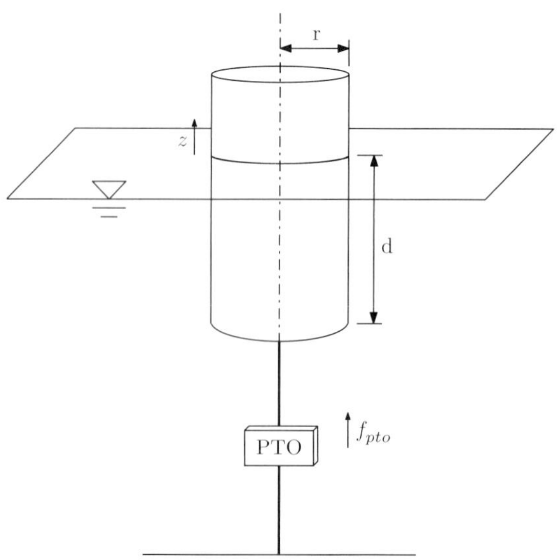

Figure 10.1 Single-body heaving WEC.

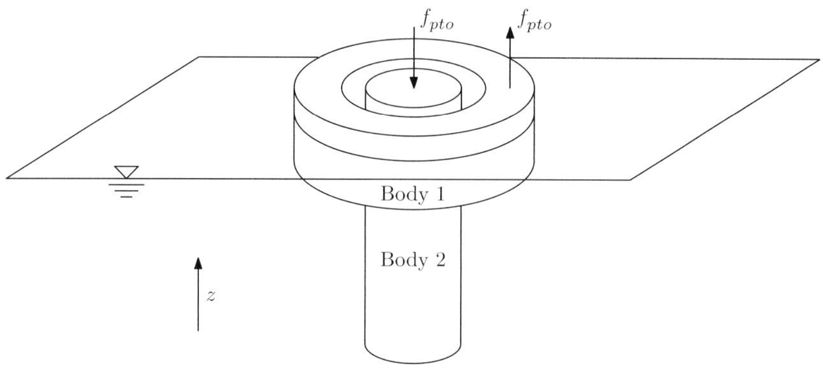

Figure 10.2 Self-reacting, two-body heaving WEC.

with the same magnitude but with opposite direction; if the system coordinates are the absolute vertical positions of the two bodies, then the absorbed energy is

$$J = \int_0^T \left(\dot{x}^B(t) - \dot{x}^A(t)\right) f_{pto}(t)\, dt$$
$$= \int_0^T [\dot{x}^B(t)\ \dot{x}^A(t)]\, \mathbf{F}_P f_{pto}(t)\, dt$$

where $\dot{x}^A(t)$ and $\dot{x}^B(t)$ are the absolute vertical velocities of bodies A and B, respectively; thus

$$\mathbf{F}_P = \begin{bmatrix} -1 \\ 1 \end{bmatrix}.$$

Additional illustrative examples, showing the procedure for the construction of the PTO configuration matrix for several WECs and arrays of WECs, are presented in [161].

10.2 Fourier–Galerkin Direct Transcription

As an illustrative example, a single-degree-of-freedom system describing the point absorber WEC depicted in Fig. 10.1 is considered [163], the motion of which is restricted to heave only; that is, the general WEC position vector $\mathbf{x}$ is now the (scalar) heave position x. In this example, the Galerkin method is used in conjunction with trigonometric polynomials as approximating functions because the trigonometric polynomials, or truncated Fourier series, seem a natural choice for an oscillating system. For such a WEC system, the time domain model of the WEC in Equation (10.2) reduces to the scalar equation

$$\mathrm{M}\ddot{x}(t) + \mathrm{B}\,\dot{x}(t) + \int_{-\infty}^{t} \mathrm{K}(t-\tau)\dot{x}(\tau)\, d\tau + \mathrm{S}\,x(t) = f_e(t) + f_{pto}(t),$$

with $\mathrm{M} = m_b + m_\infty$ and where m_b is the mass of the oscillating body and m_∞ is the added mass at infinite frequency for the heave mode. Since the PTO force is applied between the body and a fixed reference, the PTO configuration matrix is $\mathbf{F}_P = 1$, and the absorbed energy is now defined as

$$J = -\int_0^T \dot{x}(t) f_{pto}(t)\, dt. \tag{10.21}$$

For convenience of notation,[1] the state variables are renamed as $x = x_1$ and $\dot{x} = x_2$, and the PTO force, which is the control input, as $f_{pto} = u$; the resulting dynamic model is

$$\dot{x}_1 = x_2, \tag{10.22}$$

$$\mathrm{M}\dot{x}_2 = -\mathrm{B}x_2 - \int_{-\infty}^{t} \mathrm{K}(t-\tau)x_2(\tau)d\tau - \mathrm{S}x_1 + u + f_e. \tag{10.23}$$

The heave positions and velocities in Equations (10.8) and (10.9), respectively, and the PTO force in Equation (10.10), are approximated by zero-mean truncated Fourier series with N terms with N^P=N for this case, that is:

$$x_i(t) \approx \sum_{k=1}^{N/2} x_{ik}^c \cos(k\omega_0 t) + x_{ik}^s \sin(k\omega_0 t) = \Phi(t)\hat{x}_i, \tag{10.24}$$

$$u(t) \approx \sum_{k=1}^{N/2} u_k^c \cos(k\omega_0 t) + u_k^s \sin(k\omega_0 t) = \Phi(t)\hat{u}, \tag{10.25}$$

with $i = 1, 2$ and where

$$\hat{x}_i = \left[x_{i1}^c, x_{i1}^s, \ldots, x_{i\frac{N}{2}}^c + x_{i\frac{N}{2}}^s\right]^\mathsf{T},$$

$$\hat{u} = \left[u_1^c, u_1^s, \ldots, u_{\frac{N}{2}}^c + u_{\frac{N}{2}}^s\right]^\mathsf{T},$$

and

$$\Phi(t) = \left[\cos(\omega_0 t), \sin(\omega_0 t), \ldots, \cos\left(\frac{N}{2}\omega_0 t\right), \sin\left(\frac{N}{2}\omega_0 t\right)\right],$$

where the fundamental frequency is $\omega_0 = 2\pi/T$. The constant terms of the bases ($k = 0$) have been discarded because the exciting force is assumed to be zero mean; therefore, the resulting (optimal) oscillation is also expected to be zero mean.

By substituting the approximations in Equations (10.24) and (10.25) into the expression for the total absorbed energy in Equation (10.21), and by using Equation (10.11), the approximated absorbed energy J^N is

$$J^N = -\int_0^T \hat{u}^\mathsf{T}\Phi^\mathsf{T}(t)\,\Phi(t)\hat{x}_2\, dt = -\frac{T}{2}\hat{u}^\mathsf{T}\hat{x}_2. \tag{10.26}$$

In fact, the matrix Γ in Equation (10.12) is

$$\Gamma = \int_0^T \Phi^\mathsf{T}(t)\,\Phi(t)\, dt = \frac{T}{2} I_N \tag{10.27}$$

[1] In particular, for consistency with the notation used in the control theory.

where I_N is the identity matrix of size N, because the basis is orthogonal, that is,

$$\langle \phi_i, \phi_j \rangle = \int_0^T \phi_i(t)\phi_j(t)\, dt = \frac{T}{2}\delta_{ij},$$

where δ_{ij} is the Kronecker delta.

The next step of the discretization process is the construction of the linear system resulting from the minimization of the residual r^N in Equation (10.18) by applying the Galerkin method, as described in [157]. When using the Fourier series to approximate the states, the differentiation of the approximated states can be conveniently written as

$$\dot{x}_i^N = \dot{\Phi}(t)\hat{x}_i = \Phi(t)\, \mathrm{D}_\phi\, \hat{x}_i$$

where the differentiation matrix $\mathrm{D}_\phi \in \mathbb{R}^{N\times N}$ is block diagonal. Each block D_ϕ^k, for $k = 1, \ldots, N/2$, is

$$\mathrm{D}_\phi^k = \begin{bmatrix} 0 & k\,\omega_0 \\ -k\,\omega_0 & 0 \end{bmatrix}.$$

In fact, the derivative of a zero-mean Fourier series is still a Fourier series where the terms are reordered and scaled, since

$$\frac{d}{dt}\sin(\omega t) = \omega\cos(\omega t), \qquad \frac{d}{dt}\cos(\omega t) = -\omega\sin(\omega t).$$

Consequently, the approximated state equations in (10.22) and (10.23) become

$$\Phi(t)\, \mathrm{D}_\phi\, \hat{x}_1 = \Phi(t)\hat{x}_2, \tag{10.28}$$

$$\begin{aligned} \mathrm{M}\,\Phi(t)\, \mathrm{D}_\phi\, \hat{x}_2 = & -B\Phi(t)\hat{x}_2 - \int_{-\infty}^{t} \mathrm{K}(t-\tau)\Phi(t)\hat{x}_2(\tau)\, d\tau \\ & - S\Phi(t)\hat{x}_1 + \Phi(t)\hat{u} + f_e. \end{aligned} \tag{10.29}$$

The Galerkin method [157] consists of writing the dynamic equations in residual form, and then minimizing the residual by imposing its orthogonality on all the elements of the basis. However, with regard to the first state Equation (10.28), it is possible to note immediately that

$$\Phi(t)\, \mathrm{D}_\phi\, \hat{x}_1 - \Phi(t)\, \hat{x}_2 = 0 \quad \Leftrightarrow \quad \mathrm{D}_\phi\, \hat{x}_1 - \hat{x}_2 = 0, \tag{10.30}$$

since two Fourier series are equal if and only if all the corresponding coefficients are equal. The residual form of the second dynamic Equation (10.29) is

$$r_2^N = \mathrm{M}\,\Phi(t)\, \mathrm{D}_\phi\, \hat{x}_2 + \mathrm{B}\Phi(t)\hat{x}_2 + \int_{-\infty}^{t} \mathrm{K}(t-\tau)\Phi(t)\hat{x}_2\, d\tau + \mathrm{S}\Phi(t)\hat{x}_1 - \Phi(t)\hat{u} - f_e \tag{10.31}$$

and is minimized by solving

$$\langle r_2^N, \phi_i \rangle = 0, \qquad i = 1, \ldots, N, \tag{10.32}$$

which can be written, since the inner product is commutative, in a more concise form as

$$\langle \Phi^{\mathsf{T}}, r_2^N \rangle = \mathbf{0}, \tag{10.33}$$

where $\mathbf{0} \in \mathbb{R}^N$ is the zero vector (all of its elements equal to zero). Because of linearity, the inner product between ϕ_i and each term on the right-hand side of Equation (10.31) can be considered separately; thus, for the first term, the result is

$$\langle \Phi^{\mathsf{T}}, \mathrm{M}\, \Phi\, \mathrm{D}_\phi\, \hat{x}_2 \rangle = \mathrm{M} \int_0^T \Phi^{\mathsf{T}}(t)\, \Phi(t)\, dt\, \mathrm{D}_\phi\, \hat{x}_2 = \mathrm{M}\, \Gamma\, \mathrm{D}_\phi\, \hat{x}_2 = \frac{T}{2} \mathrm{M}\, \mathrm{D}_\phi\, \hat{x}_2$$

where the matrix Γ is defined in Equation (10.27). Similarly, the remaining terms of the residual in Equation (10.31) become:

$$\begin{aligned}
\langle \Phi^{\mathsf{T}}, \mathrm{B}\, \Phi\, \hat{x}_2 \rangle &= \frac{T}{2} \mathrm{B} I_N \hat{x}_2, \\
\langle \Phi^{\mathsf{T}}, \mathrm{S} \Phi \hat{x}_1 \rangle &= \frac{T}{2} \mathrm{S} I_N \hat{x}_1, \\
\langle \Phi^{\mathsf{T}}, \Phi \hat{u} \rangle &= \frac{T}{2} I_N \hat{u}, \\
\langle \Phi^{\mathsf{T}}, f_e \rangle &= \frac{T}{2} I_N \hat{e},
\end{aligned}$$

where $\hat{e}$ is the vector of Fourier coefficients of the excitation force. The product corresponding to the convolution term in Equation (10.31) requires some manipulation in order to be expressed in matrix form. Because of the linearity of the convolution integral and of the orthogonality of the basis, the reduction can be carried out one frequency at the time; thus, it is convenient to consider Φ and $\hat{x}$ as composed of two elements only, such as

$$\Phi^k = [\cos(k\omega_0 t),\ \sin(k\omega_0 t)], \qquad \hat{x}_2^k = [x_{2k}^c,\ x_{2k}^s]^{\mathsf{T}}.$$

In this case, by the commutativity property of the convolution and by the fact that the impulse response $\mathrm{K}(t) = 0$ for $t < 0$,[2] the result is

$$\int_{-\infty}^{t} \mathrm{K}(t-\tau)\Phi^k(t)\hat{x}_2^k\, d\tau = \int_0^{+\infty} \mathrm{K}(\tau)\Phi^k(t-\tau)\hat{x}_2^k\, d\tau. \tag{10.34}$$

Developing the product terms,

$$\int_0^{+\infty} \mathrm{K}(\tau)\Phi^k(t-\tau)\hat{x}_2^k\, d\tau = \int_0^{+\infty} \mathrm{K}(\tau) \left(x_{2k}^c \cos(k\omega_0(t-\tau)) + x_{2k}^s \sin(k\omega_0(t-\tau)) \right) d\tau$$

$$= x_{2k}^s \sin(k\omega_0 t) \int_0^{+\infty} \mathrm{K}(\tau) \cos(k\omega_0(\tau))\, d\tau - x_{2k}^s \cos(k\omega_0 t) \int_0^{+\infty} \mathrm{K}(\tau) \sin(k\omega_0(\tau))\, d\tau$$

$$+ x_{2k}^c \cos(k\omega_0 t) \int_0^{+\infty} \mathrm{K}(\tau) \cos(k\omega_0(\tau))\, d\tau + x_{2k}^c \sin(k\omega_0 t) \int_0^{+\infty} \mathrm{K}(\tau) \sin(k\omega_0(\tau))\, d\tau.$$

[2] Which corresponds to $\mathrm{K}(t-\tau) = 0$ for $\tau > t$.

Ogilvie's relations [164] are now applied to give:

$$-\omega(\mathrm{m}(\omega)-\mathrm{m}_\infty)=\int_0^\infty \mathrm{K}(t)\,\sin(\omega t)\,dt,$$
$$\mathrm{R}(\omega)=\int_0^\infty \mathrm{K}(t)\,\cos(\omega t)\,dt,$$
$$\mathrm{m}_\infty=\lim_{\omega\to\infty}\mathrm{m}(\omega),$$

where $\mathrm{m}(\omega)$ and $\mathrm{R}(\omega)$ are known as added mass and radiation resistance [66], respectively; the convolution integral becomes

$$\begin{aligned}&\int_0^{+\infty}\mathrm{K}(\tau)\Phi^k(t-\tau)\hat{x}_2^k\,d\tau\\ &=x_{2k}^s\Big(\sin(k\omega_0 t)\mathrm{R}(k\omega_0)-\cos(k\omega_0 t)(-k\omega_0(m(k\omega_0)-m_\infty))\Big)\\ &\quad+x_{2k}^c\Big(\cos(k\omega_0 t)\mathrm{R}(k\omega_0)+\sin(k\omega_0 t)(-k\omega_0(m(k\omega_0)-m_\infty))\Big)\\ &=\Phi^k\begin{bmatrix}\mathrm{R}(k\omega_0) & k\omega_0 m(k\omega_0)\\ -k\omega_0 m(k\omega_0) & \mathrm{R}(k\omega_0)\end{bmatrix}\hat{x}_2^k-m_\infty\Phi^k\mathrm{D}_\phi^k\hat{x}_2^k.\end{aligned}$$

Before combining all the terms, it is convenient to carry out an additional substitution: by noting that the matrix D_ϕ is invertible, and its inverse is still block diagonal with blocks

$$\mathrm{D}_\phi^{k^{-1}}=\begin{bmatrix}0 & \frac{1}{k\,\omega_0}\\ -\frac{1}{k\,\omega_0} & 0\end{bmatrix},$$

and by using Equation (10.30), the variable $\hat{x}_1$ in the residual Equation (10.31) can be substituted with

$$\hat{x}_1=\mathrm{D}_\phi^{-1}\hat{x}_2. \tag{10.35}$$

Thus, the inner product relating to the restoring force term becomes

$$\langle \mathrm{S}\Phi\mathrm{D}_\phi^{-1}\hat{x}_2,\phi_i\rangle \quad\Rightarrow\quad \frac{T}{2}\mathrm{S}I_N\mathrm{D}_\phi^{-1}\hat{x}_2.$$

Combining all the terms of the inner product in Equation (10.33), the discretized equation of motion becomes the linear system

$$\mathbf{G}\,\hat{x}_2=\hat{u}+\hat{e} \tag{10.36}$$

where the matrix $\mathbf{G}$ is block diagonal, with 2×2 blocks as

$$\mathbf{G}=\begin{bmatrix}\mathrm{D}_1 & \mathrm{M}_1 & 0 & \cdots & 0 & 0\\ -\mathrm{M}_1 & \mathrm{D}_1 & 0 & \cdots & 0 & 0\\ 0 & 0 & \ddots & & \vdots & \vdots\\ \vdots & \vdots & & \ddots & 0 & 0\\ 0 & 0 & \cdots & 0 & \mathrm{D}_{N/2} & \mathrm{M}_{N/2}\\ 0 & 0 & \cdots & 0 & -\mathrm{M}_{N/2} & \mathrm{D}_{N/2}\end{bmatrix} \tag{10.37}$$

and where

$$\mathrm{D}_k = \mathrm{R}(k\omega_0) + \mathrm{B},$$
$$\mathrm{M}_k = n\omega_0\,(m_b + m(k\omega_0)) - \mathrm{S}/(k\omega_0),$$

for $k = 1 \ldots N/2$. The matrix $\mathbf{G}$ is nonsingular, since $\mathrm{B} > 0$, thus

$$\det \mathbf{G} = \prod_{n=1}^{N/2} \mathrm{D}_n^2 + \mathrm{M}_n^2 > 0.$$

The equation of motion (10.36) can then be solved with respect to $\hat{x}_2$ as

$$\hat{x}_2 = \mathbf{G}^{-1}\hat{u} + \mathbf{G}^{-1}\hat{e}, \tag{10.38}$$

and, by substituting $\hat{x}_2$ into the approximated absorbed energy J^N in Equation (10.26), we get

$$J^N = -\frac{T}{2}\hat{u}^{\mathsf{T}}\mathbf{G}^{-1}\hat{u} - \frac{T}{2}\hat{u}^{\mathsf{T}}\mathbf{G}^{-1}\hat{e}, \tag{10.39}$$

which is a quadratic function of the variable $\hat{u}$ solely. In essence, the state variables have been eliminated by substitution, and the optimization is carried out over the control variable $\hat{u}$ only. More importantly, the constrained optimization problem, given by the cost function describing the total absorbed energy J^N in Equation (10.26) and the linear equality constraints describing the system dynamics in Equation (10.36), has been transformed into an unconstrained quadratic program.

Additionally, since radiation is a dissipative process [165], the radiation resistance is positive, that is, $\mathrm{R}(\omega) > 0$, and all the diagonal elements of the matrix $\mathbf{G}$ are positive; thus, the symmetric part of $\mathbf{G}$, $(\mathbf{G} + \mathbf{G}^{\mathsf{T}})/2$ is positive definite, and the absorbed energy function (10.39) is concave. Therefore, the optimal solution $\hat{u}^*$ for the unconstrained problem, which minimizes Equation (10.26) is then

$$\hat{u}^* = \left(\mathbf{G}^{-1} + \mathbf{G}^{-\mathsf{T}}\right)^{-1}\mathbf{G}^{-1}\hat{e}. \tag{10.40}$$

10.2.1 Force and Oscillation Amplitude Constraints

Constraints on the PTO force and on the oscillation amplitude reflect physical limitations on the device or its components. In this section, inequality constraints describing the maximum allowed force and maximum allowed oscillation amplitude are considered and described as

$$|f_{pto}(t)| \leq \mathrm{F}_{max}, \tag{10.41}$$
$$|x(t)| \leq \mathrm{Z}_{max}, \tag{10.42}$$

which correspond, for the approximated problem, to

$$|\Phi(t)\hat{u}| \leq \mathrm{F}_{max}, \tag{10.43}$$
$$|\Phi(t)\hat{x}_1| \leq \mathrm{Z}_{max}. \tag{10.44}$$

It is difficult to find the extrema of a Fourier series; thus, one possible approach to deal with the constraints in (10.43) and (10.44) is to consider their two-norm approximation, as described in [158]. An alternative approach is to enforce the constraint only at a set of specified time instants $\{t_k\}_{k=0}^{N_c}$, that is,

$$\Phi(t_k)\hat{u} \leq \mathrm{F}_{max}, \qquad -\Phi(t_k)\hat{u} \leq \mathrm{F}_{max}, \tag{10.45}$$

$$\Phi(t_k)\hat{x}_1 \leq \mathrm{Z}_{max}, \qquad -\Phi(t_k)\hat{x}_1 \leq \mathrm{Z}_{max}. \tag{10.46}$$

for $k = 0, \ldots, N_c$. By defining the vector Θ as

$$\Theta = \begin{bmatrix} \Phi_0 \\ \Phi_1 \\ \vdots \\ \Phi_{N_c} \end{bmatrix},$$

the constraints in (10.45) and (10.46) can be approximated by the linear inequalities

$$\begin{bmatrix} \Theta \\ -\Theta \end{bmatrix} \hat{u} \leq \mathbf{1}_{2(N_c+1)\times 1}\, \mathrm{F}_{max}, \tag{10.47}$$

$$\begin{bmatrix} \Theta \\ -\Theta \end{bmatrix} \hat{x}_1 \leq \mathbf{1}_{2(N_c+1)\times 1}\, \mathrm{Z}_{max}, \tag{10.48}$$

where $\mathbf{1}_{2(N_c+1)\times 1}$ is the vector of all ones of size $2(N_c+1)$. The inequality constraint, relative to the position in (10.46), can be expressed as a function of $\hat{u}$ by using Equations (10.35) and (10.38), as

$$\begin{bmatrix} \Theta \\ -\Theta \end{bmatrix} \mathrm{D}_\phi^{-1}\mathbf{G}^{-1}\hat{u} \leq \mathbf{1}_{2(N_k\times 1)}\, \mathrm{Z}_{max} - \begin{bmatrix} \Theta \\ -\Theta \end{bmatrix} \mathrm{D}_\phi^{-1}\mathbf{G}^{-1}\hat{e}. \tag{10.49}$$

In summary, the resulting constrained optimal control problem is the convex quadratic program defined by the cost function J^N in Equation (10.39), subject to the linear inequality constraints on the PTO force in (10.47) and oscillation amplitude in (10.49).

10.2.2 Simulation Results

This section presents simulation results for the heaving-body point absorber WEC depicted in Fig. 10.1, specified as a vertical cylinder of radius $r = 4$ m and draft $d = 10$ m. Both regular and irregular incident waves are considered, and PTO force and oscillation amplitude constraints are respected. Figs. 10.3–10.5 depict results for a monochromatic incident wave with period $T_r = 10$ s and amplitude $A = 5$ m. The expansions for the approximation of the state in Equation (10.24) and the control in Equation (10.25) are composed of 40 frequency components, corresponding to $N = 80$. The oscillation amplitude constraint has been set to $\mathrm{Z}_{max} = 2.5$ m, while several values of the PTO force constraints have been considered, that is, $\mathrm{F}_{max} = \{1, 1.25, 1.5, 2\} \cdot 10^6$ N. The time instants t_k have been chosen equally spaced, with a distance 10 times smaller than the smallest period of the Fourier expansion, that is, $t_{k+1} - t_k = 2T/10N$.

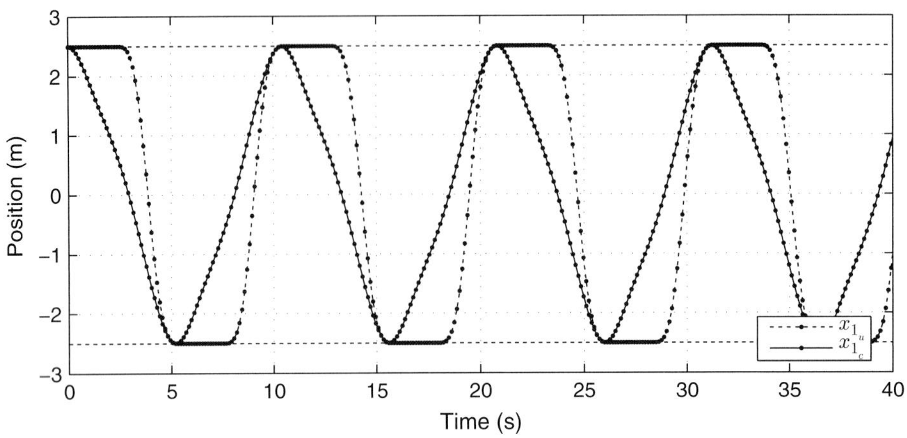

Figure 10.3 Vertical position for an incident regular wave of period $T_W = 10$ s and amplitude $A = 5$ m: x_{1_u} is the vertical position when the PTO force is not constrained; x_{1_c} is the vertical position when the PTO force is constrained.

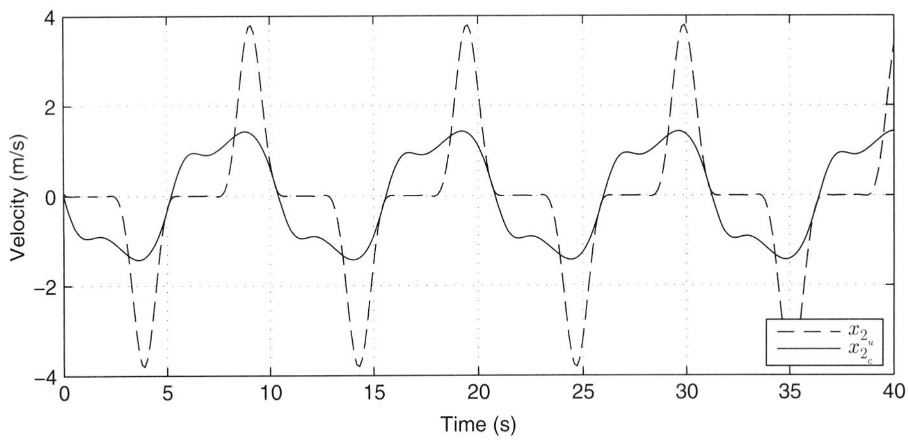

Figure 10.4 Vertical velocity for an incident regular wave of period $T_W = 10$ s and amplitude $A = 5$ m: x_{2_u} is the vertical velocity when the PTO force is not constrained; x_{2_c} is the vertical velocity when the PTO force is constrained.

Thus, for the simulation considered in this section, $Nc = 1 + T/(t_{k+1} - t_k) = 5N + 1 = 401$.

Each plot in Figs. 10.3–10.5 depicts two sets of curves: the curves associated with the variables with subscript "c" (i.e., x_{1_c}, x_{2_c}, u_c) denote the simulation results when the PTO force is constrained with the most stringent value ($F_{max} = 1 \cdot 10^6$ N), whereas the curves associated with the variables with subscript "u" (i.e., x_{1_u}, x_{2_u}, u_u) denote the simulation results when the PTO force is unconstrained. The dots superimposed on the plots of the positions (x_{1_c}, x_{1_u}) in Fig. 10.3 and the PTO forces in Fig. 10.5 denote the time points t_k at which the constraints are being enforced, via Equations (10.47) and (10.49), for the PTO force and device position, respectively.

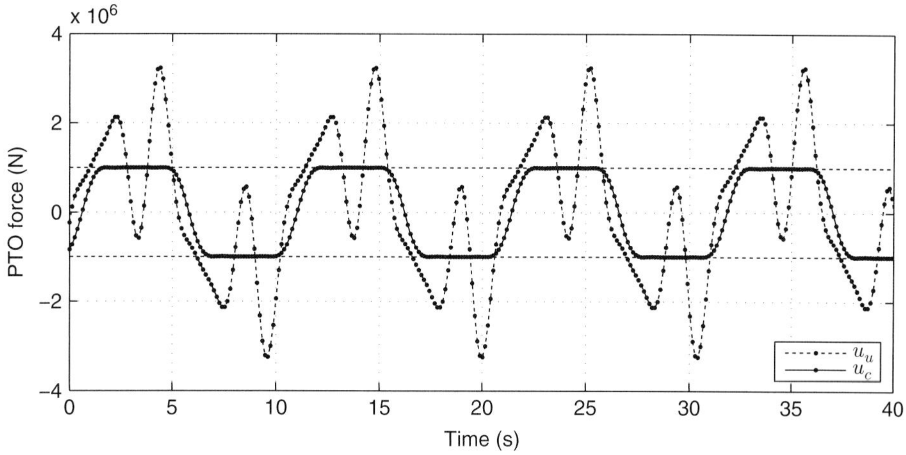

Figure 10.5 Unconstrained PTO force u_u and constrained PTO force u_c for an incident regular wave of period $T_w = 10$ s and amplitude $A = 5$ m.

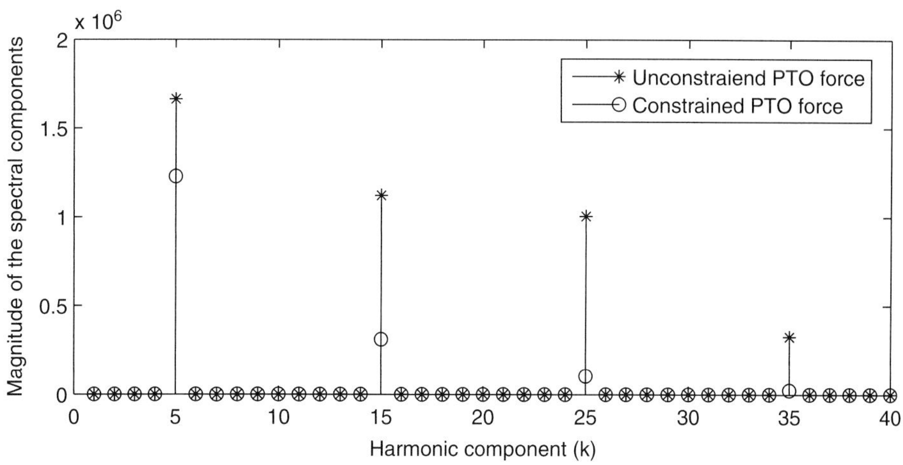

Figure 10.6 Spectral components of the PTO force.

It is interesting to note that both the optimal motion of the device and the optimal time profile of the PTO force are smoother when the PTO force is strongly constrained, thus the device is subject to less mechanical stress. Fig. 10.6 shows the frequency components of the optimal PTO force for the situation where the PTO force is unconstrained ($*$) and when the most stringent constraint is active ($\circ$), confirming the previous observation. In fact, the unconstrained PTO force is characterized by large values of higher frequency components up to the seventh harmonic, while the harmonic content of the constrained PTO force is considerably less.

Fig. 10.6 also shows that the optimal profiles of both the constrained and unconstrained PTO forces contain only odd harmonics when the position constraints are active. For the real-time implementation, it could be beneficial to reduce the dimension of the NLP problem by eliminating all the frequencies for which the energy is zero or negligible. Due to the orthogonality of the Fourier series, this could be achieved by looking at the vector of Fourier coefficients of the excitation force ($\hat{e}$) and by solving the optimization problem considering only the frequencies that contain "most" of the energy and their harmonics. The reduced problem will consist of the quadratic program with the cost function

$$J^N = -\frac{T}{2}\tilde{\mathbf{u}}^T\widetilde{\mathbf{G}}^{-1}\tilde{\mathbf{u}} - \frac{T}{2}\tilde{\mathbf{u}}^T\widetilde{\mathbf{G}}^{-1}\tilde{\mathbf{e}}$$

in place of Equation (10.39), where $\tilde{\mathbf{u}}$ and $\tilde{\mathbf{e}}$ are the "reduced" vectors of Fourier coefficients of the control signal and excitation force, respectively, which are built by extracting the elements of $\hat{u}$ and $\hat{e}$ for which the energy is nonzero, and their harmonics.

The matrix $\widetilde{\mathbf{G}}$ is derived from the matrix $\mathbf{G}$ in a similar manner by extracting the rows and columns corresponding to the frequencies that contain most of the energy, and their harmonics.

Fig. 10.7 depicts the ratio E_c/E_u of the total absorbed energy for several values of the PTO force constraints E_c and the energy absorbed in the unconstrained case E_u. With a force constraint of $1.25 \cdot 10^6$ N, which is 30% of the maximum value of the PTO force in the unconstrained case (see Fig. 10.5), it is still possible to absorb around 90% of the energy that is absorbed when the PTO force is not constrained. This analysis plays an important role in the techno-economic optimization of WECs, since absorbed energy is a major determinant of the revenue generated by the device, and the power rating is strongly related to the cost of the device.

Simulation results for irregular waves, corresponding to a Bretschneider spectrum of peak period $T_p = 10$ s and significant wave height $H_s = 1$ m, are depicted in Figs. 10.8–10.10. In this case, the approximations of the state and control in Equations (10.24) and (10.25) are composed of 80 frequencies ($N = 160$), the oscillation amplitude has been restricted to $Z_{max} = 2.5$ m, and the PTO force to $F_{max} = 5.3 \cdot 10^5$ N.

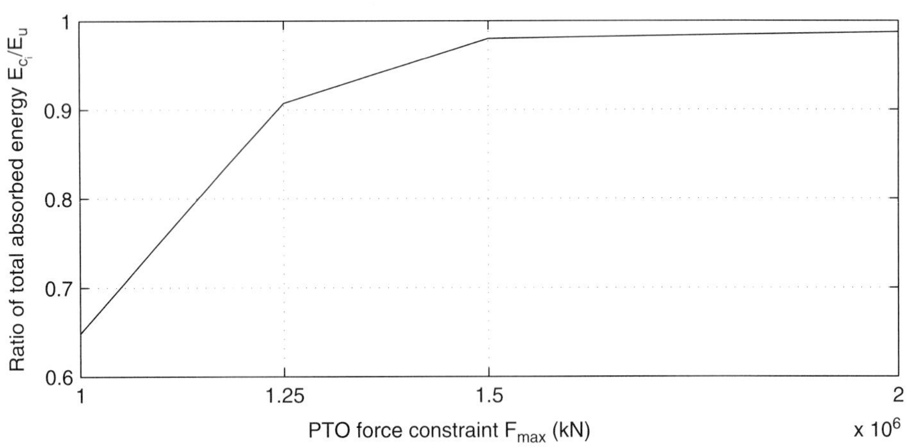

Figure 10.7 Total absorbed energy as function of the force constraint.

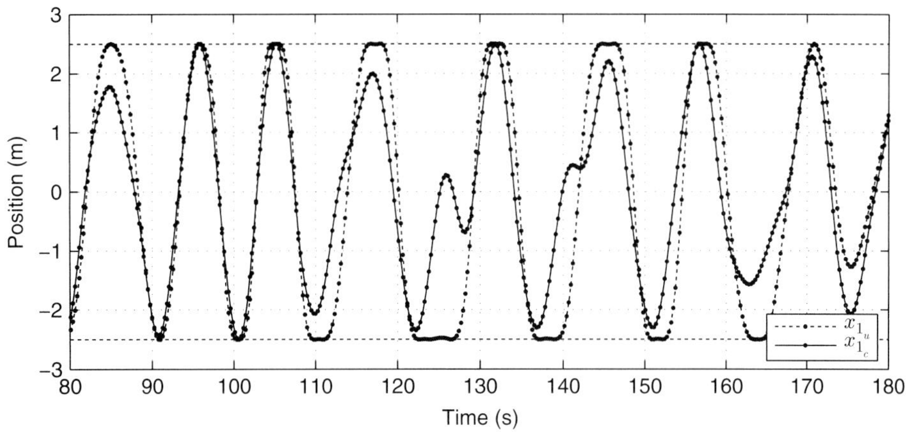

Figure 10.8 Vertical position for a Bretschneider sea state of $T_p = 10$ s and $H_s = 1$ m: x_{1_u} is the vertical position when the PTO force is not constrained; x_{1_c} is the vertical position when the PTO force is constrained.

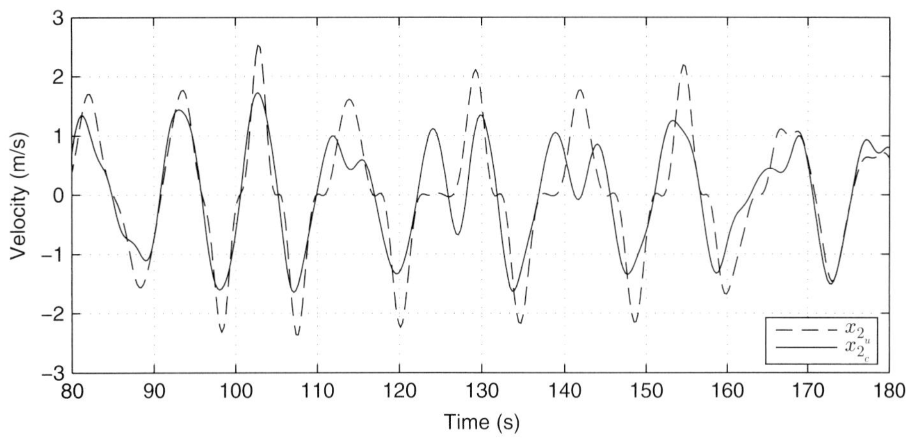

Figure 10.9 Vertical velocity for a Bretschneider sea state of $T_p = 10$ s and $H_s = 1$ m: x_{2_u} is the vertical velocity when the PTO force is not constrained; x_{2_c} is the vertical velocity when the PTO force is constrained.

With regard to the effect of constraints on the motion of the device (Figs. 10.3 and 10.4) and PTO force, the observations described for the regular incident wave still apply; namely, that a more stringent PTO force constraint results in a smoother motion and PTO force profile. Additionally, setting the PTO force constraint to 50% of the maximum unconstrained PTO force reduces the total absorbed energy by only 10%, approximately.

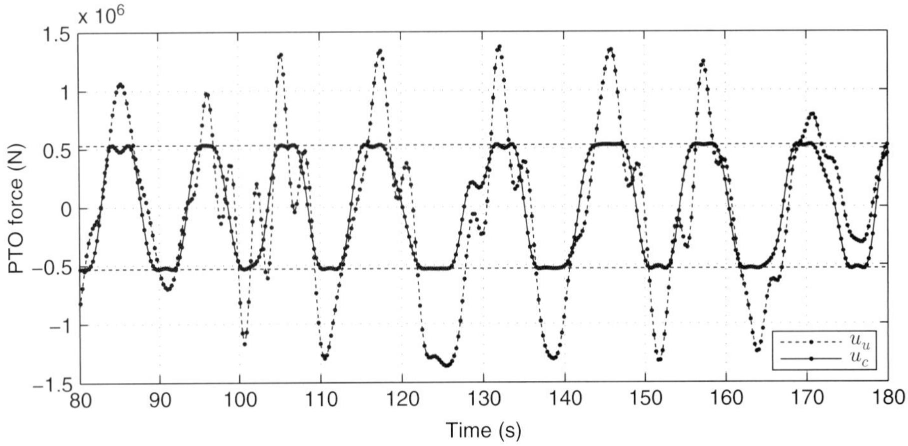

Figure 10.10 Unconstrained PTO force u_u and constrained PTO force u_c for a Bretschneider sea state of $T_p = 10$ s and $H_s = 1$ m.

10.3 Computational Considerations

In the unconstrained case, the optimal solution to the energy maximization problem is a straightforward algebraic calculation, as shown in Equation (10.40). The dimension of this matrix calculation is a function of the number of bodies, the number of degrees of freedom for each body, and the number of frequencies $N/2$ used in the discretization approximation of the system variables.

In the constrained case, an NLP problem results, with the performance function in Equation (10.39), subject to the linear inequality constraints on the PTO force in (10.47) and oscillation amplitude in (10.49). In addition to the WEC system dimension and frequencies associated with the unconstrained optimization problem, the computational load for the constrained case additionally depends on N_c, the number of time instants at which the constraints are imposed.

In order to achieve a parsimonious computational problem, N and N_c must be chosen judiciously. Application-specific experience is required to choose a minimum value of N_c so that constraints are not compromised, while there is considerable scope in pruning N by inspection, as evidenced by Fig. 10.6. In particular, the value of N should be selected in combination with the fundamental frequency ω_0 of the Fourier expansion;[3] the values of ω_0 and N can be selected, for example, by looking at the power spectrum of the wave elevation (or the excitation force). The frequency ω_0 should be smaller than the lowest frequency with significant energy content, and, conversely, N should be chosen such that $N\omega_0/2$ is larger than the larger frequency with significant energy content. Moreover, when force and amplitude constraints become active, they introduce higher harmonics in the motion and force (see, e.g., Fig. 10.6); therefore, the choice of N should also take this aspect into consideration. In practice, the smaller the value of ω_0

[3] Which is related to the period T over which the energy is being optimized as $\omega_0 = 2\pi/T$.

the greater the frequency resolution, as the Fourier components are at integer multiples of the fundamental frequencies; however, a higher frequency resolution requires a larger value of N, which corresponds to a larger dimension of the NLP problem.

A significant computational saving, however, is always achievable by reconfiguring the radiation damping convolution calculation into off-line and on-line components, as shown in following subsection.

10.3.1 Radiation Convolution Integral

In Section 10.2, it is shown that the convolution integral describing the radiation force can be solved *analytically* when the state (velocity) is approximated by using a truncated Fourier series; *thus, the numerical integration of the convolution integral can be avoided when solving the equation of motion* during the solution of the NLP problem. It should be noted that whenever the velocity is approximated by a linear combination of basis functions, the numerical integration of the convolution integral can be done off-line, for a given hydrodynamic problem and a given choice of basis functions. In fact, considering the expansion for the velocity in Equation (10.9), that is,

$$v_i(t) \approx v_i^N(t) = \sum_{k=1}^{N} x_{ik}^v,\ \phi_k(t), \qquad i = 1, \ldots, n,$$

the term of the residual of the dynamic equation involving the convolution integral in (10.16) is

$$\sum_{j=1}^{n} \int_0^t \mathrm{K}_{ij}(t-\tau) v_j^N(\tau)\, d\tau = \sum_{j=1}^{n} \sum_{k=1}^{N} x_{ik}^v \int_0^t \mathrm{K}_{ij}(t-\tau)\, \phi_k(\tau)\, d\tau.$$

The minimization of the residual by means of the mean weighted residual method results in the inner products $\langle r_i^v, \psi_j \rangle$ in (10.18); because of linearity, that is $\langle f+g, h\rangle = \langle f, h\rangle + \langle g, h\rangle$, each of the inner products corresponding to the convolution integral can be considered separately, and they are

$$\left\langle \int_0^t \mathrm{K}_{ij}(t-\tau)\, \phi_k(\tau)\, d\tau, \psi_l(t) \right\rangle = \int_0^T \int_0^t \mathrm{K}_{ij}(t-\tau)\, \phi_k(\tau)\, d\tau\ \psi_l(t)\, dt = \kappa_{ijkl}.$$

The elements κ_{ijkl} are not functions of time; they only depend on the basis functions ϕ_k, the test functions ψ_k and the radiation impulse responses K_{ij}. Consequently, the terms of the inner products $\langle r_i^v, \psi_j \rangle$ in Equation (10.18) can be written as

$$\left\langle \sum_{j=1}^{n} \int_0^t \mathrm{K}_{ij}(t-\tau) v_j^N(\tau)\, d\tau, \psi_l(t) \right\rangle = \sum_{j=1}^{n} \sum_{k=1}^{N} x_{ik}^v \kappa_{ijkl}.$$

The coefficients κ_{ijkl} can be calculated either analytically, if possible, or by numerical integration once the basis functions, test functions, and the hydrodynamics have been chosen. In other words, by applying the mean weighted residual method for the approximation of the equation of motion, *all the computation involving the integration of the convolution integral can be carried out off-line*, reducing the computational load when solving the NLP problem of the approximated optimal control problem.

10.4 Conclusion

The chapter presents a general mathematical framework for the solution of the WEC optimal control problem. The flexibility of the approach permits a wide variety of basis functions for variable (motion and PTO force) parameterization to be employed, while the degree of approximation (choice of N) provides a trade-off between approximation accuracy and computational complexity. Such complexity scaling makes the algorithm ideally suited to real-time control, and the chapter shows how inspection of the PTO spectrum can result in considerable computational savings with minimal reduction in approximation fidelity. Other key features include the efficient representation of the radiation-damping convolution, which further improves the computational attractiveness of the method. While the simulation results included are for the single-body case, multibody WEC systems and WEC arrays (see Chapter 11) are easily accommodated using an appropriate model and PTO configuration matrix, as outlined in Section 10.1.1, with simulation examples for the two-body case shown in [161] and an application to the control of WEC arrays shown in [166].

11 Control Strategies for WEC Arrays

11.1 Introduction

Just like wind farms, it is clear that commercial wave energy farms will contain multiple wave energy devices, relying on an economy of scale in reducing overall costs due to sea area leasing, moorings, electrical connections, and maintenance. However, while wind turbines interact only destructively in a collective (due to shadowing effects, etc.), wave energy devices in a farm have the possibility to interact *constructively*, since radiated waves from one device can reinforce the motion of another device.

However, the design of such a positively interacting collective depends on a number of factors, including the wave farm layout. In addition, the optimality of the layout (spacing, orientation, etc.), with respect to positive reinforcement, depends on sea state and incident wave direction, which, in general, are variable.

For a given wave farm layout, the geometry and motion of each device determines the radiated wave patterns, where the motion is strongly influenced by the control strategy employed on each device. This opens up the possibility that devices can, via inter device communication, coordinate their motion to the mutual benefit of all devices in the farm, with maximum farm power production as an objective. Such a structure, for a simple two-device array, is shown in Fig. 11.1.

Wave energy arrays include both large devices (circa 1 MW), individually moored, and closely packed arrays of devices, which are contained within a larger superstructure. Examples of larger individual devices include Pelamis [167], Oyster [168], and Powerbuoy [169], while closely spaced arrays include the Wavestar [45], Fred Olsen Lifesaver [170], and the Manchester Bobber [171] devices, and the device developed by Trident Technologies [172]. In particular, compact circular arrays have been shown to be capable of extracting more energy than a single device of the same volume, and offer a larger spectral bandwidth [173].

Original work in the area of global control of arrays was carried out by Evans [30] and Falnes [174], who developed optimal control for a number of oscillating bodies. The focus was on extending complex-conjugate control to the multi-body case. Since then, however, little progress has been made on optimal array control, though a lot of research has focused on using differential (linear) damping on individual array elements. In particular, the benefits of optimal global array control over independent control are relatively unclear. In some cases [45], independent (optimal) control of array elements is evaluated, but there is no comparison with the global control case.

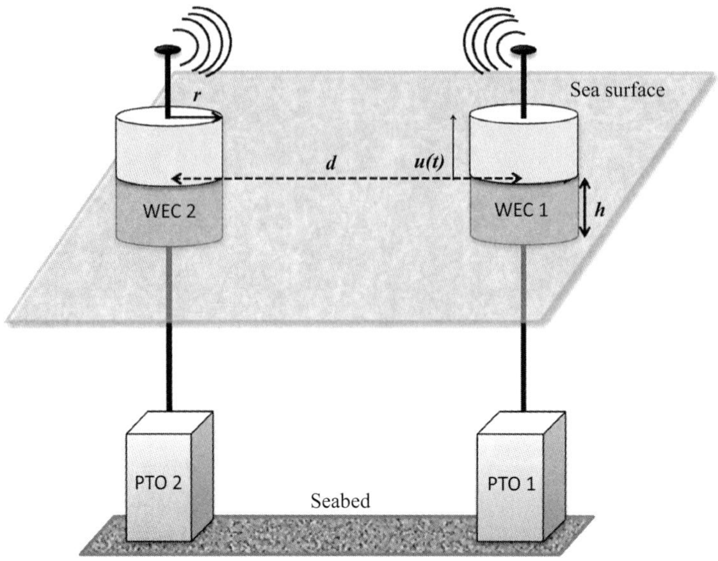

Figure 11.1 Layout for a two-body array. Motion and power take-off (PTO) force is restricted to heave only.

A number of studies have looked at the sub-optimal case where a diagonal radiation damping matrix (considering the change in the boundary value problem but not inter device dynamic interactions) is employed. However, these are essentially "independent" controllers, with no inter-device communication:

- Justino and Clement [175] show that the optimal diagonal "controller" (ignoring radiation interactions) is simply the complex conjugate of the diagonal radiation impedance matrix, while a similar, but simplified controller parameterization is considered in [176]. Comparable approaches are also taken in [177] and [172].
- A number of researchers have considered linear damping, which is optimized for each array device [178–181].
- A number of studies consider linear damping, where a consistent level of damping is determined for each device, but is designed to maximize total array energy [177–179].

In this chapter, two control strategies are compared for the control of wave energy converter (WEC) arrays, namely global control (GC) and independent control (IC), in order to determine the potential benefits of co operative global control, facilitated by inter-device communication. Both GC and IC are model-based control strategies; however, GC is based on a centralized control algorithm that uses the *complete hydrodynamic model* of the array and features full sharing of device motion information among the array members, whereas, with IC, each device is controlled independently using the hydrodynamic model of a single isolated device. For comparison purposes, it is assumed that the total local hydrodynamic force on each WEC can be estimated for both GC and IC cases, where the total hydrodynamic force is intended to be the sum of the force exerted by the incoming wave, diffraction, and radiation. While several layouts, separations, and body geometries are considered, it is the objective of this study not to investigate the optimal shapes or layouts, but rather to achieve some consistency of conclusion regarding the effectiveness of array control strategies. However, it is appreciated that WEC array performance is sensitive to both layout and device geometry [89], which is further explored in Chapter 14.

In the literature, the predominant metric used to evaluate the performance of a WEC array is the q-factor [89] (see Equation (6.126)), which indicates the *average* device performance of an array element compared with an isolated device. Here, however, the ratio of the total array (energy) performance for the GC and IC cases is considered. The optimal GC and IC problems are formulated in the time domain and are then discretized and solved as nonlinear programs. While a time domain formulation is not necessary for the fundamental unconstrained case considered in this chapter (and essentially the same results can be obtained by using the frequency domain approach originally developed by Evans [30] and Falnes [174]), it provides the possibility to extend the comparison to the constrained case. A preliminary analysis of the constrained GC problem is reported in [182] with a more complete treatment given in [183].

Some useful background for the models and measures associated with WEC arrays in given in Section 6.10. In general, the mathematical framework in this current chapter follows that for Chapter 10, bearing in mind that the quantities (parameters and variables) used to describe arrays are generally of higher dimension.

11.2 Hydrodynamic Model Formulation

In order to get an overview of the value of various control strategies in an array context, the control systems are implemented on different array layouts and for several geometries of the WECs composing the arrays. The WECs are vertical cylinders of radius r and draft h, and the distance between their vertical axes is denoted by d, as depicted in Fig. 11.1 for the particular case of a two-body array.

11.2.1 Equations of Motion

For this work, we assume linear wave theory and non-compressible irrotational flow. Therefore, the motion of the device can be described, in the time domain, by Cummins' equation [63] as originally presented in Equations (4.27) and (10.2) and repeated here for convenience:

$$\mathrm{M}_t\,\ddot{x}(t) + \mathrm{B}\,\dot{x}(t) + \int_{t_0}^{t} \mathrm{K}(t-\tau)\dot{x}(\tau)\,d\tau + \mathrm{S}\,x(t) = f(t) \tag{11.1}$$

where $x(t) \in \mathbb{R}^n$ is the vector of the vertical positions of the WECs, and n is the number of degrees of freedom of the system, which corresponds to the number of WECs composing the array, because the motion of each device is assumed to be restricted to heave only. $\mathrm{M}_t = \mathrm{M} + \mathrm{m}_\infty$ where $\mathrm{M} \in \mathbb{R}^{n\times n}$ is the generalized mass matrix and

$\mathrm{m}_\infty \in \mathbb{R}^{n\times n}$ is the asymptotic values of the added mass at infinite frequency; $\mathrm{B} \in \mathbb{R}^{n\times n}$ is the viscous damping term; $\mathrm{S} \in \mathbb{R}^{n\times n}$ is the hydrodynamic stiffness, while $\mathrm{K}(t) \in \mathbb{R}^{n\times n}$ is the matrix of the radiation impulse responses. The vector of external forces $f(t) \in \mathbb{R}^n$ is given by $f(t) = f_e(t) + f_{pto}(t)$ where $f_e(t)$ is the exciting force and $f_{pto}(t)$ is the power take-off (PTO) force. The excitation force is calculated as $f_e(t) = \mathcal{F}^{-1}\{\mathrm{K_e}(\omega)\eta(\omega)\}$, where $\eta(\omega)$ is the Fourier transform of the wave elevation and $K_e(\omega)$ is the frequency domain exciting force transfer function.

11.2.2 Numerical Modeling

The hydrodynamic coefficients m_∞ and $\mathrm{K}(t)$ are calculated from the frequency domain radiation impedance matrix $\mathrm{Z}(\omega)$ by applying Ogilvie's relations [164, 121]. The matrices $\mathrm{K_e}(\omega)$, $\mathrm{Z}(\omega)$, and S are then computed in the boundary element solver WAMIT [120]. This package is widely applied in studies of WEC arrays, as in [178, 180, 181]. In WAMIT, the high-order method is used and computations are performed for 160 frequencies equally spaced at intervals of 0.0151 rad/s. All control simulations are implemented in Matlab and are performed with the same frequency resolution as the WAMIT output. The software simulates the motion of the devices, the PTO forces, the instantaneous converted power, and the vertical velocities and displacements of the WECs for a representative surface elevation time history of a given wave spectrum.

11.2.3 Viscous Damping Approximation

The viscous effect of a fluid on a body is usually described, under certain conditions [184], as a force proportional to the square of the relative velocity between the body and the fluid surrounding the body; that is, $f_d = (1/2)C_d\rho\dot{x}|\dot{x}|A$, where ρ is the fluid density, A is the area of the body projected onto the plane orthogonal to the velocity $\dot{x}$, and C_d is the drag coefficient, which is obtained experimentally (see, for e.g., appendix 2 in [184]). Several recent studies have considered the effects of viscosity on WEC motion, such as [172, 185, 186].

The objective of the approximation is to find a force $\hat{f}_d$ that is linearly proportional to the velocity and that dissipates the same amount of energy as the force f_d. This procedure is known as Lorentz linearization [187], and it has been used in the case of WECs by Folley et al. [188]. The approximation is carried out by equating the work of the nonlinear force f_d with the work of the linear approximation $\hat{f}_d = B_v\dot{x}$, which results in $\int_0^T f_d\dot{x}dt = \int_0^T \hat{f}_d\dot{x}dt$. By means of simple manipulations, the damping coefficient B_v can be expressed as

$$B_v = \frac{C_d\rho A\int_0^T \dot{x}^2(t)|\dot{x}(t)|\,dt}{2\int_0^T \dot{x}^2(t)\,dt}. \tag{11.2}$$

For the special case of a vertical cylinder of radius r oscillating sinusoidally in heave with velocity $\dot{x}(t) = V_o\cos(\omega t)$, the coefficient B_v is

$$B_v = 4C_d\rho r^2\omega V_o/3. \tag{11.3}$$

Knowledge of the velocity is required to calculate the coefficient B_v, but the velocity itself depends on B_v; therefore, an iterative procedure has been implemented. The procedure is initiated by setting B_v to an initial value B_v^0, which is not critical for the convergence because the relation between B_v and the velocity $\dot{u}$ is monotonic ($\dot{u}$ decreases when B_v increases). The ith iteration of the procedure is composed of two steps:

1. Calculate the velocity $\dot{x}^i$ using the value of B_v^{i-1}.
2. Calculate B_v^i using the velocity $\dot{x}^i$ and Equation (11.2), or (11.3) in the case of regular waves.

The procedure is stopped when the difference between two subsequent values of B_v is smaller than a threshold δ, that is, $|B_v^i - B_v^{i-1}| < \delta$. For any given geometry, controller type and sea state, the coefficient B_v is calculated by simulating an isolated device.

11.3 Energy-Maximizing Control

The control problem is defined as follows: find the optimal profile for the PTO forces that maximizes the total energy absorbed by the array described by the equation of motion (11.1) over a time interval of length T. The total energy absorbed by the array is considered to be the sum of the mechanical work performed by each of the PTO forces, as

$$W = -\sum_{k=1}^{n_d} \int_0^T \dot{x}^k(t)\, f_{pto}^k(t)\, dt \tag{11.4}$$

where $\dot{x}^k(t)$ and $f_{pto}^k(t)$ are, respectively, the heave velocity and the PTO force of the kth device, similar to Equation (10.1), and n_d denotes the number of devices.

In the ideal case, assuming the wave excitation is known completely into the future, the optimization is performed over an infinitely long time interval, that is, for $T \to +\infty$; however, it has been shown [189] that a limited horizon suffices to achieve close to optimal energy absorption. As a consequence, the real-time implementation of the control algorithm can be performed in a receding horizon fashion, as described in [158], where future knowledge of either wave elevation or excitation force, up to time T ahead, is obtained by prediction. Since the focus of this chapter is on the control of arrays of WECs, we neglect the effects of prediction by applying the separation principle [190], which is a standard approach in control engineering, allowing the design of the optimal control to be separated from the design of the predictor; thus, perfect knowledge of future wave elevation is assumed, in order to isolate the effect of control on the total energy produced.

11.3.1 Discretization

As in Chapter 10, the control problem is discretized by approximating the velocity and the PTO force with a linear combination of basis functions, resulting in a finite-dimension optimization problem. Again, trigonometric functions are chosen as basis functions, so the device velocities and PTO forces, respectively, are approximated with the truncated zero-mean Fourier series

$$\dot{x}_k(t) \approx \sum_{n_f=1}^{N/2} v_k^{n_f,c} \cos(n_f\omega_0 t) + v_k^{nn_f,s} \sin(n_f\omega_0 t), \tag{11.5}$$

$$f_{pto_k}(t) \approx \sum_{n_f=1}^{N/2} \phi_k^{n_f,c} \cos(n_f\omega_0 t) + \phi_k^{n_f,s} \sin(n_f\omega_0 t). \tag{11.6}$$

The best approximation of the solution for the equation of motion (11.1) is sought by applying the Galerkin method, the details of which are presented in [158] and Section 10.2, and the result is the linear system

$$\mathrm{G}X = P + \mathrm{E}, \tag{11.7}$$

where X, P, E, and G are defined as

$$X = \begin{bmatrix} X_1 \\ X_2 \end{bmatrix}, \quad P = \begin{bmatrix} P_1 \\ P_2 \end{bmatrix}, \quad \mathrm{E} = \begin{bmatrix} \mathrm{E}_1 \\ \mathrm{E}_2 \end{bmatrix}, \quad \mathrm{G} = \begin{bmatrix} \mathrm{G}_{11} & \mathrm{G}_{12} \\ \mathrm{G}_{21} & \mathrm{G}_{22} \end{bmatrix}. \tag{11.8}$$

The vectors X_k and P_k, for $k = 1, 2$ (for e.g.), are the vectors of the Fourier coefficients of the velocity and PTO force of the kth device, respectively, and are arranged as

$$X_k = \left[v_k^{1,c}, v_k^{1,s}, v_k^{2,c}, v_k^{2,s}, \ldots, v_k^{\frac{N}{2},c}, v_k^{\frac{N}{2},s} \right]^T, \qquad \text{for } k = 1, 2.$$
$$P_k = \left[\phi_k^{1,c}, \phi_k^{1,s}, \phi_k^{2,c}, \phi_k^{2,s}, \ldots, \phi_k^{\frac{N}{2},c}, \phi_k^{\frac{N}{2},s} \right]^T,$$

The elements of the vectors E_k are the Fourier coefficients of the excitation force on the kth device and are arranged in the same manner as the vectors X_k and P_k. The matrices $\mathrm{G}_{ij} \in \mathbb{R}^{N\times N}$ composing the matrix G are block diagonal, where each of the $N/2$ square blocks is of size two and the lth block is defined as

$$\begin{aligned} \mathrm{G}_{ij}^l &= \begin{bmatrix} \mathrm{D}_{ij}^l & \mathrm{M}_{ij}^l \\ -\mathrm{M}_{ij}^l & \mathrm{D}_{ij}^l \end{bmatrix}, \qquad \text{for } l = 1, \ldots, N/2 \\ \mathrm{D}_{ij}^l &= \mathrm{R}_{ij}(l\omega_0) + \mathrm{B}_{ij}, \\ \mathrm{M}_{ij}^l &= l\omega_0 \left(\mathrm{M}_{ij} + \mathrm{m}_{ij}(l\omega_0) \right) - \mathrm{S}_{ij}/(l\omega_0). \end{aligned} \tag{11.9}$$

B_{ij}, M_{ij}, and S_{ij} are, respectively, the elements of the matrices B, M, and S, while $\mathrm{R}_{ij}(\omega)$ and $\mathrm{m}_{ij}(\omega)$ are the elements of the radiation impedance matrix $Z(\omega)$, which is computed by WAMIT and is defined as $\mathrm{Z}(\omega) = \mathrm{R}(\omega) + j\omega\mathrm{m}(\omega)$.

11.3.2 Global Control

The control system of the GC strategy has information relating to the whole configuration of the array; the resulting optimization problem is defined by the cost function $\mathrm{W} = -(T/2)P^T X$, which is obtained by substituting (11.5) and (11.6) into the definition of the total absorbed energy in (11.4). If G is nonsingular, the cost function can be expressed as a function of P by solving Equation (11.7) with respect to X, and the coefficients $P^\star$ of the optimal PTO forces that maximize the absorbed energy for the array are obtained by solving the optimization problem[1]

$$P^\star = \arg\max_P -P^T \mathrm{G}^{-1} P - P^T \mathrm{G}^{-1} \mathrm{E}. \tag{11.10}$$

11.3.3 Independent Control

For the IC case, it is assumed that each device is equipped with its own controller and an excitation force estimator. It is also assumed that no communication occurs between the devices, and each controller uses the model of a single isolated device. That is, the control system of each device uses the model (11.1) where the variables $u(t)$ and $f(t)$ are scalars and where M_t, B, K(t), and S are also scalars and are the parameters of a single isolated device with the same shape and dimensions of the devices composing the corresponding array. For the example of an array composed of two WECs (as shown in Fig. 11.1), the controllers of devices 1 and 2 use, respectively, the discretized models

$$\mathrm{G}_s X_1 = P_1 + \bar{\mathrm{E}}_1, \tag{11.11}$$

$$\mathrm{G}_s X_2 = P_2 + \bar{\mathrm{E}}_2, \tag{11.12}$$

where $\mathrm{G}_s \in \mathbb{R}^{N\times N}$ is analogous to the matrix G in the approximated equation of motion of the array in (11.7). However, in this case, G_s is calculated using the hydrodynamic coefficients of a single isolated device. In particular, G_s is a block diagonal matrix with square blocks of size two and is constructed in the same manner as the diagonal blocks G_{ii} of the matrix G, as described in Equations (11.9) for $i = j$, with the difference being that the elements D^l_{ij} and M^l_{ij} of G_s are calculated using the radiation impedance matrix of a single isolated device.

$\bar{\mathrm{E}}_k$ is the excitation force measured by the estimator on WEC k; more specifically, $\bar{\mathrm{E}}_1 = \mathrm{E}_1 - \mathrm{G}_{12}X_2$ and $\bar{\mathrm{E}}_2 = \mathrm{E}_2 - \mathrm{G}_{21}X_1$. In fact, it is assumed that the excitation force estimator on each device is not capable of discerning the excitation force due to incoming waves from the radiation generated by other bodies; therefore, the estimator provides a signal that is the sum of the radiation force caused by other bodies ($\mathrm{G}_{12}X_2$ and $\mathrm{G}_{21}X_1$) and excitation from incoming waves (E_1 and E_2).

Each of the independent controllers calculates the optimal PTO force that maximizes the energy absorbed by the corresponding WEC using the models in Equations (11.11)

[1] The term $T/2$ is a positive scaling factor that can be neglected for the calculation of the optimal PTO force because the vector $P^\star$ that maximizes $\mathrm{W} = -(T/2)P^T X$ also maximizes the cost function $\overline{W} = -P^T X$.

and (11.12). The optimal PTO forces (P_1^* and P_2^*) are the solutions of the optimization problems

$$P_1^* = \arg\max_{P_1} W_1 = \arg\max_{P_1} -P_1^T \mathrm{G}_s^{-1} P_1 - P_1^T \mathrm{G}_s^{-1} \bar{E}_1, \quad (11.13)$$

$$P_2^* = \arg\max_{P_2} W_2 = \arg\max_{P_2} -P_2^T \mathrm{G}_s^{-1} P_2 - P_2^T \mathrm{G}_s^{-1} \bar{E}_2. \quad (11.14)$$

the cost functions of which are the energy absorbed by each device, that is, $W_1 = -P_1^T X_1$ and $W_2 = -P_2^T X_2$. However, the solutions of the optimization problems in Equations (11.13) and (11.14) are coupled because P_1^* depends on $\bar{E}_1$, which is a function of the velocity of body 2 (X_2), and vice versa; the problem is then effectively solved iteratively. The initial condition is considered to be with the PTOs switched off ($P_1 = P_2 = 0$). The velocities X_1 and X_2 are then calculated by means of the equation of motion (14.21), and the controller calculates the PTO forces by solving the optimization problems (11.13) and (11.14). When the PTO forces are applied to the WECs, the new velocities are calculated again, using Equation (14.21), and the process is repeated. The computations performed by the controller at the ith step of the iteration are

$$X^i = \mathrm{G}^{-1}(\hat{P}^{i-1} + E),$$

$$\hat{P}_1^i = \arg\max_{P_1} -P_1^T \mathrm{G}_s^{-1} P_1 - P_1^T \mathrm{G}_s^{-1} \left(E_1 - \mathrm{G}_{12} X_2^i\right),$$

$$\hat{P}_2^i = \arg\max_{P_2} -P_2^T \mathrm{G}_s^{-1} P_2 - P_2^T \mathrm{G}_s^{-1} \left(E_2 - \mathrm{G}_{21} X_1^i\right),$$

where $i > 1$, $\hat{P}^0 = 0$, and $\hat{P}^i$, $\hat{P}_1^i$, $\hat{P}_2^i$, X^i, X_1^i, X_2^i are defined according to Equations (11.8). The iteration stops when the PTO forces approach to their asymptotic values, which is implemented by the condition $\|P_k^i - P_k^{i-1}\| < \varepsilon \ \forall k = 1, 2$.

11.4 Results

Three possible array layouts of heaving cylinders are considered, as shown in Fig. 11.2. Layout 1 is composed of two devices; layout 2 is a linear array of three WECs, while layout 3 is composed of three WECs placed at the vertices of an equilateral triangle. For each layout, 24 values of the inter body spacing d are chosen, logarithmically spaced between $2.2r$ and $10r$. Three device geometries for the WECs composing each array are simulated, each of which has approximately the same volume ($\approx 160\pi$ m^3) but different resonant periods; Table 11.1 lists the radius r, draft h, and resonant period T_r for each device.

Table 11.1

WEC Geometry	1	2	3
Radius r (m)	4	5	6.25
Draft h (m)	10	6	4
Resonance T_r (s)	7.1	5.9	5.4

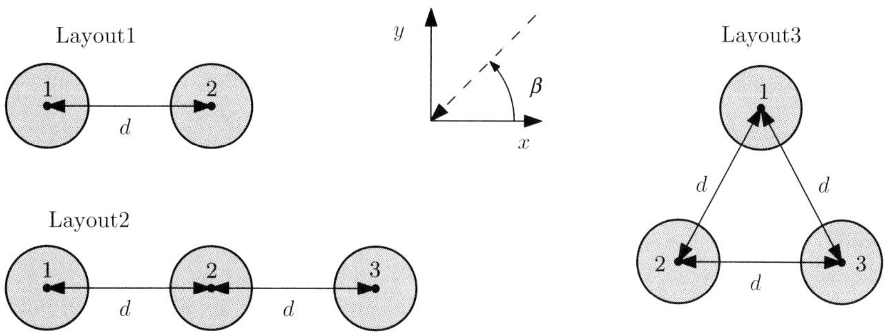

Figure 11.2 Top view of array layouts and of incident wave angle β.

The optimal control laws for each WEC in the case of IC are obtained by iteration, and it is assumed that, between consecutive iterations, there is enough time for the waves radiated from a device to reach all the other devices and enough time for the estimator and predictor on each WEC to build a reliable forecast of the incoming waves (to reach quasi–steady state). While this is not feasible in practice, it provides a best-case scenario for IC, providing an upper performance bound for the IC case. However, the comparison between the GC and the IC cases is also carried out by considering only the first iteration of the ICs, which start from an initial condition where each PTO is switched off. The comparison of the GC case with first iteration of the IC case is interesting from a practical point of view, since it highlights the effect of the PTO (and therefore the control system) on the interaction between devices.

Each of the three array layouts depicted in Fig. 11.2 is simulated for all three WEC geometries in Table 11.1. Both GC and IC cases are computed for each of the resulting nine possible arrays with inter-body spacing ranging between $2.2r$ and $10r$, and considering four Bretschneider spectra with $H_s = 1$ m and $T_p = \{6, 8, 10, 12\}$ s. The parameters ω_0 and N for the discretization of the control problem are $N = 160$ and $\omega_0 = 2\pi/200$ rad/s, respectively, while the thresholds δ and β for the adaptive approximation of the viscous damping (Section 11.2.3) and the IC case (Section 11.3.3), respectively, are $\delta = 1$ and $\varepsilon = 10$.

The patterns exhibited by the results are illustrated by the representative cases depicted in Figs. 11.3–11.5, which show the ratio between the energy absorbed using IC and GC (E_i/E_g). In addition, Tables 11.2–11.4 summarize the totality of the results for the different geometries, with the table entries representing the integral of the difference between the (ideal) IC and GC cases ($\int_{2.2}^{10}(1 - E_i/E_g)d(d/r)$), as highlighted by the shaded area in Fig. 11.3. In Figs. 11.3–11.5, the asterisks denote curves corresponding to the asymptotic value of the IC, that is, when the control algorithm reaches (the ideal situation of) convergence in the calculation of the control law, while the circles denote the curves corresponding to the first iteration.

As expected, GC always performs better than IC in terms of absorbed energy, with the difference decreasing monotonically with an increase in the separation distance.

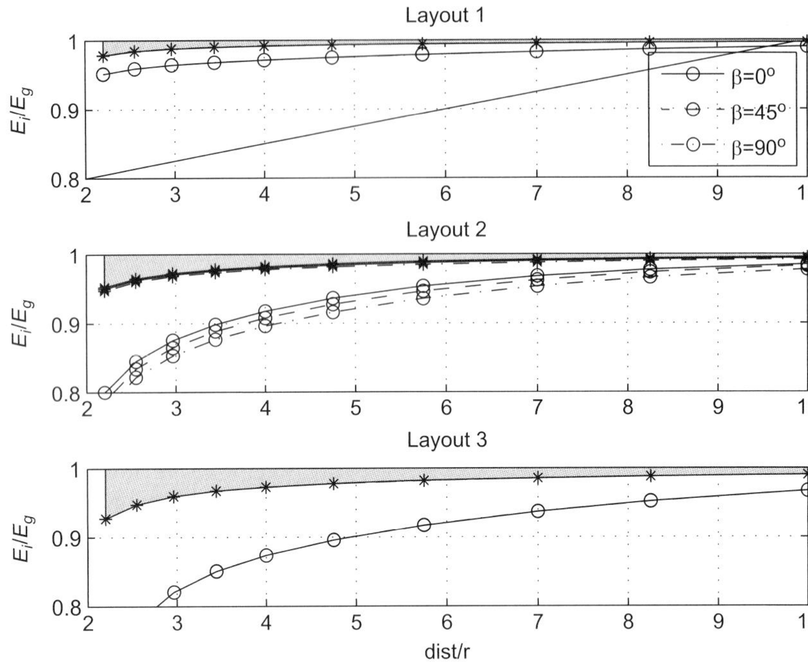

Figure 11.3 Relative performance of IC and GC for geometry 1, $T_p = 12$ s ($\lambda = 225$ m). The mark $*$ identifies the asymptotic values of E_i/E_g; the mark $\circ$ identifies the first iteration E_i^1/E_g.

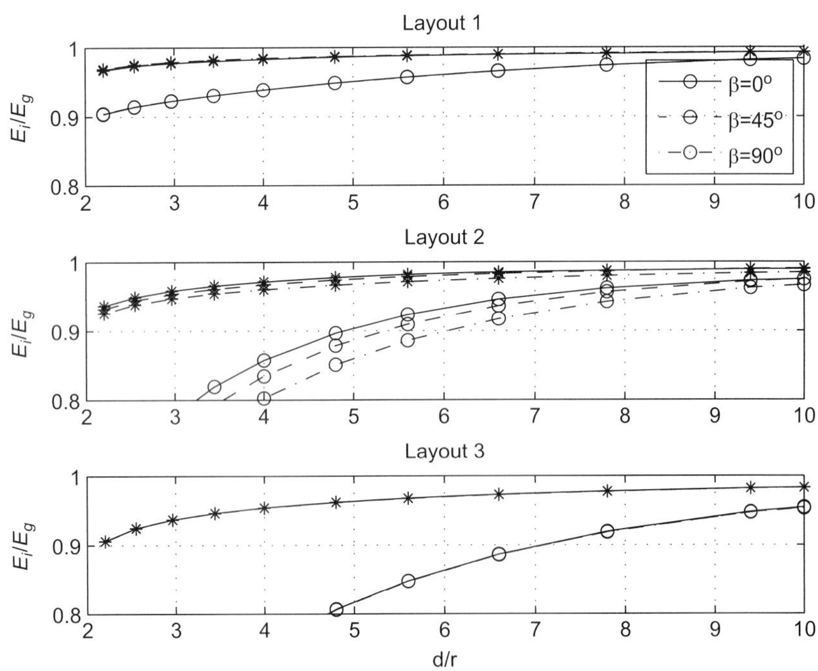

Figure 11.4 Relative performance of IC and GC for geometry 2, $T_p = 12$ s ($\lambda = 225$ m).

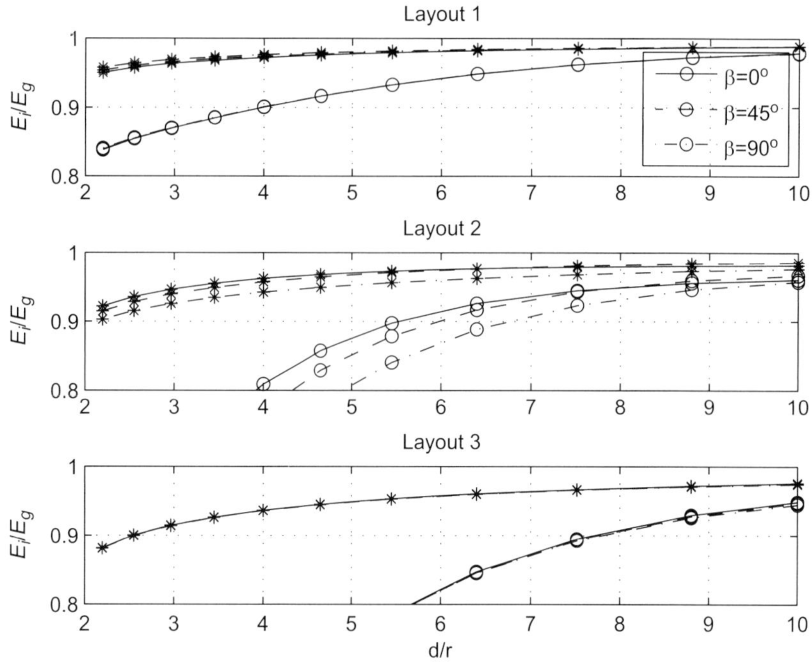

Figure 11.5 Relative performance of IC and GC for geometry 3, $T_p = 12$ s ($\lambda = 225$ m).

Although the difference between GC and IC, for the ideal IC case, is generally smaller than 5–10%, the relative performance is considerably larger when comparison is made with the first iteration of the IC. In addition, the performance of the IC degrades when the number of bodies in the array increases.

Comparing Figs. 11.3 and 11.5, it is evident that the IC cases suffer a degradation in performance for devices with stronger radiative properties. In fact, at any distance d, the ratio E_i/E_g for geometry 3 (Fig. 11.5) is smaller than E_i/E_g for geometry 2 (Fig. 11.4), which is a flatter device compared with geometry 3. In turn, E_i/E_g for geometry 2 is smaller than E_i/E_g for geometry 1 (Fig. 11.3), which is flatter than geometry 2. By "flatter," it is meant that r/h is larger, with resulting stronger radiative properties. Tables 11.2–11.4 show how the relative behaviors of the GC and IC cases are affected by the sea state, for different geometries. For layout 1, composed of two WECs, the relative performance of the asymptotic value of the IC case with respect to the GC case is not significantly affected by sea state, while more significant variations exist for layouts 2 and 3.

For layouts 2 and 3, the patterns in the E_i/E_g curves are still similar to the case of layout 1; however, a dependency on the heading angle emerges for layout 2, especially when considering the first iteration of the IC. This phenomenon shows that the performance of the ICs is affected by the wave heading angle only in the case of the extended linear array (layout 2).

Fig. 11.8 shows the comparison of performance between GC and IC, for geometry 3, with $T_p = 6$ s. GC performs progressively worse than IC with increasing device displacement, as the curves E_i/E_g, for any heading angle β, always reach their minima for $\Delta x = 0$ and $\Delta y = 0$.

Similarly to the case where the estimation of the excitation force is correct, the degradation in performance of GC with respect to IC is smaller for longer wavelengths (see Tables 11.5 – 11.7) and for devices with smaller radiation characteristics (compare Table 11.5 with Table 11.7).

When the estimation of the excitation force is affected by the error due to the position offset of one device, the relative performance of GC with respect to the IC case degrades significantly when the offset takes place in the direction of the incoming waves. Fig. 11.8, for example, shows that when $\beta = 0$ degrees, E_i/E_g is not affected by offsets along y but is obviously sensitive to displacements along x. Conversely, when $\beta = 90$ degrees, E_i/E_g is strongly affected by variations in position along y but is not sensitive to variations along the x direction. The explanation for this effect is related to the fact that a displacement of the body in the direction of the incoming waves corresponds to a phase shift in the excitation force, while a displacement in the direction orthogonal to the incoming waves affects only the diffraction component. In particular, the Froude–Krylov force is constant in a direction orthogonal to the wave direction (plane wave case), while the diffraction force depends on the layout of the array and on the geometry of the devices. The effect of a small displacement along y on the diffraction component is negligible because the distance d between the bodies remains unaltered, that is, $d \approx \bar{d}$. However, a displacement along x may affect the diffraction component because the variation in the distance between the devices is equal to the magnitude of the offset Δx, that is, $|d - \bar{d}| = |\Delta x|$.

Fig. 11.8 allows us to discern between the effects of the diffraction and Froude–Krylov forces on the relative performance between GC and IC. In particular, the curve corresponding to $\beta = 0$ degrees is perfectly horizontal when the displacement occurs along y, where variations of both diffraction and Froude–Krylov forces are negligible. For $\beta = 90$ degrees, the curve corresponding to displacements along y is not constant, and its variation is due to diffraction only because, in this case, the Froude–Krylov component is constant along y.

11.6 Discussion

One striking feature of the results is the characteristic "asymptotic" convergence of E_i/E_g to 1, as d/r becomes large. This can be explained by the reduction in interaction effects as the separation distance between array elements increases.

In [192], it is shown that, for a two-body array of heaving or surging WECs aligned with the direction of the incoming waves and large d, *when radiation is negligible*, the alteration of the energy absorption due to wave interaction effects decreases asymptotically with the square root of the separating distance. A similar discussion is carried out in [191], where further development accounts for the effects of radiation at closer distances; although the authors provide a model for the approximation of the interaction factor as a function of the separating distance, the effect of radiation on the interaction between WECs was not isolated from the effect of diffraction. Separation between the effects of diffraction and radiation on the energy absorption have been considered in Babarit et al. [193], where the problem has been studied by considering a two-body array and keeping one device fixed, inhibiting radiation.

Consider a two-body array with no incoming waves, with the first device (WEC 1) induced to oscillate by means of a pre specified PTO force (P_1), and the second device (WEC 2) optimally controlled so as to absorb the maximum amount of energy radiated by the first device. As expected, from analytical considerations, the maximum amount of energy transferred from the first device to the second is inversely proportional to the distance d. The first step to show this effect is to define H, if G is nonsingular, as

$$\mathrm{H}=\mathrm{G}^{-1}=\begin{bmatrix}\mathrm{H}_{11} & \mathrm{H}_{12}\\ \mathrm{H}_{21} & \mathrm{H}_{22}\end{bmatrix}.$$

From the discretized equation of motion (11.7), the energy absorbed by WEC 2, denoted as J_2, is

$$J_2 = P_2^T X_2 = P_2^T \mathbf{H}_{21} P_1 + P_2^T \mathbf{H}_{22} P_2. \tag{11.16}$$

For this two-body case, it can be seen that the symmetric part of $\mathbf{G}$ (the symmetric part of $\mathbf{G}$ is $(\mathbf{G}+\mathbf{G}^T)/2$) is positive definite, thus also the symmetric parts of H and, given the structure of $\mathbf{G}$, H_{12} are positive definite, and Equation (11.16) is a convex quadratic function of P_2. Therefore, for any given value of the PTO force P_1, the optimal value of the PTO force P_2 that maximizes J_2 is $P_2^* = (\mathbf{G}+\mathbf{G}^T)^{-1}\mathbf{H}_{21}P_1$.

The "radiation interaction factor," w_{21}, defined as the ratio of the energy delivered to the system by means of the mechanical work of the PTO on WEC 1 and the energy recovered by means of the optimally controlled WEC 2, is $w_{21} = J_2/J_1 = P_2^T X_2 / P_1^T X_1$. Using Equations (11.7) and (11.8), and some further manipulation, w_{21} can be evaluated as

$$w_{21} = \frac{P_1^T \mathbf{H}_{21}^T \left(\bar{\mathbf{H}}_2^T \mathbf{H}_{22} \bar{\mathbf{H}}_2 + \bar{\mathbf{H}}_2^T\right) \mathbf{H}_{21} P_1}{P_1^T \left(\mathbf{H}_{11} + \mathbf{H}_{12} \bar{\mathbf{H}}_2 \mathbf{H}_{21}\right) P_1} \tag{11.17}$$

where $\bar{\mathbf{H}}_2 = \left(\mathbf{H}_{22} + \mathbf{H}_{22}^T\right)^{-1}$.

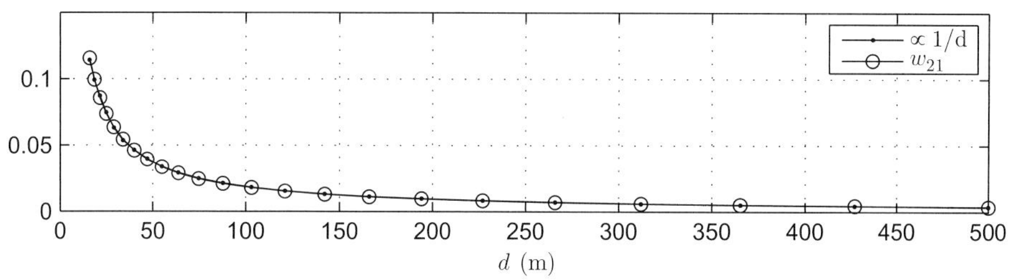

Figure 11.9 Interaction factor w_{21} for geometry 3 as a function of d.

Fig. 11.9 shows that w_{21} is inversely proportional to the inter body distance d. In particular, the curve marked with circles ($\circ$) depicts the values of w_{21} for an array composed of two devices of geometry 3 as a function of the inter body distance d, where P_1 has been chosen as a random vector. The curve marked with dots ($\cdot$) is obtained by fitting the values of w_{21} with a function $f(d) = \alpha/d$, where the value of α is obtained using least squares.

11.7 Conclusions

The results in Section 11.4 clearly indicate that a significant performance improvement (up to 10% or more) can be obtained using global control of arrays, compared with independent control. Considering the more realistic case, where the IC cases will not have time to fully converge on a quasi–steady state (which assumes that the sea state is strictly stationary), significantly greater benefits of GC are indicated, with improvements of greater than 20% possible for small d. The benefits of GC are greatest for layouts that have maximum device interaction (e.g., layout 3 in our analysis) and devices that have strong radiation properties (e.g., geometry 3, which has a large radius/draft ratio).

The degree of benefit of GC over IC, as expected, diminishes as the interaction between devices decreases. Therefore, the greatest benefit of GC is likely to be manifest for closely packed arrays of WECs, such as Wavestar [45], Fred Olsen Lifesaver [170], Manchester Bobber [171], and Trident Technologies [172] systems. In addition, since the individual devices within these systems are held within a relatively rigid framework, GC is not likely to suffer from performance degradation due to incorrect specification of the position of the individual devices, as detailed in the analysis in Section 11.5.

For GC, some extra complexity and computation are required. However, the addition of hardware components is minimal, though some inter-device communication is required (which could be achieved wirelessly or performed in tandem with the electrical connections). The main additional complexity is in the form of the hydrodynamic model and, consequently, the control equations. Given a potential 10% improvement in energy converted, this would seem to be a small price to pay, given the considerable capital cost of a wave energy array system, and should be considered as a solution component in the challenge to make wave energy economic.

The study presented in this chapter does not consider force or displacement constraints on the WECs. Such constraints have the potential to impact the co-ordinated movement between the devices in an array under GC, in particular. The interested reader is referred to [183], which shows a decrease in the gap between GC and IC for active PTO force constraints, while constraints on WEC motion amplitude have a relatively minor effect.

12 Wave Forecasting for Control

12.1 Introduction

Energy conversion in most wave energy converters (WECs) is based either on relative oscillation between bodies or on oscillating pressure distributions within fixed or moving chambers. Oscillators generally have pronounced resonances, which enable efficient power absorption over only a restricted range of frequencies. As has been shown in previous chapters, to cope with the variations of wave spectra, a control system can be designed to alter the oscillator dynamics such that the efficient energy conversion occurs over a wide range of wave conditions [194].

The control approach, in the early stages of wave energy research, consisted of frequency domain relationships regulating the dynamics of the system to be tuned for maximum energy absorption at different peak frequencies corresponding to different incoming wave spectra [66, 194]. Although being an advantageous approach for real sea spectra, frequency domain techniques do not generally allow real-time control on a wave-by-wave basis, which can significantly raise the device's productivity and therefore its economical viability. Real-time optimal control can be directly derived from the aforementioned optimal frequency relationships [66, 194]. The main difficulties arise from the fact that the transformation into the time domain results in *noncausal* transfer functions, so that the conditions for *optimal* power absorption can be realized only if future motion of the device, or of the future incident wave profile, is known [135, 194, 195], as shown in Chapter 7. However, a variety of suboptimal causal control schemes are also possible, for example, as reported in [112] and Chapter 8.

The problem of short-term wave prediction of the actual wave elevation profile at a specific point of the sea surface, for some seconds into the future, is central to the more general issue of time domain control of WECs. As an example, Fig. 12.1 illustrates a possible digital control scheme for a generic oscillating body, where the controlled variable is its oscillation velocity, $u(k)$, and the control action is performed through a control force $F(k)$. The estimate of the optimal reference, $\hat{u}_{opt}(k)$, is computed by an algorithm that requires future values, $\hat{\eta}(k+l|k)$, of the incident wave elevation, $\eta(k)$. Depending on the operating principle of the specific device, as well as the control strategy adopted and the available instrumentation, the quantities involved may vary, but the logic can always be described by a framework such as that proposed in Fig. 12.1. In particular, the quantity to be predicted may be the wave excitation force, the oscillation velocity for an oscillating body, or the air pressure inside the chamber

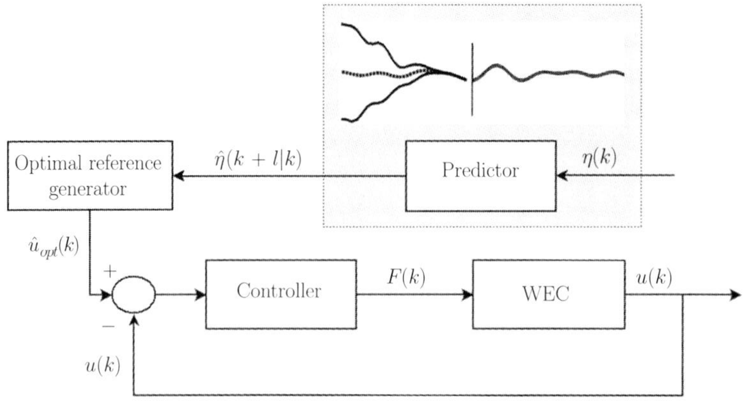

Figure 12.1 Wave predictions are required in order to generate an optimal reference for a generic time domain control of a WEC.

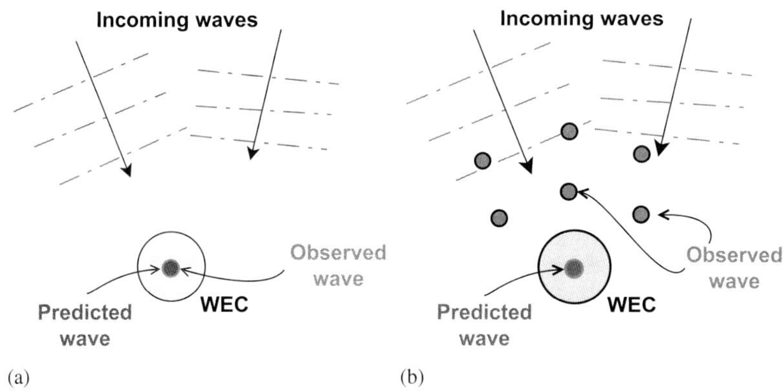

Figure 12.2 The two main approaches to wave forecasting. (a) Prediction based only on local single-point measurements. (b) Prediction based on reconstruction of wave field from array of distant measurements.

of an oscillating water column, and they all depend, ultimately, on the incident wave elevation.

The main approach followed in the literature is based on a spatial prediction of the wave elevation, as in Fig. 12.2(b), meaning that the wave field at a certain location is reconstructed from one or more observations at nearby locations [98, 124, 196, 197] (see also Chapter 7 and Section 12.8). The forecasting model, in this situation, requires an array of spatial measurements and can become very complex, because it has to take into account the possible multidirectionality of waves [98], the presence of radiated and diffracted waves [198], and eventual nonlinearities in the wave propagation (refer to [199] for a more detailed literature survey).

The solution proposed in this chapter is the prediction of the wave elevation based only on its past history at the same point of the sea surface, as in Fig. 12.2(a). Such an alternative approach introduces significant advantages with respect to the spatial prediction, in terms of complexity of the models (multidirectionality and radiation do not need to be considered unless the device itself requires it, e.g., directional WECs) and the amount of instrumentation required (measurements at only one point are required). Some preliminary work, following this purely time-series approach, may be found in [194, 200], where autoregressive (AR) models are proposed for prediction of future oscillation velocity of a generic device, and in [201], where AR and hybrid Kautz/AR models are utilized to predict the wave elevation time series. Both of them present only results with data numerically generated from standard wave spectra.

This chapter presents a variety of forecasting models and validates them against real observations. Initially, the prediction requirements of WEC control schemes are examined in Section 12.2. The forecasting models are then presented in Section 12.4 and a methodology to evaluate the confidence intervals of the predictions is outlined in Section 12.5. For the case studies described in Section 12.6, sample forecasting results are presented in Section 12.7, while the benefits of up-wave measurements are assessed in Section 12.8.

12.2 Prediction Requirements

The present section outlines a methodology for the quantification of the prediction requirements concerning reactive control, structured as in Fig. 12.3. After a fundamental discussion of the noncausality of the control in Section 12.2.1, the prediction requirements are quantified in Sections 12.2.2, while the possibility of manipulating the control to achieve a reduction in the prediction requirements is also discussed, in Section 12.2.3.

12.2.1 Analysis of Noncausality

The origin of noncausality in the control of WECs was articulated in Chapter 7 and is repeated here, in the context of quantifying the prediction requirements. The noncausality of the control strategy described in Section 8.2, and shown in Fig. 12.3, is encapsulated by the following relationship:

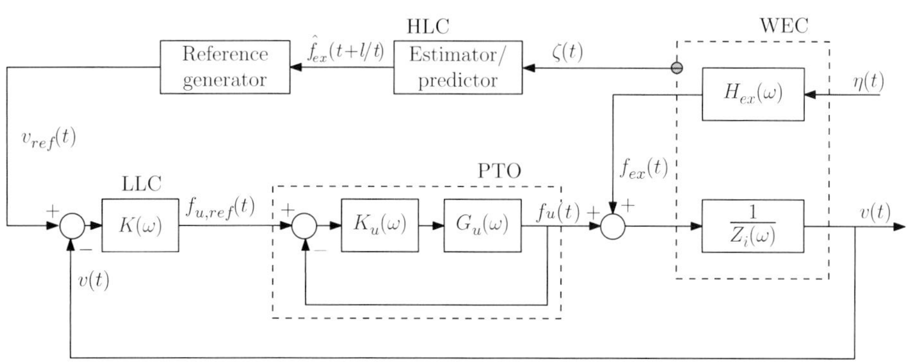

Figure 12.3 General structure for a real-time controller of a WEC.

$$V_{opt}(\omega) = \frac{1}{2B(\omega) + 2K_f} F_{ex}(\omega) \triangleq H_{opt}(\omega) F_{ex}(\omega), \tag{12.1}$$

which can be equivalently written in the time domain as

$$v_{opt}(t) = \int_{-\infty}^{t} h_{opt}(\tau) f_{ex}(t - \tau) d\tau \tag{12.2}$$

where $h_{opt}(t)$ is the inverse Fourier of $H_{opt}(\omega)$:

$$h_{opt}(t) = \mathcal{F}^{-1}\left\{\frac{1}{2B(\omega) + 2K_f}\right\} = \mathcal{F}^{-1}\left\{H_{opt}(\omega)\right\}. \tag{12.3}$$

When $\omega \to \infty$, the radiation resistance $B(\omega)$ goes to zero and therefore $H_{opt}(\omega)$ tends to the constant value $1/2K_f$. This means that the transform relationship in Equation (12.3) is valid only in a generalized sense and, more correctly,

$$h_{opt}(t) = \frac{1}{2K_f}\delta(t) + \mathcal{F}^{-1}\left\{H_{opt}(\omega) - \frac{1}{2K_f}\right\} = \frac{1}{2K_f}\delta(t) + k_{opt}(t) \tag{12.4}$$

where $\delta(t)$ is the Dirac delta function. The inverse Fourier transform in the right-hand term of Equation (12.4), yielding the function $k_{opt}(t)$, is now well defined. Note, as far as noncausality is concerned, $h_{opt}(t)$ or $k_{opt}(t)$ have the same characteristics, as they differ only at time $t = 0$.

The impulse response $h_{opt}(t)$ is real and even (the Fourier transform is real and even) and tends to zero for infinite time (the Fourier transform tends to a constant value for $\omega = 0$). The decay to zero is dominated, in general, by an exponential with a time constant, τ_1, which is the reciprocal of the lowest frequency pole of the corresponding transfer function.

Due to the presence of the radiation resistance $B(\omega)$, $H_{opt}(\omega)$ is known numerically only for some frequencies, and an explicit finite-order representation is not immediately available. From the behavior shown in Fig. 12.4, valid for systems with a single-peaked radiation characteristic, it is however straightforward to identify the lowest frequency

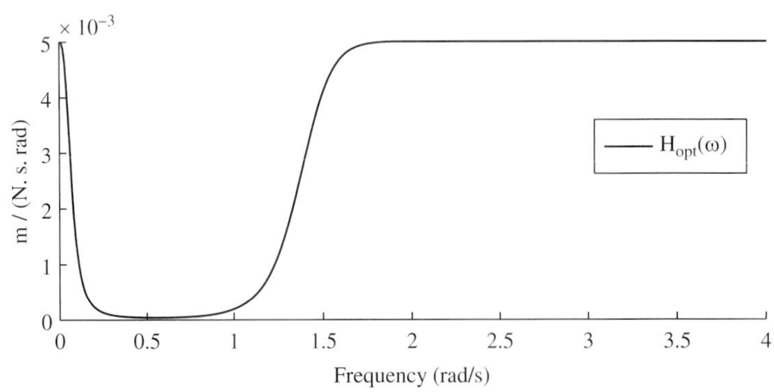

Figure 12.4 Noncausal reference-generator transfer function between excitation force and optimal velocity. It is a real and even function of the frequency.

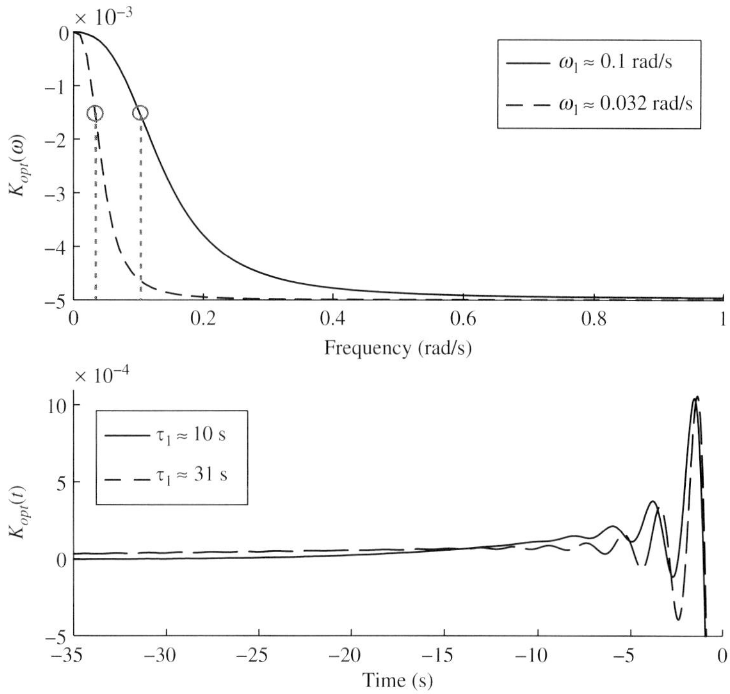

Figure 12.5 Relation between lowest frequency pole and noncausality of the impulse response. $K_{opt}(\omega)$ and $k_{opt}(t)$ are calculated for two heaving buoys.

pole. The function $H_{opt}(\omega)$ can, in fact, be seen as a low-pass filter cascaded with a high-pass filter, where the attenuation band corresponds to the resonance band of the system's radiation function. The lowest frequency pole, namely, ω_l, can be estimated as the cut-off frequency (at -3 dB) of the low-pass filter.

The time constant $\tau_l = 1/\omega_l$ is the dominant parameter that influences the decay to zero of $h_{opt}(t)$, and consequently the noncausality of the control action. As an example, Fig. 12.5 shows the relation between τ_l and ω_l for two different floating systems.

12.2.2 Prediction Requirements

In Section 12.2.1, a parameter strictly connected to the radiation of the floating system, τ_l, was identified as a measure of the noncausality of its control system. It would be interesting to see if a direct proportionality exists between τ_l and the actual forecasting horizon required for an effective approximation of the convolution in Equation (12.2), for the generation of the optimal velocity reference.

The required prediction horizon is determined from a range of suboptimal oscillation velocities, calculated from the integral in Equation (12.2), truncated at different times $\tau = -L$:

$$v_{opt}^{L}(t) = \int_{-L}^{L} h_{opt}(\tau) f_{ex}(t-\tau) d\tau, \tag{12.5}$$

so that future values for the excitation force only up to $f_{ex}(t+L)$ (supposed to be perfectly known) are included. Note that the causal part of the integral is also truncated to maintain the evenness of the impulse response. In the extreme case of $L = 0$, the integral in Equation (12.5) becomes causal and no future values of the excitation force are utilized.

The velocity $v^L_{opt}(t)$ is then ideally imposed onto the system, using a load force that can be directly derived from Equation (10.2). The relative capture width (RCW), obtained with the velocity thus calculated (and with the required power take-off (PTO) force), is

$$\text{RCW}(L) = \frac{P_u(L)}{P_w \cdot D} \tag{12.6}$$

where $P_u(L)$ is the time-averaged absorbed power, D is the width of the device (the diameter in the case of a cylinder or sphere), and P_w is the wave power per unit width [66]:

$$P_w = \frac{1}{2}\rho g^2 \int_0^{+\infty} \frac{\Xi(\omega)\Xi^*(\omega)}{\omega} d\omega. \tag{12.7}$$

In Equation (12.7), g is the acceleration due to gravity and ρ is the water density. Also, deep water is assumed, that is, $kh >> 1$ [66], where k is the wave number and h is the water depth.

Intuitively, it is expected that RCW(L) increases when larger values of L are considered, but only up to a critical future horizon, beyond which the power gain from considering extra future information of the excitation force is negligible. Such a critical future horizon will be proportional to the time constant τ_1.

12.2.3 Reduction of Prediction Requirements

In some situations, it may happen that the prediction requirements, determined from Section 12.2.2, are beyond the ability of the prediction model. One solution could be the adoption of complex forecasting solutions, making use of expensive instrumentation and of more computationally intensive models [202]. A critical evaluation of the noncausality of $H_{opt}(\omega)$, which also takes into account the typical frequency distribution of the excitation force, offers a much simpler alternative.

For a number of reasons, the wave excitation force is always contained within the stop band of $H_{opt}(\omega)$:

- Waves do not appear at frequencies below 0.2–0.3 rad/s (30–40 s).
- Waves at frequencies higher than the attenuation band of $H_{opt}(\omega)$ are filtered out by the excitation transfer function, $H_{ex}(\omega)$. There exists, in fact, a correspondence between the radiation and excitation bandwidth (based on the Haskind relation) [66, 203].
- If the system is well suited to the location of deployment, its radiation response will roughly match most of the sea states, and this corresponds to the attenuation band of $H_{opt}(\omega)$, as discussed in Section 12.2.2.

Based on these considerations, it may be argued that the reference generator will mostly work at frequencies within the attenuation band of $H_{opt}(\omega)$ (0.4–1 rad/s in Fig. 12.4). As a consequence, there exists the possibility of manipulating the shape of $H_{opt}(\omega)$ in order to reduce its noncausality, while at the same time maintaining its effectiveness in most sea conditions.

In Section 12.2.2, the noncausality is related to the time constant τ_1, which is the reciprocal of the pole at the lowest frequency of $H_{opt}(\omega)$. A possible intuitive manipulation may therefore consist of lowering the gain of $H_{opt}(\omega)$ outside the attenuation band, where its input (the excitation force) will not usually appear. This would move the first pole towards higher frequencies, thus reducing the time constant τ_1. As an example, consider Fig. 12.6(a), where the original $H_{opt}(\omega)$ is modified to a new function, termed $H_{red}(\omega)$. As a result, the impulse response $h_{red}(t)$, which is still real and even, decays much more quickly to zero than $h_{opt}(t)$, as shown in Fig. 12.6(b).

Note that the modification proposed here is not rigorous, in any sense, and is based only on qualitative considerations. It is not within the scope of this chapter to find

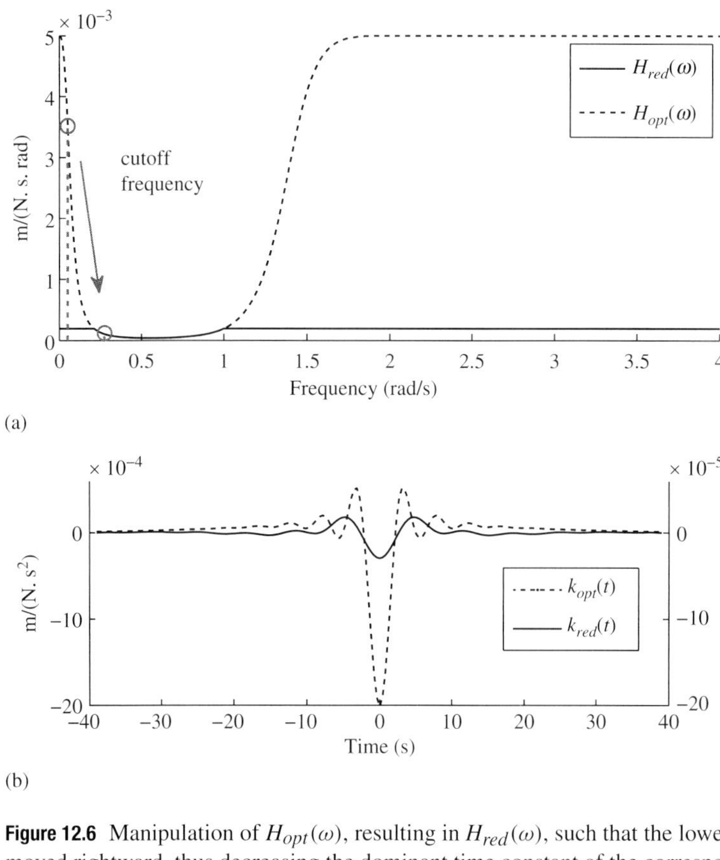

Figure 12.6 Manipulation of $H_{opt}(\omega)$, resulting in $H_{red}(\omega)$, such that the lowest frequency pole is moved rightward, thus decreasing the dominant time constant of the correspondent impulse response. Note that the impulse response $k_{opt}(t)$ instead of $h_{opt}(t)$ is shown, as from Equation (12.4).

the best reshaping of $H_{opt}(\omega)$ that minimises the noncausality and at the same time maximizes the approximation of the optimality condition.

A very important remark should also be mentioned. From Figs. 12.4 and 12.6(a), it may be noted that, over its attenuation band, the function $H_{opt}(\omega)$ is quite flat. This observation may suggest that a constant approximation of $H_{opt}(\omega)$ could be close to optimal in most sea states. The possibility of approximating $H_{opt}(\omega)$ with a constant value is investigated in more detail in [122].

12.3 Sample Prediction Requirements Results

Consider a wave energy conversion system consisting of a bottom-referenced floating body constrained to move in the heave direction only. Both cylinders and spheres are considered, with a range of values for the radius and the height. The detail of the geometries is specified in Table 12.1, where R is the radius, h is the draft at rest, M the submerged mass at rest, ω_0 the resonance frequency, $\Delta\omega$ the bandwidth of the radiation resistance $B(\omega)$, and τ_1 the main time constant of the control $h_{opt}(t)$. As the radius of the sphere increases, the bandwidth and the resonance frequency get smaller, while the dominant time constant of the noncausal control, τ_1, increases. Regarding the cylinders, the time constant τ_1 is mainly affected by the radius, while it changes only slightly for different heights.

Table 12.1 Geometry of Heaving Buoys.

	R (m)	h (m)	M (kg$\times 10^5$)	ω_0 (rad/s)	$\Delta\omega$ (rad/s)	τ_1 (s)
S_1	4	4	1.38	1.63	1.60	10.87
S_2	5	5	2.70	1.43	1.45	14.16
S_3	6	6	4.66	1.32	1.37	18.62
C_1	3	4	0.93	1.40	1.03	10.00
C_2	3	8	1.86	1.08	0.79	9.97
C_3	3	12	2.80	0.92	0.66	9.94
C_4	3	16	3.73	0.82	0.59	9.91
C_5	3	20	4.66	0.74	0.53	9.89
C_6	5	4	2.59	1.23	0.98	19.80
C_7	5	8	5.18	1.02	0.75	19.80
C_8	5	12	7.77	0.89	0.63	19.76
C_9	5	16	10.40	0.79	0.56	19.7
C_{10}	5	20	12.90	0.72	0.51	19.76
C_{11}	7	4	5.07	1.14	0.95	31.06
C_{12}	7	8	10.10	0.98	7.2	30.96
C_{13}	7	12	15.20	0.85	0.62	30.96
C_{14}	7	16	25.40	0.70	0.54	30.96
C_{15}	7	20	31.71	0.64	0.50	30.9

C: cylinder; S: sphere.

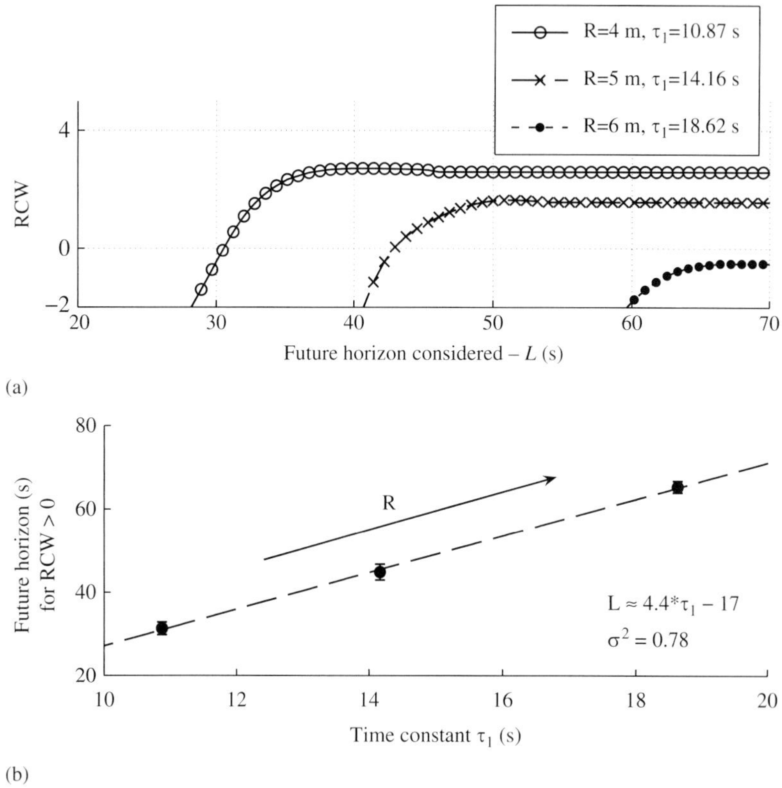

Figure 12.7 Performance of reactive control, applied to heaving spheres, when only limited future information of the excitation force, up to L seconds in the future, is considered. (a) Detail for a specific sea state. (b) Required horizon averaged over all sea states.

Based on the methodology proposed in Section 12.2.2, reactive control is implemented with the (truncated) noncausal law in Equation (12.5), where only future knowledge of the excitation force up to L seconds into the future is included. The resulting performance, RCW(L), is calculated for a range of values of the lead time L. Fig. 12.7(a) shows the behavior of RCW(L) for the three spheres of Table 12.1, in a particular sea state. As expected, the performance improves if more future information about the excitation force is included in the control. At some point, however, the optimum is approached and no further benefit is gained from extra knowledge of the future. This critical future horizon depends on the geometry and increases with τ_l, as expected. Note that the optimal value of RCW changes with the size of the device, in accordance with the general theory of axisymmetric heaving buoys [204].

In Fig. 12.7(b), the average lead time L, for which RCW(L) > 0, over all the considered sea states (about 200), is plotted against the time constant τ_l corresponding to the case of the three spheres. Interestingly, the relation between the future horizon required and τ_l is quite linear (the variance of the residuals of a linear fit ~ 0.78). In addition, the standard deviation related to each mean value (dot) in Fig. 12.7(b) is very

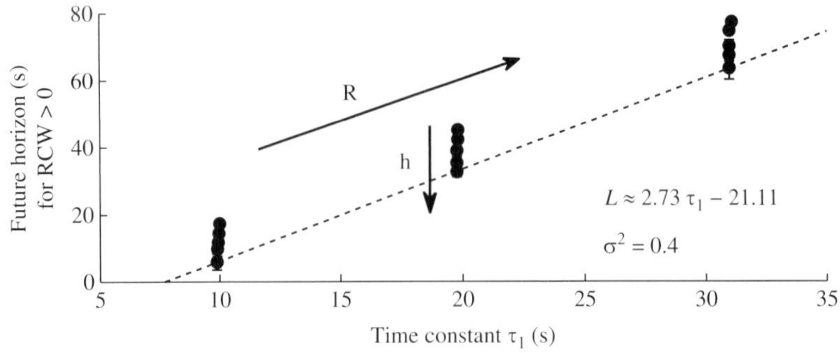

Figure 12.8 Future prediction horizon, L, required for effective implementation of reactive control, applied to heaving cylinders.

small (bar centered around each dot), which confirms the independence of the prediction requirements from the specific wave conditions.

In the case of the cylinders, there are two parameters to be considered: the height and the radius. The time constant τ_1 is mainly influenced by the radius, so that the prediction requirements are not expected to be significantly affected by variations in height. Fig. 12.8 shows the required prediction horizon L, such that $\text{RCW}(L) > 0$, averaged over all the considered sea states, against the time constant τ_1, for each cylinder. As for the spheres, the forecasting horizon required increases with τ_1 (and with the radius). In addition, for a given height, the relation is linear (variance of the residuals of a linear fit ~ 0.04). To a lesser extent, changes in height also affect the prediction requirements, and the trend is consistent: the cylinders are less demanding in terms of prediction when their height increases. The parameter τ_1 remains, though, the most critical in relation to the prediction requirements.

The possibility of manipulating the control such that the noncausality is reduced was proposed in Section 12.2.3. Intuitive modifications to the reference generator $H_{opt}(\omega)$ are applied, as shown in the example of Fig. 12.6(a). As a result of implementing reactive control based on the modified $H_{red}(\omega)$, Fig. 12.9(a) shows the behavior of $\text{RCW}(L)$ for two different systems, cylinders C_1 and C_{10} of Table 12.1. A stark reduction of the prediction horizon required for an effective implementation of reactive control is achieved. In addition, the performance drop is minor in most cases. The deviation from the optimal performance, in particular, will be affected by the bandwidth of the system (flatness of $H_{opt}(\omega)$) and by the excitation force spectral distribution, as noted in Section 12.2.3. Fig. 12.9(b) shows the distribution, as a function of the bandwidth of the system, of the maximum RCW achievable, when $H_{red}(\omega)$ is adopted, as percentage of the RCW obtained from ideal reactive control. The average (dot) performance is usually well above 80% and increases with the bandwidth of the system. The standard deviation (bar centered around dot) is quite large only for very narrow-banded systems. Note that $H_{red}(\omega)$ was only qualitatively derived from $H_{opt}(\omega)$, and lower performance losses could be achieved through an optimization that takes into account the spectral distribution of the sea state, as also mentioned in Section 12.2.3.

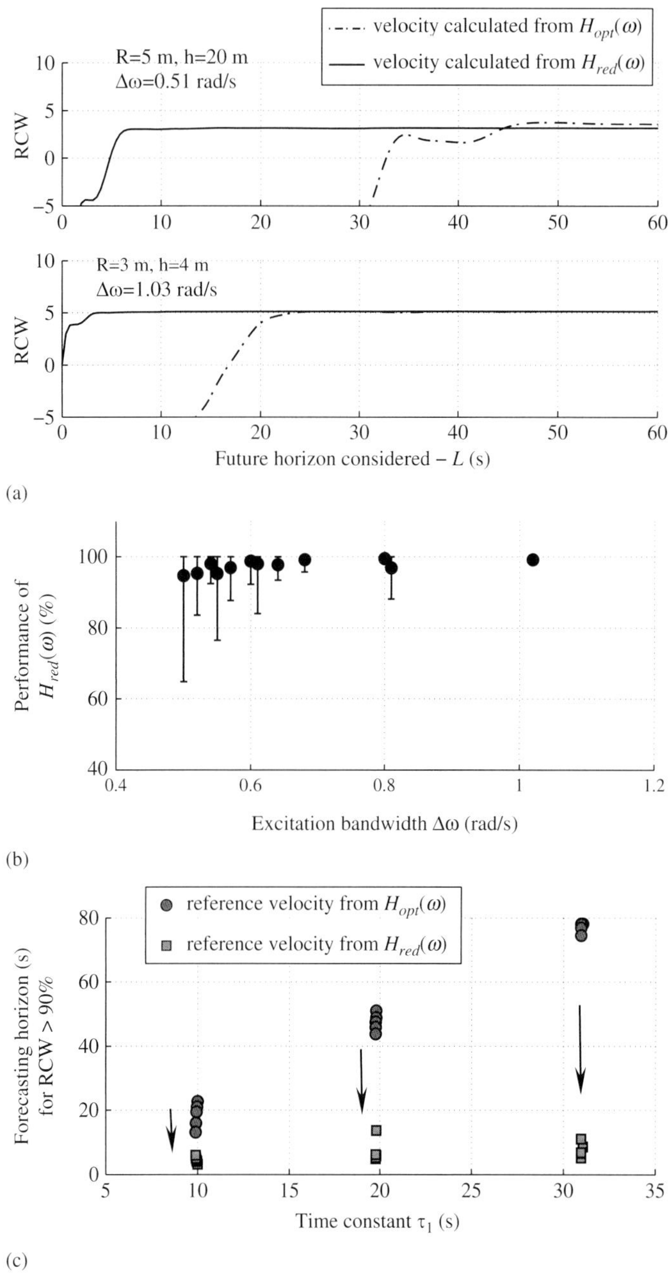

Figure 12.9 Reduction of prediction requirements through manipulation of the noncausal control law. (a) Detail for a specific sea state. (b) Max RCW, when using $H_{red}(\omega)$, as percentage of ideal RCW, obtained with $H_{opt}(\omega)$. (c) Average over all sea states.

By applying the procedure to all the cylinders, and calculating the prediction requirements in all the sea states, the results of Fig. 12.9(c) are obtained. The prediction requirements are significantly reduced to a horizon shorter than 10 seconds, which is well within the ability of stochastic wave forecasting models [115]. Also, it is interesting to note that the differences between the cylinders are minimal, so that no geometry seems to be particularly advantageous in terms of required forecasting horizon.

12.4 Forecasting Models

12.4.1 Cyclical Models

The most straightforward forecasting model is a cyclical model, where the wave elevation, $\eta(k)$, is expressed as a superposition of a number m of linear harmonic components:

$$\eta(k) = \sum_{i=1}^{m} \alpha_i \cos(\omega_i k) + \beta_i \sin(\omega_i k) + \zeta(k). \tag{12.8}$$

An error term, $\zeta(k)$, is also introduced, while the phase and amplitude information for each harmonic component is contained in the parameters α_i and β_i [205, 206].

The model in Equation (12.8) is completely characterized by the parameters α_i,β_i and by the frequencies ω_i. It can fitted to the data through a nonlinear estimation procedure (the model is nonlinear in the frequencies) and utilized to predict the future wave elevation. It needs, however, to be adapted to the time variations of the wave spectrum (amplitudes, phases, and frequencies are nonconstant), so that a first solution has been considered [206], where the frequencies are kept constant and the model becomes perfectly linear in the parameters α_i,β_i.

The choice of the frequencies is a crucial one. In [206], it was pointed out how, while the range of frequencies is easy to determine, the distribution of the ω_i inside this range is more problematic. A robust choice would be a constant spacing over the entire range, with the spacing as small as possible in order to give an accurate coverage of the spectrum (appropriate constant frequency intervals are also discussed in [207]). Alternative nonhomogeneous solutions may be considered, for example, with the frequencies more densely distributed around the energy peaks. Such more efficient solutions, however, would be unreliable in the context of a model with fixed frequencies, due to the time variations of the energy distribution in wave spectra (which can be verified through the wavelet transform [207, 208]).

Once the frequencies are determined, a model for the amplitude variations has to be chosen. Initially, the cyclical structural model proposed by Harvey [209] is adopted:

$$\eta(k) = \sum_{i=1}^{m} \psi_i(k) + \zeta(k) \tag{12.9}$$

and

$$\begin{bmatrix} \psi_i(k+1) \\ \psi_i^*(k+1) \end{bmatrix} = \begin{bmatrix} \cos(\omega_i T_s) & \sin(\omega_i T_s) \\ -\sin(\omega_i T_s) & \cos(\omega_i T_s) \end{bmatrix} \begin{bmatrix} \psi_i(k) \\ \psi_i^*(k) \end{bmatrix} + \begin{bmatrix} w_i(k) \\ w_i^*(k) \end{bmatrix}, \quad i = 1, \ldots, m \tag{12.10}$$

where T_s is the sampling period; vectors $[\psi_i(k) \quad \psi_i^*(k)]^T$ model the m cyclical components; and it can be verified that $\psi_i(0) = \alpha_i$ and $\psi_i^*(0) = \beta_i$. The Gaussian white disturbance $[w_i(k) \quad w_i^*(k)]^T$ models the amplitude and phase variation. In [205] and [206], an alternative stochastic model for the amplitudes $\alpha_i(k)$ and $\beta_i(k)$ is proposed: the dynamic harmonic regression (DHR) model [210]. However, the higher complexity required (four state variables for each harmonic against two of Harvey's model) is not justified by an improvement of the modeling ability, so the DHR model is not further considered.

From Equations (12.9) and (12.10), the following state space form is derived:

$$\begin{aligned} \mathbf{x}(k+1) &= \mathbf{A}\mathbf{x}(k) + \mathbf{w}(k), \\ \eta(k) &= \mathbf{C}\mathbf{x}(k) + \zeta(k), \end{aligned} \tag{12.11}$$

where

$$\mathbf{x}(k) \triangleq \left[\psi_1(k)\ \psi_1^*(k)\ \ldots\ \psi_m(k)\ \psi_m^*(k)\right]^T \in \mathbb{R}^{2m\times 1}, \tag{12.12}$$

$$\mathbf{w}(k) \triangleq \left[w_1(k)\ w_1^*(k)\ ...\ w_m(k)\ w_m^*(k)\right]^T \in \mathbb{R}^{2m\times 1}, \tag{12.13}$$

$$\mathbf{A} \triangleq diag\left\{\begin{bmatrix} \cos(\omega_i T_s) & \sin(\omega_i T_s) \\ -\sin(\omega_i T_s) & \cos(\omega_i T_s) \end{bmatrix}\right\} \in \mathbb{R}^{2m\times 2m}, \tag{12.14}$$

$$\mathbf{C} \triangleq [1 \quad 0 \quad 1 \quad 0 \quad \ldots \quad 1 \quad 0] \in \mathbb{R}^{1\times 2m}. \tag{12.15}$$

The state space form is suited to the application of the Kalman filter for recursive on-line adaptation. The initialization is provided by regular least-squares estimates on a number of batch observations, and then the Kalman filter is applied on-line, once a covariance matrix for the state and output disturbances is available. When the estimate of the model's parameters, $\hat{\mathbf{x}}(k|k)$, is available at any instant k, the l-step-ahead prediction $\hat{\eta}(k+l|k)$ is obtained through the free evolution of the model in Equations (12.11) as

$$\hat{\eta}(k+l|k) = \mathbf{C}\mathbf{A}^l\hat{\mathbf{x}}(k|k). \tag{12.16}$$

12.4.2 Sinusoidal Extrapolation with the Extended Kalman Filter

A step further from the cyclical models with fixed frequencies would be to consider *variable frequencies* that are updated on-line, along with the amplitudes and phases, on the basis of free surface wave elevation measurements.

We can, therefore, attempt to model the wave elevation as a single cyclical component of the type in Equation (12.10), but with a time-varying frequency $\omega(k)$:

$$\eta(k) = \psi(k) + \zeta(k), \tag{12.17}$$

with

$$\begin{bmatrix} \psi(k+1) \\ \psi^*(k+1) \\ \omega(k+1) \end{bmatrix} = \begin{bmatrix} \cos(\omega(k)T_s) & \sin(\omega(k)T_s) & 0 \\ -\sin(\omega(k)T_s) & \cos(\omega(k)T_s) & 0 \\ 0 & 0 & 1 \end{bmatrix} \cdot \begin{bmatrix} \psi(k) \\ \psi^*(k) \\ \omega(k) \end{bmatrix} + \begin{bmatrix} \varepsilon(k) \\ \varepsilon^*(k) \\ \kappa(k) \end{bmatrix}, \tag{12.18}$$

where $\varepsilon(k)$, $\varepsilon^*(k)$, and $\zeta(k)$ are random disturbances; $\eta(k)$ is the wave elevation; and a model for the variability of $\omega(k)$ has been introduced, assuming a simple random walk driven by the additional white noise $\kappa(k)$. The model in Equation (12.18), of course, is *nonlinear* in $\omega(k)$, and an explicit linear state-space structure cannot be formulated.

As a consequence, a linear recursive estimator cannot be directly applied. It is possible, however, to utilize an extension of the Kalman filter to nonlinear models, namely, the extended Kalman filter (EKF), assuming that the discrete time step, T_s, is sufficiently small to permit the prediction equations to be approximated by a linearized form, based on the truncation of the Taylor expansion of Equation (12.18) at the first order [123]. The estimate of the state vector at each time step is, therefore, given by linear recursive equations, while the prediction $\hat{\eta}(k+l|k) = \hat{\psi}(k+l|k)$ is obtained from the free evolution of the nonlinear model in Equation (12.18).

The extension to a model with $m > 1$ variable frequencies is not straightforward. It was found that a superposition of m models of the type expressed in Equation (12.18) does not offer any appreciable advantage [208]. For the case of a single frequency, the model can be considered as a *sinusoidal extrapolation*.

12.4.3 AR Models

The wave elevation $\eta(k)$ is assumed to be linearly dependent on a number, n, of its past values, through the parameters a_i:

$$\eta(k) = \sum_{i=1}^{n} a_i \eta(k-i) + \zeta(k) \tag{12.19}$$

where a disturbance term $\zeta(k)$ is included. If an estimate of the parameters at instant k, $\hat{a}_i(k)$, is computed and the noise is assumed to be Gaussian and white, the best prediction of the future wave elevation $\hat{\eta}(k+l|k)$ at instant k can be derived from Equation (12.19) as

$$\hat{\eta}(k+l|k) = \sum_{i=1}^{n} \hat{a}_i(k)\hat{\eta}(k+l-i|k) \tag{12.20}$$

where, obviously, $\hat{\eta}(k+l-i|k) \equiv \eta(k)$ if $k+l-i \le k$ (i.e., information has been already acquired and there is no need of prediction).

The general shape of the prediction function $\hat{\eta}(k+l|k)$ is completely determined by the poles, p_i, of the AR model in Equation (12.19). In the case of $m/2$ (when m is even) couples of complex-conjugate poles, p_i and p_i^* [211],

$$\hat{\eta}(k+l|k) = \sum_{i=1}^{m/2} c_i(k)|p_i|^l \sin(\angle p_i k + \varphi_i(k)) \tag{12.21}$$

where the coefficients $c_i(k)$ depend on the last n observations $\eta(k)$, $\eta(k-1), \ldots, \eta(k-n+1)$. Thus, an AR model with only complex-conjugate poles is *implicitly* a cyclical model, where the frequencies are related to the phase, $\angle p_i$, of each pole, and the amplitudes and phases of the harmonic components are related to the last n measurements of each time instant k, so that they adapt to the observations.

The AR coefficients, a_i, are estimated from a number N of batch observations through the minimization of a multistep-ahead cost functional, referred to as long-range predictive identification (LPRI) [212]:

$$J_{LPRI} = \sum_{k=1}^{N} \sum_{j=1}^{N_l} \left[\eta(k) - \hat{\eta}(k|k-j)\right]^2 \tag{12.22}$$

where N_l is the forecasting horizon over which the AR model is to be optimized. The function J_{LPRI} is minimized with a standard algorithm for nonlinear least-squares problems, the Gauss-Newton algorithm, initialized with the estimates from regular least squares [212].

Note that, from Equation (12.21), the implicit frequencies are related to the poles of the AR model, so if the AR parameters are kept constant, the frequencies will be constant as well. An adaptivity mechanism based on the LPRI function (12.22) could be implemented, as proposed in [212]. However, as will also be shown in the results of Section 13.5, a static AR model maintains its accuracy for a long time after being estimated (more than 2 hours), in spite of spectral variations (for more details, refer to [208]).

12.4.4 Neural Networks

In spite of the nonlinear modeling capability, neural networks (NN) have the great disadvantage of offering a model completely enclosed in a black box, where inherent characteristics cannot be analyzed by inspection or analytical calculation. So, whereas in the cyclical and AR models, an analysis of the estimated parameters and frequencies and their variations in an adaptive structure can provide indications about the real process behavior and its main characteristics, this would not be possible with neural networks.

For the problem under study, a nonlinear relationship is created through a multilayer perceptron [213]:

$$\eta(k) = \mathrm{NN}(\eta(k-1), \eta(k-2), \ldots, \eta(k-n)). \tag{12.23}$$

Only structures with one linear output neuron and two hidden layers, consisting of a number of nonlinear neurons, varying between three and seven each, are considered. Several orders of regression n were also considered. The model is trained using the Levenberg–Marquard algorithm [214] on a set of batch data and utilized for multistep-ahead prediction.

The structure in Equation (12.23) is, of course, not the only possibility, and many others could be considered. For example, a priori knowledge about the process (harmonic nature) may be considered as follows:

$$\eta(k) = \mathrm{NN}(\cos(\omega_1 T_s k + \varphi_1), \ldots, \cos(\omega_n T_s k + \varphi_n)), \tag{12.24}$$

though such an input structure would retain some of the limitations of cyclical models with fixed frequencies.

12.4.5 Other Possibilities

Other possible wave forecasting models were analyzed but were found unsuitable and therefore discarded (refer to [208] for a detailed discussion).

A candidate alternative technique is Gaussian processes (GPs), whose basic idea is to place a Gaussian prior directly on the space of functions [215] underlying the data, without assuming any particular function parametrization but only specifying the prior's mean and covariance. In order to model the cyclical characteristics of the sea, the harmonic frequencies of the covariance function have to be assigned permanently in the initial estimation of the model. This leads to the issue, concerning the choice of frequencies, that emerged from the discussion of cyclical models with fixed frequencies; see Section 12.4.1.

Also, the introduction of a moving average (MA) term in the AR model, giving an ARMA model, has been analyzed, and it was found that no real change in the forecasting function's shape, with respect to AR models, can be obtained, so that no significant improvement in the results can be expected. ARMA models can, however, be utilized in order to obtain more parsimonious forecasting models, but this is not considered here.

Particle filters [216] may be considered for the on-line state estimation of the nonlinear model (12.18) and can deal directly with nonlinear models. However, while they do not assume a Gaussian distribution of the state variable, and therefore may give better estimation results than the EKF, they still allow only a single frequency to be tracked, and would therefore be subject to same main performance limitation as the EKF.

12.5 Confidence Intervals

The wave elevation predictions alone, as computed by any of the models presented, do not give sufficiently complete information about the future of the signal, as they are inevitably affected by an error. It is fundamental to have an indication about the extent of this error and about the confidence in the forecasts computed by the prediction algorithm. If the l-step-ahead prediction error is Gaussian,

$$\hat{e}(k+l|k) = \eta(k+l) - \hat{\eta}(k+l|k) \approx \aleph\left(0, \sigma_l^2\right), \tag{12.25}$$

then the variance σ_l^2 is all we need in order to define its probability distribution. We can assume that the error is contained within a *confidence interval*, with probability δ, as follows:

$$-z_{\frac{\delta}{2}} \leq \hat{e}(k+l|k) \leq +z_{\frac{\delta}{2}}. \tag{12.26}$$

In (12.26) $z_{\frac{\delta}{2}}$ is the value of the probability distribution such that

$$P\{-z_{\frac{\delta}{2}} \leq \hat{e}(k+l|k) \leq +z_{\frac{\delta}{2}}\} \triangleq \int_{-z_{\frac{\delta}{2}}}^{+z_{\frac{\delta}{2}}} p(y)dy = \delta \tag{12.27}$$

where the probability density function $p(\cdot)$, with the distribution of the forecasting error considered to be zero-mean Gaussian, assumes the following structure:

$$p(\hat{e}(k+l|k)) = \frac{1}{\sqrt{2\pi}\sigma_l} e^{-\frac{\hat{e}(k+l|k)^2}{2\sigma_l^2}}. \tag{12.28}$$

The estimate of the variance σ_l^2 could be calculated from the specific model parameters and from the statistics of the parameter estimation algorithm, which is not straightforward and also could be misleading if the model is not sufficiently accurate. A more straightforward alternative, however, is adopted, where the estimate of the variance of the forecasting error is based purely on the past history of the prediction errors,

$$\hat{\sigma}_l^2 = \frac{1}{N-1} \sum_{k=1}^{N} \hat{e}(k+l|k)^2, \tag{12.29}$$

where N is the number of past observations available.

The estimate of σ_l^2 can also be recursively updated as soon as new observations become available [217], via

$$\hat{\sigma}_l^2(k) = \frac{k-2}{k-1}\hat{\sigma}_l(k-1) + \frac{1}{k}\left[\eta(k) - \hat{\eta}(k|k-l)\right]^2, \; k \geq 2. \tag{12.30}$$

12.6 Case Studies

The data utilized for the case studies come from two locations.

1. The Irish Marine Institute provided real observations from a data buoy located in Galway Bay, on the west coast of Ireland, at approximately 53° 13′ N, 9° 18′ W (water depth nearly 20 m). Data consist of 20-min record sets for each hour, collected at a sampling frequency of 16.08 rad/s (2.56 Hz). The location is sheltered from the Atlantic Ocean so that the wave height magnitude, in general, is relatively small.
2. Wave elevation time series are also available from the Atlantic Ocean at Pico Island, in the Azores Archipelago, at approximately 38° 33′ N, 28° 34′ W (water depth around 40 m). The Pico data are collected in the form of two contiguous 30-min record sets for each hour, with a sampling frequency of 8.04 rad/s (1.28 Hz) [218–220].

The wave climate at the Galway Bay site is dominated by relatively low energy sea states (significant wave height less than 2 m) that, most of the time, have a broad spectral distribution with no clear energy peak, due to the superposition of low-frequency swell(s) and high-frequency wind waves of similar energy content. Wave systems off the coast of Pico, on the other hand, usually have a more defined low-frequency peak (around 0.7 rad/s), and their significant wave height ranges, usually, from 1 to 5 m. A more detailed analysis of the wave data at the two locations can be found in [208].

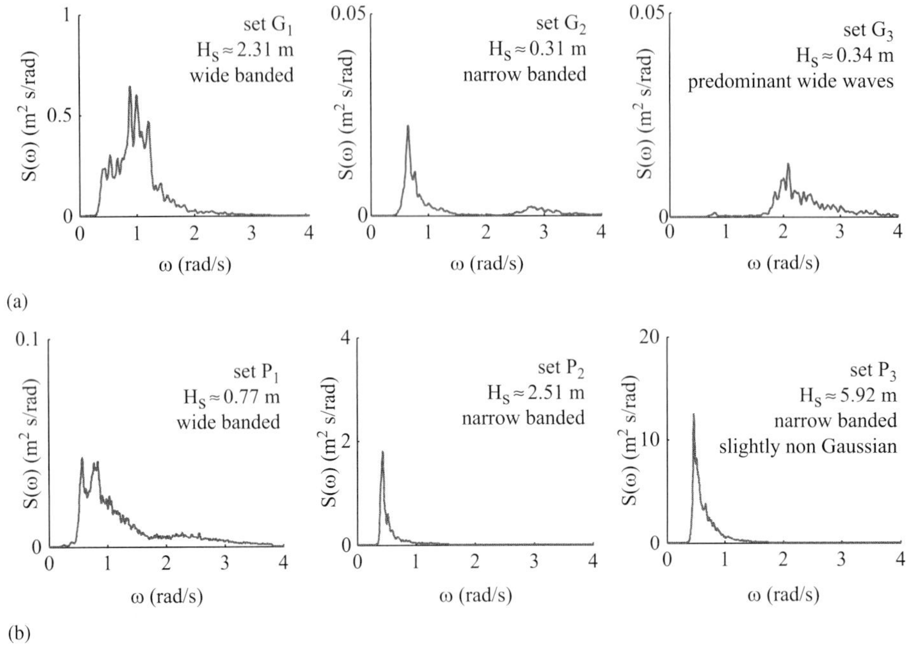

Figure 12.10 Wave spectra of sample data sets for the two locations: (a) Galway Bay and (b) Pico Island.

The wave spectra of three significant data sets at the two locations, shown in Fig. 12.10, will be utilized to test the forecasting models of Section 12.4. In particular, one wide-banded and one narrow-banded sea state from each of the two sites is considered. Then, a situation where wind waves predominate is selected from the Galway Bay data, and a very high energy wave system, where the sea bottom slightly affects the wave symmetry (this was analyzed through higher order spectral analysis and skewness and kurtosis indices [206]), is chosen from the Pico Island data.

One main characteristic that emerges from the wave data analysis, which is well known, is that a significant portion of the wave energy is usually concentrated at low frequencies. In contrast, high frequencies dominate the spectrum only for very low energy sea states, which are of diminished interest for wave energy conversion. It is also known that low-frequency waves (the swell) are more regular and less affected by nonlinearities (which can be verified by means of the bispectrum [206, 208, 221]). It is reasonable, therefore, to low-pass filter the wave elevation and focus the prediction only on the low frequencies, which, intuitively, might improve the accuracy of the forecasts and the length of the forecasting horizon.

A quantification of the possible benefits of low-pass filtering on the prediction can be obtained by measuring the *predictability* of the wave elevation time series. It is argued that, for predictability analysis, it is not necessary to design any predictors; we just have to know how much information about future signal values can be obtained from the past [222]. A simpler measure of predictability than the very general approach proposed in

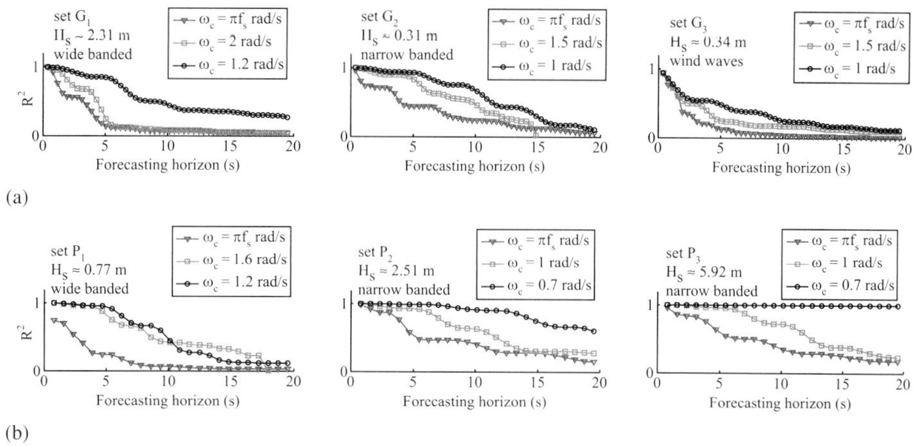

Figure 12.11 Predictability of wave elevation time series, low-pass filtered with different cut-off frequencies ω_c: (a) Galway Bay and (b) Pico Island.

[222, 223] (based on the mutual information notion) will be adopted here, as previously proposed in [206]. It assumes that a linear relationship exists between the future values of the wave elevation and its past values. A predictability index, $R^2(l)$, is estimated as

$$R^2(l) \triangleq \frac{E\{\hat{\eta}(k+l|k)^2\}}{E\{\eta(k)^2\}} \tag{12.31}$$

where $\eta(k)$ is the wave elevation, supposed to have a zero mean; $E\{\cdot\}$ is the expectation operator; and $\hat{\eta}(k+l|k)$ is the optimal l-step-ahead prediction. A very efficient algorithm for the estimation of $R(k)^2$, which is utilized in this study, was proposed in [224].

In Fig. 12.11 it is shown that the improvement of predictability of the wave data detailed in Fig. 12.10, when only low-frequency components are considered, is quite significant, particularly in the case of narrow-banded sea states. The suggestion that it is reasonable to focus the wave forecasting algorithms on only the low-frequency waves is, therefore, numerically confirmed. Note that the choice of the cut-off frequencies ω_c is based on visual inspection of the spectral shape of each data set, such that the amount of energy discarded is not significant (apart from the data set G_3, where the wind waves dominate the spectrum). The cut-off frequency will ultimately represent a significant parameter at the design stage of any specific wave prediction algorithm, and it will be determined as a compromise between improvement in the prediction accuracy and amount of discarded wave components/energy.

12.7 Sample Autoregressive Forecasting Results

Each of the data sets of Fig. 12.10 was split up into training and validation sets. For the Galway Bay data, the training and validation sets consist of two consecutive data sets of 3072 samples (20 min each at a sampling frequency of 16.08 rad/s, or 2.56 Hz). In the case of the Pico Island data, because the consecutive data sets are contiguous in time, training and validation sets have been chosen as four consecutive segments each (9216 samples per set, equivalent to 2 hours at a sampling frequency of 8.04 rad/s, or 1.28 Hz).

The prediction accuracy is measured with the following goodness-of-fit index, for each forecasting horizon l:

$$\mathcal{F}(l) = \left(1 - \frac{\sqrt{\sum_k \left[\eta(k+l) - \hat{\eta}(k+l|k)\right]^2}}{\sqrt{\sum_k \eta(k)^2}}\right) \cdot 100. \tag{12.32}$$

Here $\eta(k+l)$ is the wave elevation and $\hat{\eta}(k+l|k)$ is its prediction based on the information up to instant k. A 100% value for $\mathcal{F}(l)$ means that the wave elevation time series is perfectly predicted l steps into the future. Note that the quantity $\mathcal{F}(l)$ has a direct correspondence with the variance of the prediction error, which, as discussed in Section 12.5, is utilized to characterize the confidence interval of the forecasts. In particular, if $\hat{\sigma}_l^2$ is an estimate of the variance of the l-step, ahead prediction error, then.

$$\hat{\sigma}_l^2 = \left(1 - \frac{\mathcal{F}(l)}{100}\right)^2 \sum_k \eta(k)^2. \tag{12.33}$$

The focus is on the prediction of the low-frequency components, which contain the most energy, as discussed in Section 12.6. Ideal zero-phase low-pass filters are approximated with off-line forward and backward filtering through type I Chebyshev filters (order 15, maximum error in the pass band 10^{-3}) discretized with the bilinear transform. The cut-off frequency, ω_c, has been chosen according to the specific shape of the wave spectrum (Fig. 12.10). Fig. 12.12 shows the prediction accuracy of different configurations of the models outlined in Section 12.4, for two wave systems, over a forecasting horizon of about 20 s (50 samples for the Galway data and 25 samples for the Pico data).

The first characteristic that emerges from Fig. 12.12 is that all the selected forecasting models perform better for the narrow-banded sea state P_2, as shown in Fig. 12.12(b), than for the broader-banded G_1, shown in Fig. 12.12(a), and this was expected after the predictability analysis carried out on the wave data, shown in Fig. 12.11. In particular, the Harvey cyclical model and the sinusoidal extrapolation through the EKF are the most dramatically affected by the bandwidth of the wave signal, and their $\mathcal{F}(l)$ for the data set G_1 dies out relatively quickly, going below 0% after only 5 s. Note also that a cyclical model with a single variable frequency offers an accuracy comparable to that of the cyclical models with many fixed frequencies, and at the same time represents a much more efficient approach (three system states compared with hundreds for the Harvey model).

AR models offer much more accurate predictions further into the future than the cyclical models for all the sea states. In the case of P_2 (Fig. 12.12(b)) an accuracy of $\mathcal{F}(l) > 90\%$ is maintained for more than 20 s into the future, about two wave periods of a wave with period 10.47 s (corresponding to a peak frequency of 0.6 rad/s, as from Fig. 12.10). AR model performance is still affected by the bandwidth of the sea state,

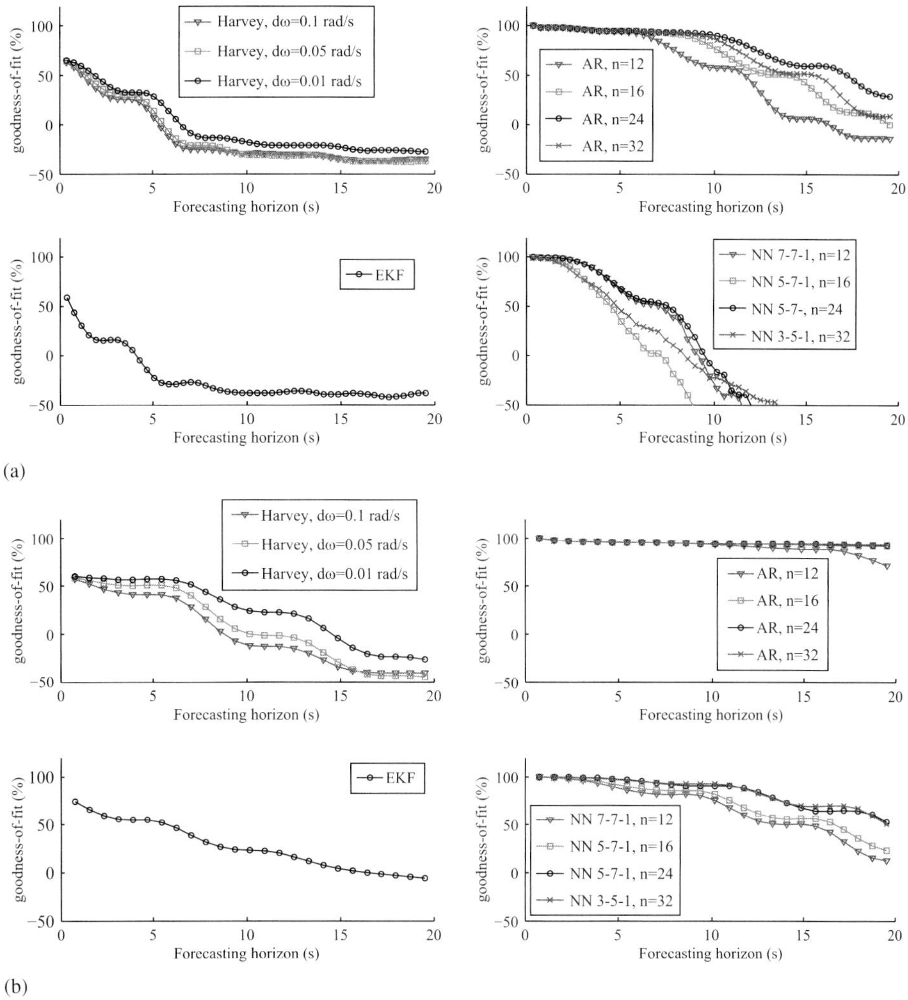

Figure 12.12 Goodness-of-fit $\mathcal{F}(l)$ for different forecasting models on two data sets: (a) Galway Bay set G_1 and (b) Pico set P_2.

but relatively accurate predictions, $\mathcal{F}(l) > 90\%$ for more than 10 s in the future, are still obtained for the set G_1, although higher order models, for example, $n = 24, 32$, are required.

For the neural networks, the results of some of the more promising configurations show an accuracy similar to that of AR models for the case of the narrow-banded sea state P_2, while the prediction deteriorates fairly quickly in the case of the data set G_1. The search of possible network configurations was not exhaustive, but was broad enough to conclude that neural networks are not able to offer any significant improvement in the short-term prediction of the wave elevation, with respect to simple AR models, to justify the higher complexity and the lack of any physical meaning. Note that high-energy sea states with a well-defined low-frequency swell like P_2 present relatively weak nonlinearities (from analysis of the bispectrum and of skewness and

Table 12.2 Goodness-of-Fit $\mathcal{F}(l)$ for Different Forecasting Models on the Galway Bay Data Sets of Fig. 12.10(a)

		$G_1, \omega_c = 1.2$ rad/s			$G_2, \omega_c = 1$ rad/s			$G_3, \omega_c = 1$ rad/s		
	Model	$l = 10$	$l = 25$	$l = 50$	$l = 10$	$l = 25$	$l = 50$	$l = 10$	$l = 25$	$l = 50$
Harvey	$d\omega = 0.1$	25.6%	–	–	30.6%	–	–	–	–	–
	$d\omega = 0.05$	30.7%	–	–	38.3%	–	–	–	–	–
	$d\omega = 0.01$	32.4%	–	–	47.6%	–	–	–	–	–
EKF	–	5.4%	–	–	46.0%	–	–	25.3%	–	–
AR	$n = 12$	94.9%	58.2%	–	98.1%	87.1%	37.2%	72.2%	57.6%	12.2%
	$n = 24$	94.9%	88.3%	9.1%	98.1%	94.5%	66.3%	72.2%	68.8%	28.4%
	$n = 32$	94.9%	91.4%	29.3%	98.1%	94.6%	69.9%	72.2%	68.8%	41.8%
NN	7-7-1, $n = 12$	84.3%	–	–	89.4%	48.9%	–	86.1%	24.1%	–
	5-7-1, $n = 24$	84.4%	–	–	91.1%	59.6%	–	89.3%	39.8%	–
	3-5-1, $n = 32$	68.0%	–	–	85.3%	30.1%	–	93.8%	20.1%	–

Table 12.3 Goodness-of-Fit $\mathcal{F}(l)$ for Different Forecasting Models on the Pico Island Data Sets of Fig. 12.10(b)

		$P_1, \omega_c = 1$ rad/s			$P_2, \omega_c = 0.7$ rad/s			$P_3, \omega_c = 0.7$ rad/s		
	Model	$l = 5$	$l = 12$	$l = 25$	$l = 5$	$l = 12$	$l = 25$	$l = 5$	$l = 12$	$l = 25$
Harvey	$d\omega = 0.1$	15.0%	–	–	41.2%	–	–	38.0%	–	–
	$d\omega = 0.05$	20.4%	–	–	50.5%	5.6%	–	43.5%	0.3%	–
	$d\omega = 0.01$	28.8%	–	–	56.7%	28.5%	–	53.1%	27.5%	–
EKF	–	8.3%	–	–	55.2%	24.2%	–	56.3%	20.8%	–
AR	$n = 12$	96.9%	69.6%	–	95.9%	94.1%	71.6%	98.7%	96.4%	81.1%
	$n = 24$	99.0%	95.7%	42.9%	95.9%	94.5%	92.6%	98.7%	96.7%	93.4%
	$n = 32$	99.0%	96.5%	52.3%	95.9%	94.5%	93.1%	98.7%	96.7%	93.7%
NN	7-7-1, $n = 12$	87.4%	20.9%	–	93.1%	80.8%	12.4%	97.1%	86.0%	39.9%
	5-7-1, $n = 24$	73.8%	–	–	98.9%	90.4%	52.8%	99.7%	95.5%	71.8%
	3-5-1, $n = 32$	74.7%	–	–	98.7%	92.5%	50.6%	97.3%	91.5%	48.8%

kurtosis indices [208]), unless shallow water conditions apply (which is not the case here).

More detailed results over the six sea states considered in this study are shown in Table 12.2, regarding the Galway Bay wave data, and in Table 12.3, for the data from Pico. Negative values of $\mathcal{F}(l)$ were omitted, as they imply a prediction error comparable with the forecast value itself. The results regarding the sea states G_2 (narrow banded) and P_1 (broad banded) do not add any significant information to the discussion outlined so far, which is focused on the sea states G_1 and P_2 and on Fig. 12.12. A further interesting point, however, concerns the situation for the sea state dominated by wind waves, G_3. In this case, the neural networks retain a good accuracy in the short term, with $\mathcal{F}(l)$ around 90% for $l = 10$, outperforming the AR models, which are significantly

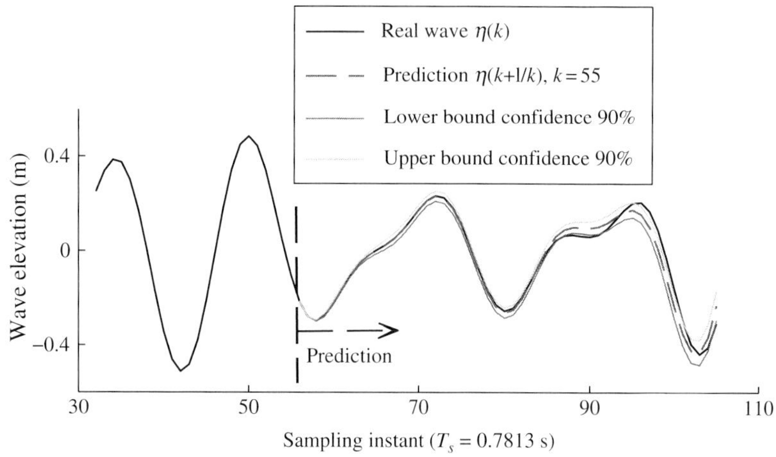

Figure 12.13 Confidence interval and predictions $\hat{\eta}(k+l|k)$, for $l=1$ to $l=50$, at a specific time instant k, calculated with an AR model of order $n=24$ on the data set P_2, filtered with cut-off frequency $\omega_C = 0.7$ rad/s.

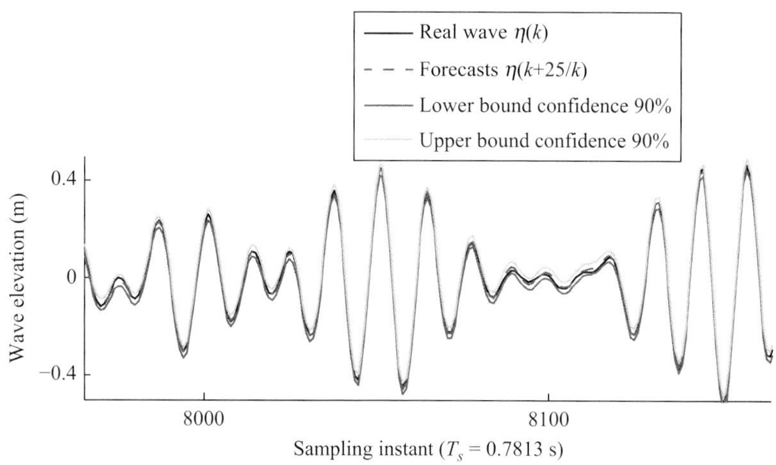

Figure 12.14 Confidence interval and receding horizon 25-step-ahead predictions $\hat{\eta}(k+25|k)$, for some k, calculated with an AR model of order $n=24$ on the data set P_2, filtered with cut-off frequency $\omega_C = 0.7$ rad/s.

affected by the nonlinearities in the wave system (detectable through the bispectrum; refer to [208] for a detailed analysis).

The detail of the prediction $\hat{\eta}(k+l|k)$ with an AR model on the data set P_2, at a specific instant $k=55$ and for a forecasting horizon $l=1$ to 50 samples (nearly 39 s), is shown in Fig. 12.13, along with the 90% confidence interval, estimated according to the methodology described in Section 12.5. In Fig. 12.14 the comparison between the real wave elevation and the receding horizon 25-step-ahead prediction $\hat{\eta}(k+25|k)$ is shown, computed with an AR model of order $n=24$, along with the corresponding 90% confidence interval.

12.8 Use of Up-Wave Measurements in Wave Forecasting

As suggested in Chapter 7 and in Fig. 12.2, up-wave measurements can also be utilized to estimate the future wave profile incident on a WEC. In this section, a case study is used to illustrate an approach and to draw some conclusions on the potential benefit of up-wave measurements, made using time series models, on wave forecasting systems. However, it should be borne in mind that this case study is for a particular site and specific sensor types, and therefore general conclusions are difficult to draw.

Regarding the spatial approach for wave prediction, a number of previous studies have documented various solutions. In [225], digital filters were deployed for the real-time prediction of waves incident upon a wave energy device, while in [102], the remote wave profile is measured by means of a nautical radar and is propagated in time and space in order to predict the wave field and estimate the motion of a vessel. In [226], a shallow-angle LIDAR was adopted in order to measure the time evolution of spatial profiles of sea waves over an extended region of several hundred meters. In [227], the fast Fourier transform is employed to decompose the wave elevation into individual frequencies at a given point, propagate each individual frequency component using the dispersion relation, and reconstruct the wave elevation at a different temporal and spatial point. In [228], an overview of a deterministic sea-wave prediction model is presented, which is used to estimate the wave elevation at the point of interest, given the measurement of the free surface elevation at a point at some distance from the prediction site. Finally, in [229], a linear harmonic model fitted to distant wave elevation measurements is presented, while in [230] the hydrodynamics of the chamber of an oscillating water column (OWC) is predicted by means of a neural network based on the measurements of the incident wave elevation.

Here, the comparative benefits to wave prediction accuracy using a sensor for the measurement of the free surface elevation up-wave of the WEC, compared with using historical measurements alone, are examined. In particular, an autoregressive with exogenous input (ARX) model, which extends the AR model with the inclusion of the additional information coming from the measurement of the free surface elevation up-wave of the WEC, is employed. The ARX models proposed are validated against real observations coming from the Pico OWC WEC plant located on Pico Island [230]. Two different types of sensors for the up-wave elevation measurement are tested. The objective is to estimate future variations in the displacement of the water column inside the OWC. A more complete exposure of this case study is given in [231].

12.8.1 Case Study Characteristics

A sonar sensor for recording the chamber wave elevation is mounted on the roof of the OWC chamber, while two different sensors for the measurement of the free surface elevation around 60 m in front of the Pico OWC chamber wall are deployed [230]. During the first deployment period, the “Aquadopp” hydrostatic pressure sensor and acoustic doppler current meter unit was utilized to measure the hydrodynamics of the free surface elevation up-wave of the WEC. Due to ongoing technical difficulties with the sensor cable, an alternative sensor was developed and used at the same location as the Aquadopp during the second deployment period. This alternative sensor, in the following referred to as the "pneumatic sensor," consists of a small steel box fixed on the sea floor with an open bottom. The variation of the pressure of the entrapped air inside the box, due to the change of the water column above, is measured by a pressure transducer. From the pressure variation of the air inside the box, the water surface elevation can be obtained [230] as

$$\eta_w = \frac{p_{hyd}}{\rho g} \tag{12.34}$$

where p_{hyd} represents the dynamic water pressure at 60 m in front of the OWC, ρ is the density of the sea water, and g is the gravitational acceleration. The data available from the Aquadopp sensor consist of 25 hours of records, while the data provided by the pneumatic sensor consist of 10 hours of records. The sampling frequency for both of the deployment periods is 2 Hz.

The spectra of the chamber wave elevation variations measured during the first and second deployment periods are shown in Fig. 12.15(a) and 12.15(b), respectively, while the corresponding spectra of the up-wave elevation measured with the Aquadopp and pneumatic sensor are shown in Fig. 12.15(c) and 12.15(d), respectively. The spectra of

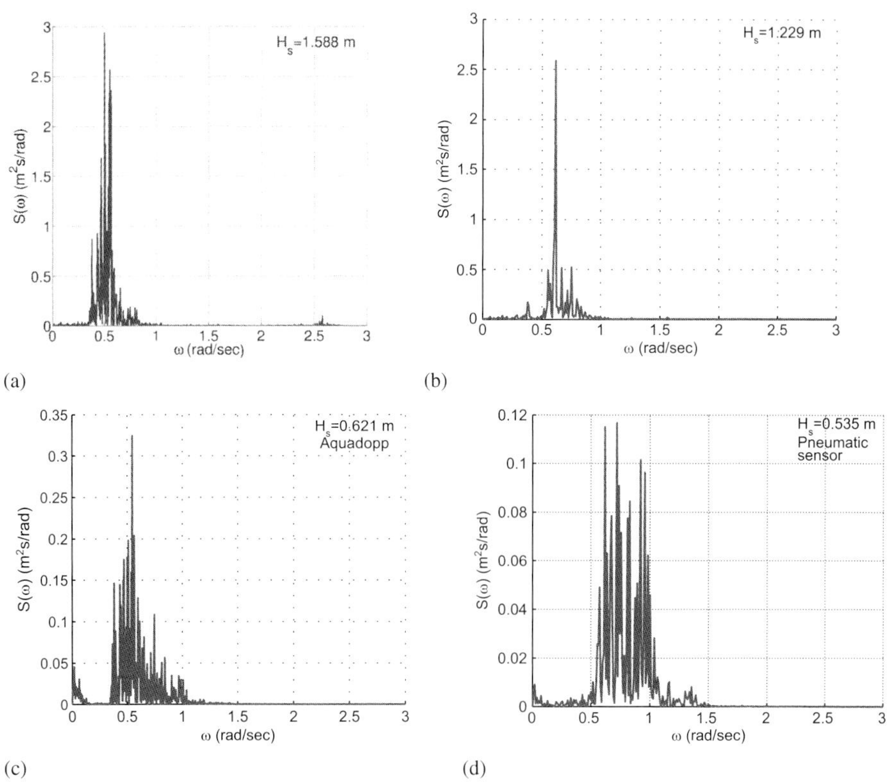

Figure 12.15 Chamber wave elevation spectra for first (a) and second (b) deployment periods and up-wave elevation spectra from Aquadopp sensor (c) and from pneumatic sensor (d).

the chamber wave elevation show a resonance around 0.62 rad/s for both deployment periods, which corresponds to a dominant wave period of approximately 10 s. However, the spectra of the free surface elevation up-wave of the WEC show that, in addition to the dominant wave frequency, other components exist at higher frequency. Because the incident waves are reflected by the wall of the OWC, frequency components of shorter period can be present in the spectra of up-wave elevation [230].

Since the low-frequency components carry most of the wave energy, it is a reasonable assumption to predict only the low-frequency components of the chamber wave elevation [232]. In fact, the high frequencies are representative of low-energy sea states, which are not relevant for wave energy conversion. Also, neglecting the sea spectral components at high frequency yields a more regular and more predictable wave elevation, and accurate predictions can be also obtained for long forecasting horizons. Therefore, the data are filtered with an ideal zero-phase low-pass filter with cut-off frequency ω_C. The filter is implemented with off-line forward and backward filtering through Chebyshev filters of type I, discretized with the bilinear transform. The cut-off frequency was selected in order to preserve the dominant wave period of the sea state.

If the free surface elevation up-wave of the WEC propagates only in the direction toward the chamber, with a time delay t_{prop}, termed the "propagation time," the incident wave elevation causes the change of the chamber wave elevation. The propagation time can be inferred from considerations of wave propagation in a finite water depth. Assuming that there are no attenuation effects on the travelling wave (deep water is assumed), the wave elevation η propagates in the direction x, defined as positive from the location of the up-wave sensor to the chamber, according to the following expression [233]:

$$\eta(t,x) = A\cos(\omega t - k(\omega)x) = \mathrm{Re}(Ae^{j(\omega t - k(\omega)x)}) \tag{12.35}$$

where $k(\omega)$ is the wave number, which is a function of the frequency ω and water depth h. The wave propagates with a phase velocity given by [66]

$$v_p = \frac{\omega}{k(\omega)}. \tag{12.36}$$

In shallow water with depth h, the waves can reach a maximum velocity $v_{max} = \sqrt{gh}$, where g is the gravitation acceleration. Since the up-wave sensor is placed at a location with a water depth $h = 8\,\mathrm{m}$, a velocity $v_{max} = 8.85\,\mathrm{m/s}$ is obtained. If it is assumed that all the waves travel at the same phase velocity equal to the maximum velocity, then the free surface elevation up-wave of the WEC reaches the chamber without any phase distortion. If d denotes the spatial distance between the point of measurement of the free surface elevation up-wave of the WEC and the point of measurement of the chamber elevation, the propagation time t_{prop} of the wave is as follows [233]:

$$t_{prop} = \frac{d}{v_{max}}. \tag{12.37}$$

Since the point of measurement of the up-wave elevation is at a distance $d = 60\,\mathrm{m}$ from the chamber, using Equation 12.37, a propagation time around 7 s is obtained.

If there are no nonlinearities or attenuation effects on the wave propagation, the use of the up-wave elevation measurement provides advance knowledge of the future wave elevation inside the chamber for a forecasting horizon equal to the propagation time. In reality, the presence of eventual nonlinearities on the wave propagation and multidirectionality of the waves distort the effect of the incoming wave on the chamber wave elevation. As a result, the measurement from the up-wave sensor gives partial information about the future wave elevation inside the chamber. As shown in Equation (12.37), the distance of the up-wave sensor from the chamber determines the maximum forecasting horizon. The measurements of the up-wave elevation given by a sensor located at a long distance from the chamber provides less accurate information about the future chamber wave elevation, due to the increasing effect of nonlinearities and multidirectional waves. Also, in order to acquire the measurements from a sensor more distant from the site where the prediction is required, greater costs for the data cable are involved.

12.8.2 Forecasting Models

Up-Wave-Only Models

Here, a model for forecasting the chamber wave elevation based *only* on the measurement of the free surface elevation up-wave of the WEC is presented. In particular, a finite impulse response (FIR) model is designed that assumes that the chamber wave elevation at time instant k is a linear combination of n_b past values of the free surface elevation up-wave of the WEC. The up-wave measurement is considered to be the input of the model and is denoted by u. The model considered is of the following form:

$$\eta_{ch}(k) = \sum_{i=1}^{n_b} b_i u(k - i + 1 - n_k) + \xi(k) \tag{12.38}$$

where ξ is considered to be a white noise with zero mean and variance σ^2. The term n_k is the delay (in number of sample periods) that occurs before the output is affected by the input. If the data coming from the up-wave sensor are delayed by the propagation time, then, at time instant k, the water level in the chamber is influenced by the up-wave elevation at the same time instant. Therefore, the (additional) delay n_k is considered to be zero. The use of the FIR filter takes into account the unmodeled dynamics between the up-wave elevation and the chamber wave elevation, which was not considered in the calculation of the propagation time alone. Also, the FIR filter models the dynamics of the sensor itself. In fact, since the up-wave sensor deployed approximates the free surface elevation based on a pressure signal, the FIR filter also models any dynamic relationship between the pressure and the free surface elevation. Given a set of parameters b_i, from Equation (12.38), the l-step-ahead prediction is given as follows:

$$\hat{\eta}_{ch}(k + l|k) = \sum_{i=1}^{n_b} b_i u(k + l - i + 1). \tag{12.39}$$

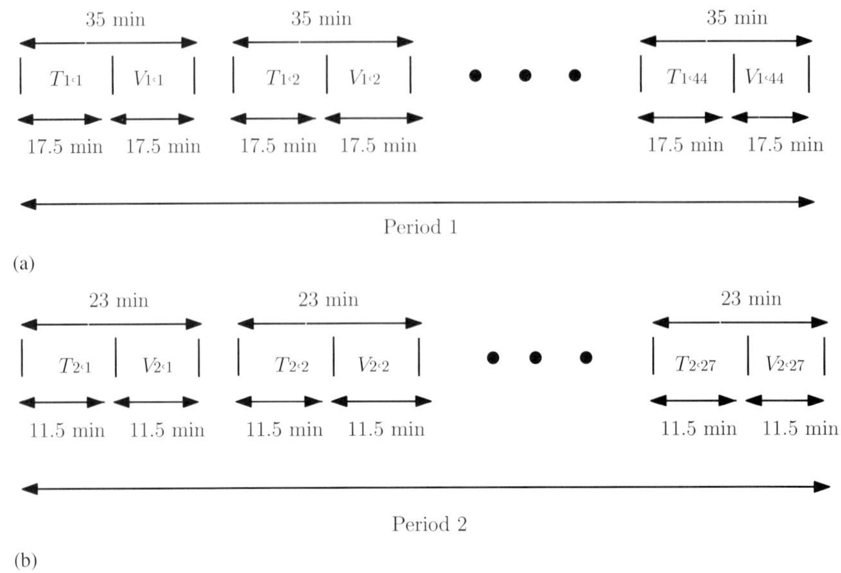

Figure 12.16 Time-series data segmentation for the first (a) and second (b) deployment period.

In Equation (12.39), the time series of the up-wave elevation is known up to a certain prediction horizon, equal to the propagation time of the wave from the location of the sensor.

As shown in Fig. 12.16, the data for each deployment period were divided into training and validation sets. The coefficients b_i of the FIR model, at an instant k, are estimated using a batch of training data $T_{i,j}$ of dimension N_1 consisting of past observations of the water level in the chamber and measurements of the free surface elevation up-wave of the WEC. The cost function to be minimized, for the estimation of the model parameters, is the variance of the one-step-ahead prediction error, which is given as follows:

$$J_{LS} = \sum_{k=1}^{N_1} (\eta_{ch}(k+1) - \hat{\eta}_{ch}(k+1|k))^2, \tag{12.40}$$

which is a linear least squares (LS) problem. For the estimation of the model parameters, a cost function corresponding to long-range predictive identification (LPRI) [234] was considered, with the initial value of the parameters given by the minimization of Equation (12.40). However, since the minimization of the LPRI cost function for different forecasting horizons did not yield a significantly different value of the parameters compared with their initial estimates, the LPRI cost function was not further considered.

An important element in the identification of the FIR model with cost function (12.40) is the choice of the order n_b. The objective is to identify a model that guarantees an accurate fitting of the training data without overfitting or, rather, the identification of a overparametrized model that also interpolates the stochastic noise present in the measurements. Given a previously unseen batch of validation data $V_{i,j}$ of dimension N_2,

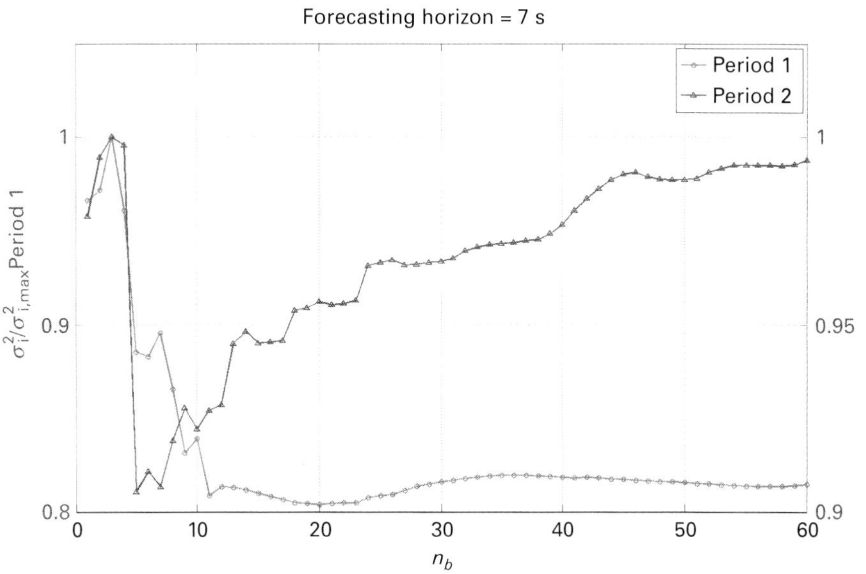

Figure 12.17 Normalized prediction error variance for a forecasting horizon of 7 s as a function of the order n_b averaged across validation data sets for the first and second deployment periods

consisting of observations of the chamber and up-wave elevations, the accuracy of the identified model can be validated by means of the variance of the prediction error σ_l^2 for a forecasting horizon l computed as follows:

$$\sigma_l^2 = \frac{1}{N_2}\sum_{k=1}^{N_2} \hat{e}(k+l|k)^2 = \frac{1}{N_2}\sum_{k=1}^{N_2} (\eta_{ch}(k+l) - \hat{\eta}_{ch}(k+l|k))^2. \tag{12.41}$$

In order to select the best order $\overline{n}_b$ that avoids overfitting the data, FIR filters of different orders n_b were trained and validated by means of the data coming from the two deployment periods. In Fig. 12.17, the variance σ_l^2 for a forecasting horizon equal to the propagation time of 7 s is computed for FIR filters of different order n_b. The variance is averaged across the validation data sets coming from both of the deployment periods, normalized by the appropriate maximum value. An order n_b equal to 10 is chosen as point of trade-off between the minima of the two curves.

Autoregressive Models

In this section, a model for forecasting the chamber wave elevation based *only on its past history* is presented. In particular, an AR model, as presented in Section 12.4.3, is proposed that assumes that the chamber wave elevation at time instant k is linearly dependent on a number n_a of its past values. Thus, the model considered is of the following form:

$$\eta_{ch}(k) = \sum_{i=1}^{n_a} \alpha_i \eta_{ch}(k-i) + \xi(k). \tag{12.42}$$

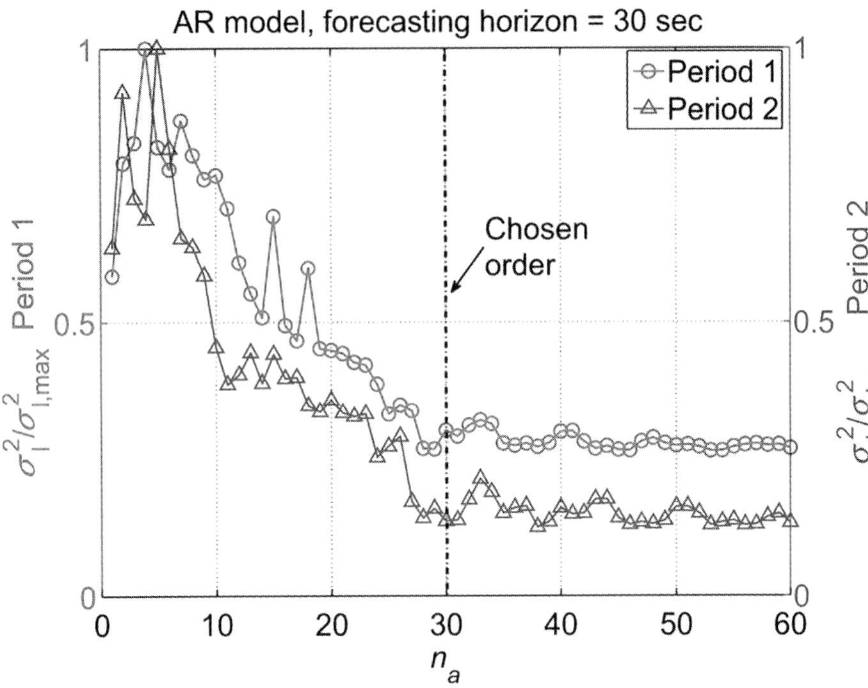

Figure 12.18 Normalized prediction error variance for a forecasting horizon of 30 s as a function of the order n_a averaged across validation data sets for the first and second deployment periods.

Given a set of parameters α_i, from Equation (12.42) the l-step-ahead prediction is given as follows:

$$\hat{\eta}_{ch}(k+l|k) = \sum_{i=1}^{n_a} \alpha_i \hat{\eta}_{ch}(k+l-i|k) \tag{12.43}$$

where $\hat{\eta}_{ch}(k+l-i|k) \equiv \eta_{ch}(k+l-i)$ if $k+l-i \leq k$, since the information is already available and there is no need to have a prediction [232]. The coefficients α_i of the AR model are estimated using the cost function given by Equation (12.40), and the identified models can be validated according to the variance of the prediction error over a forecasting horizon l, given by Equation (12.41). AR models of different orders n_a were trained and validated by means of the measurements of the wave elevation inside the chamber for both of the deployment periods. In Fig. 12.18, the variance σ_l^2 for a forecasting horizon equal to 30 s, normalized by its maximum value and averaged across the validation sets, is plotted against different order n_a for both of the deployment periods. An order n_a equal to 30 is chosen since it represents a minimum point for both of the 2 curves.

Combination Up-Wave/Autoregressive Models

In this section, a model for forecasting the chamber wave elevation based on a combination of its past history *and* measurements of the free surface elevation up-wave of the WEC is presented. In particular, an ARX model is proposed that assumes that the chamber wave elevation at time instant k is linearly dependent on a number n_a of its past values and on n_b up-wave measurements. Thus, the model considered is of the following form:

$$\eta_{ch}(k) = \sum_{i=1}^{n_a} a_i \eta_{ch}(k-i) + \sum_{i=1}^{n_b} b_i u(k-i+1) + \xi(k). \tag{12.44}$$

Given a set of parameters a_i and b_i, from Equation (12.44), the l-step-ahead prediction is given as follows:

$$\hat{\eta}_{ch}(k+l|k) = \sum_{i=1}^{n_a} a_i \hat{\eta}_{ch}(k+l-i|k) + \sum_{i=1}^{n_b} b_i u(k+l-i+1). \tag{12.45}$$

In equation (12.45), no prediction of the input is made until a prediction horizon, equal to the propagation time of the wave from the location of the sensor, is reached. The limit of the prediction horizon, in time steps, is denoted as l_{max}, and given as

$$l_{max} = \frac{t_{prop}}{t_{samp}} = \frac{dv_p}{t_{samp}} \tag{12.46}$$

where t_{samp} is the sampling period for the chamber and up-wave elevation data. Therefore, the prediction of $\hat{\eta}_{ch}$ can be made with an ARX model only until l_{max} steps, after which an AR-only model is used to predict the water level in the chamber. Thus, the complete prediction of the wave elevation using an ARX model is given as follows:

$$\hat{\eta}_{ch}(k+l|k) = \begin{cases} \sum_{i=1}^{n_a} a_i \hat{\eta}_{ch}(k+l-i|k) \\ \quad + \cdots \sum_{i=1}^{n_b} b_i u(k+l-i+1), & \text{if } l \leq l_{max}, \\ \sum_{i=1}^{n_a} \alpha_i \hat{\eta}_{ch}(k+l-i|k), & \text{else ,} \end{cases} \tag{12.47}$$

The coefficients a_i, b_i of the ARX model and α_i of the AR model are estimated individually using the cost function given by Equation (12.40), and the identified models were validated according to the variance of the prediction error, over a forecasting horizon l, given by Equation (12.41).

Given an order $n_a = 30$ for the AR model obtained in the previous section, ARX models of different orders n_b were trained and validated for both of the deployment periods. The optimal order n_b selected for the FIR model cannot be used directly for the ARX model, since, in Equation 12.44, the chamber wave elevation is not exclusively determined by the up-wave measurement but is also determined by its past values. In Fig. 12.19, the normalized variance of the prediction error, averaged across the validation data sets, is plotted against different orders n_b for both of the deployment

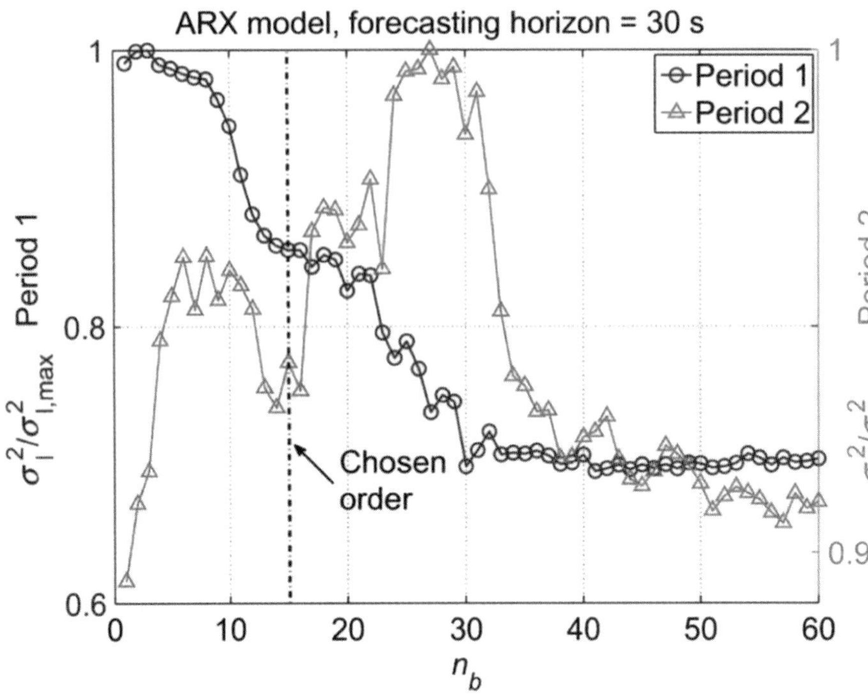

Figure 12.19 Normalized prediction error variance for a forecasting horizon of 30 s as a function of the order n_b with $n_a = 30$ averaged across validation data sets for the first and second deployment periods.

periods. By inspection of Fig. 12.19, an order $n_b = 15$ is selected as a trade-off point between the minima of the two curves. The order n_b, chosen for the ARX model, is different from the order n_b of the FIR model, since the ARX models were trained and validated for a fixed choice of $n_a = 30$.

12.8.3 Comparative Results

Here, the forecasts of the water level in the chamber given by the FIR and ARX models is compared with the forecasting given by the AR model over a range of prediction horizons. The index used for the prediction accuracy for every forecasting horizon l is the goodness-of-fit (GOF) index, which is defined as follows:

$$\text{GOF}(l) = \left(1 - \frac{\sqrt{\sum_k (\eta(k+l) - \hat{\eta}(k+l|k))^2}}{\sqrt{\sum_k \eta(k)^2}} \right) \cdot 100. \tag{12.48}$$

In order to compare the performance between the AR model and the FIR or ARX model, the relative goodness of fit (RGOF) is introduced:

$$\begin{aligned} \text{RGOF}_{FIR}(l) &= \text{GOF}_{FIR}(l) - \text{GOF}_{AR}(l), \\ \text{RGOF}_{ARX}(l) &= \text{GOF}_{ARX}(l) - \text{GOF}_{AR}(l). \end{aligned} \tag{12.49}$$

The RGOF between an AR model of order $n_a = 30$ and an FIR model of order $n_b = 10$, averaged across the validation data sets, is plotted against the forecasting horizon for the first and second deployment period in Fig. 12.20. As shown in Fig. 12.20, the AR model is able to provide more accurate predictions of the wave elevation in the chamber for all forecasting horizons up to 7 s. In Fig. 12.21, the RGOF between an AR model of order $n_a = 30$ and an ARX model of order $n_a = 30$ and $n_b = 15$, averaged across the validation data sets of the first deployment period, is plotted against the prediction horizon for different values of the cut-off frequency ω_c. As Fig. 12.21 shows, the AR model performs better than the ARX model for prediction horizons that range from 10 up to 22 or 27 s, depending on ω_c. The choice of the cut-off frequency ω_c depends on frequency components of the chamber wave elevation that need to be predicted. Since the resonance of the chamber is around 0.6 rad/s, a cut-off frequency $w_c = 0.7$ rad/s is selected for filtering the chamber and up-wave elevation. Note that the improvement in the accuracy of the prediction of the chamber wave elevation for long horizons obtained with the ARX model has to be evaluated with respect to the prediction requirements in the control of the WEC. In fact, because of the noncausal transfer function involved in the realization of the real-time optimal control of the WEC, the future wave elevation may be needed for a time horizon less than 30 s [235]. Regarding the second deployment period, in Fig. 12.22, the RGOF between an AR model of order $n_a = 30$ and an ARX model of order $n_a = 30$ and $n_b = 15$, averaged

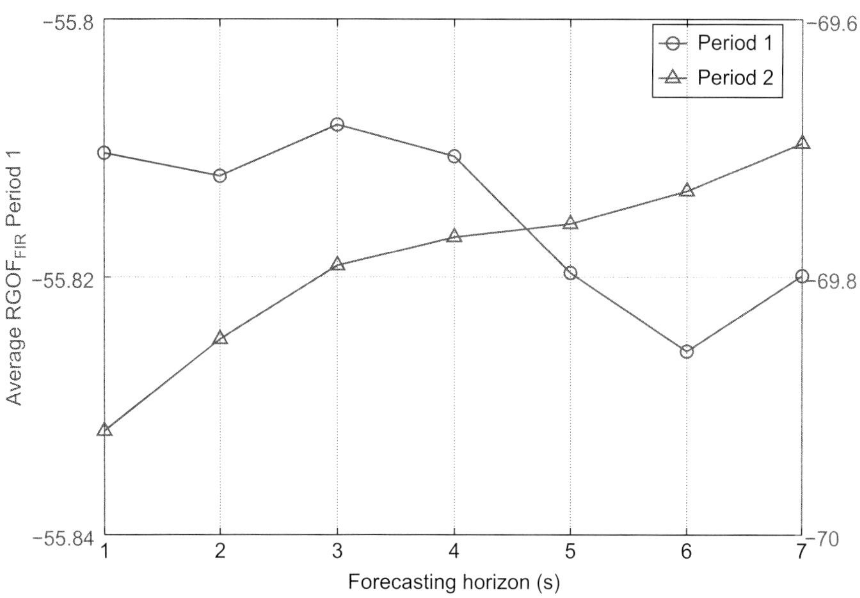

Figure 12.20 Relative goodness of fit (RGOF) between AR model ($n_a = 30$) and FIR model ($n_b = 10$) averaged across the validation data sets of the first and second deployment periods.

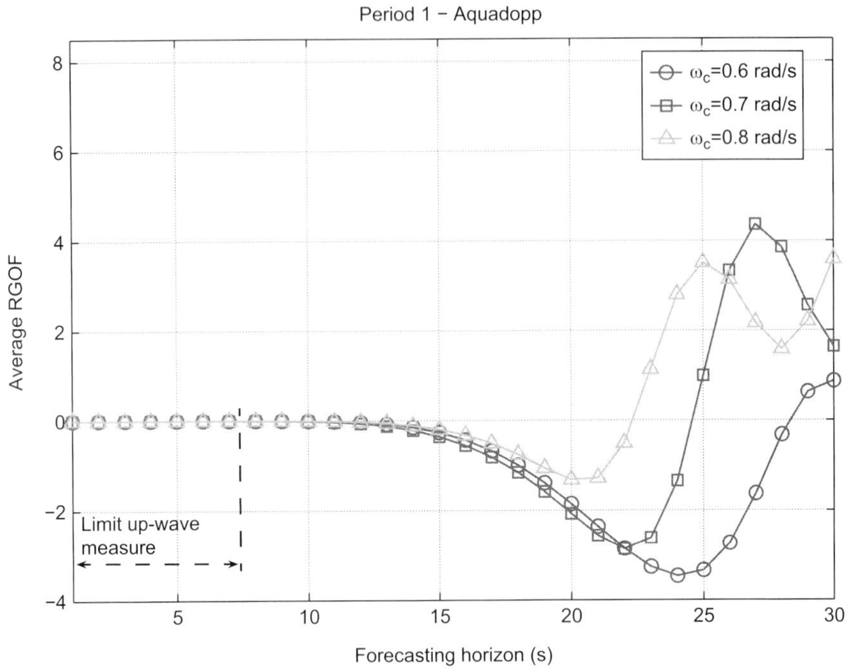

Figure 12.21 RGOF between AR model ($n_a = 30$) and ARX model ($n_a = 30, n_b = 15$) averaged across the validation data sets of the first deployment period.

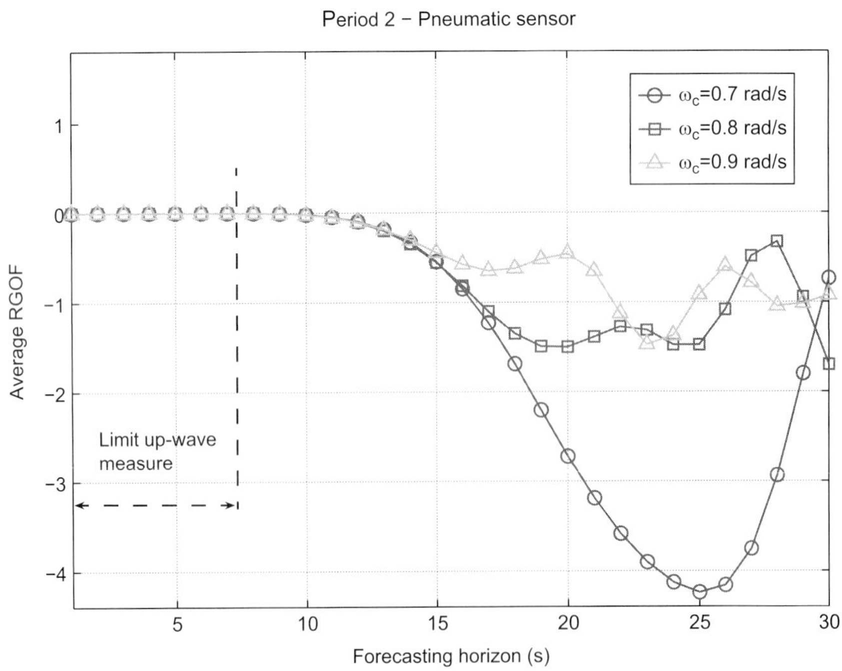

Figure 12.22 Relative Goodness of Fit (RGOF) between AR model ($n_a = 30$) and ARX model ($n_a = 30, n_b = 15$) averaged across the validation data sets of the second deployment period.

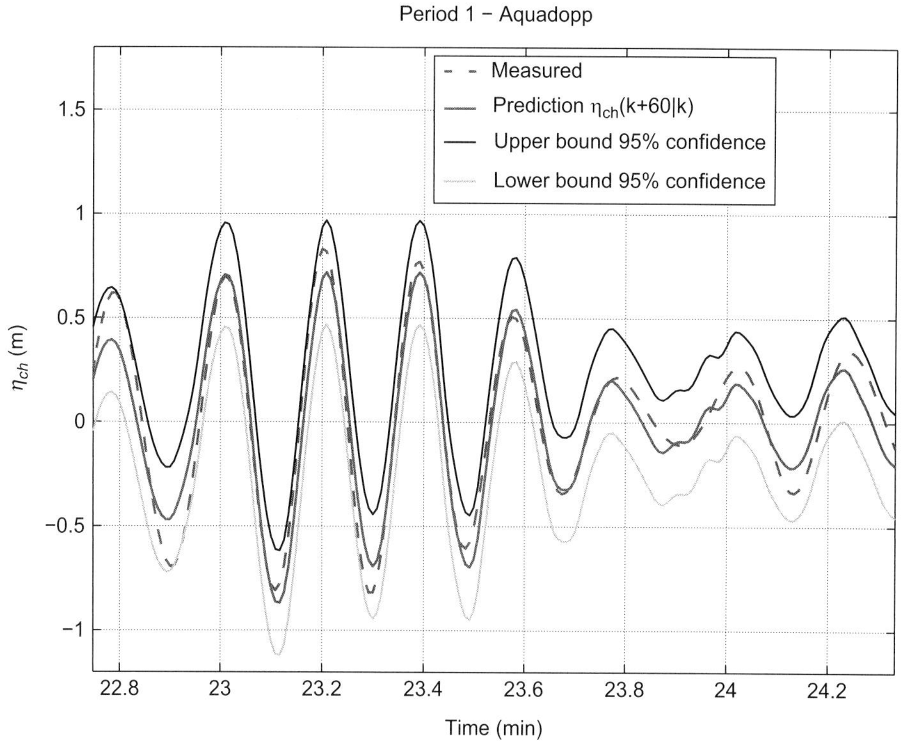

Figure 12.23 Comparison between measured chamber wave elevation and prediction $\eta_{ch}(k+60|k)$ computed with an ARX model of order $n_a = 30$ and $n_b = 15$ with data set corresponding to the first deployment period, filtered with a cut-off frequency of $\omega_c = 0.7$ rad/s.

across the validation data sets, is plotted against the prediction horizon for different values of the cut-off frequency ω_c. As Fig. 12.22 shows, the ARX model provides less accurate predictions of the wave elevation inside the chamber than the AR model for every prediction horizon.

A comparison between predicted and actual chamber wave elevations for a forecasting horizon of 30 s, for an ARX model of order $n_a = 30$ and $n_b = 15$, is made. The ARX model utilizes the free surface elevation up-wave of the WEC provided by the Aquadopp and pneumatic sensors, shown in Figs. 12.23 and 12.24, respectively. The prediction of the chamber wave elevation is plotted together with confidence intervals, which can be assumed to be a Gaussian process with zero mean and variance σ_l^2 given as follows [232]:

$$\hat{e}(k+l|k) = \eta(k+l) - \hat{\eta}(k+l|k) \cong \mathcal{N}(0, \sigma_l^2). \tag{12.50}$$

Finally, it may be noted that there appears to be little advantage, at least for the current case study, of employing nonlinear ARX models, since the results reported in [236] show an inferior performance for artificial neural network models compared with linear ARX models.

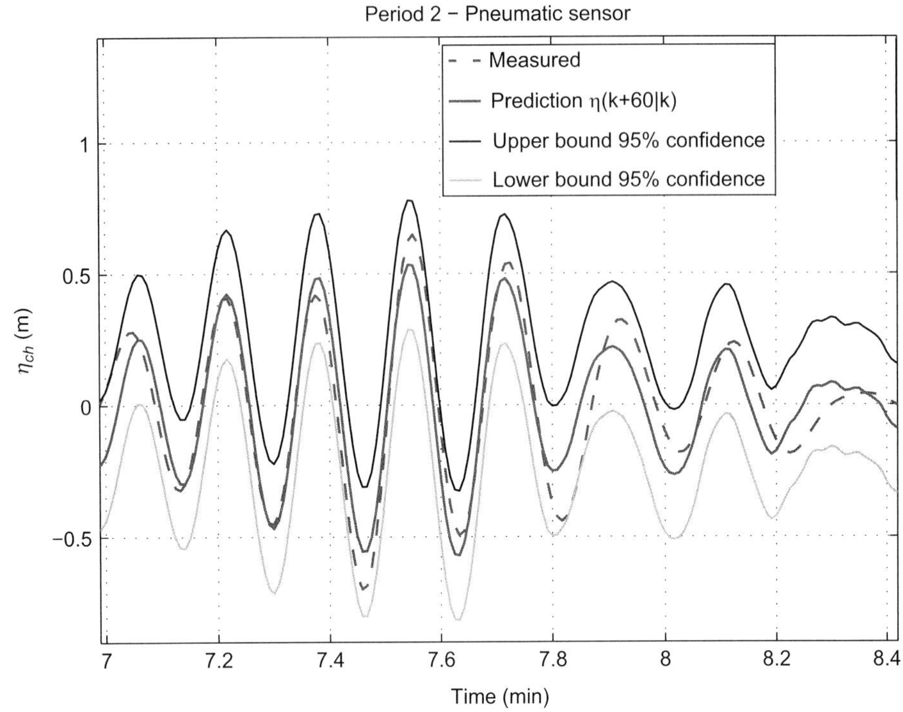

Figure 12.24 Comparison between measured chamber wave elevation and prediction $\eta_{ch}(k+60|k)$ computed with an ARX model of order $n_a = 30$ and $n_b = 15$ with data set corresponding to the second deployment period, filtered with a cut-off frequency of $\omega_c = 0.7$ rad/s.

12.9 Conclusion

This chapter is focused on the problem of short-term wave prediction, which is a central topic in the wave energy field, in order to improve the economic viability of a WEC. As opposed to the classic reconstruction of the wave field from distant measurements, an approach is proposed where the wave elevation is treated as a univariate time series.

To put the wave forecasting problem in context, a study of the prediction requirements in real-time control of WECs was presented. Focus was put on unconstrained reactive control, which allowed the characterisation of the relation between such requirements and some fundamental properties of the floating system.

A time constant, denoted τ_1, related to the shape of the radiation resistance of a floating system, was identified as the most critical parameter influencing the noncausality of the control law. From time domain simulations of different heaving buoys (spheres and cylinders), over a wide variety of real wave conditions, it was verified that the relation between τ_1 and the future horizon required for an effective implementation of reactive control is practically linear.

Simple manipulation of the control, based on prior knowledge of the excitation force spectral distribution, was also proposed for the reduction of the prediction requirements.

Such reshaping of the noncausal control law allows a significant decrease in the prediction requirements, so that they fall within the range of predictability achievable with simple stochastic predictors. Minor performance losses, in terms of energy capture, are verified in most sea states. The possibility of making the control law causal was also proposed, and this is investigated in more detail in [122].

Wave forecasting results on real wave elevation data from Galway Bay and Pico Island showed how a relatively simple AR model, which implicitly models the cyclical behavior of the waves, can offer a very accurate prediction of the low-frequency swell waves for up to two typical wave periods into the future. It is also shown that no significant benefit can be expected in using nonlinear forecasting models, such as neural networks, unless highly nonlinear sea states are encountered. In deep water locations, however, nonlinearities due to interactions between different wave components occur only for very low energy wind wave systems, which are of poor interest in a wave energy context (an analysis of nonlinearities in waves is provided in [208]).

Concerning the use of the wave prediction algorithms in real-time WEC control, the real-time filter implementation needs to be given further attention, since a zero-phase filter was assumed in the current analysis. With a focus on achievable prediction, such an assumption can be justified. In any case, as mentioned in Section 12.2.3, the controller of a WEC may require the prediction of the *effect* of the wave elevation on the device, for example, the wave excitation force, which is effectively a low-pass-filtered version of the wave elevation, with the filtering provided by the device itself.

One further effect of importance, relating to wave forecasting for WEC control systems, is the degree to which the WEC control quality might be adversely affected by wave forecasting errors. A general treatment of this is currently not available, and therefore beyond the scope of this book, but the interested reader may find some clues in [237], which looks at some particular cases.

Finally, the employment of up-wave measurement systems for wave or excitation force prediction needs further study, since the case study results presented here, using time series models, show no clear benefit in employing up-wave measurement, compared with the use of purely AR models. In addition to any dubious performance improvements over AR models, the not insignificant capital and maintenance costs of up-wave measurement systems are important factors. However, depending on the nature of the wave spectrum, particularly in the case of broader wave spectra, up-wave measurements may have greater benefit.

Part VI

Toward Overall WEC System Hydrodynamic Optimization

The final part of this book examines some broader aspects of optimum wave energy system design and how various design decisions can be impacted by the control strategy employed. Chapter 13 begins with a focus on force and amplitude constraints and how they can be optimally traded off, from a control perspective, to maximize energy capture. The first part of Chapter 14 examines broader aspects of the WEC geometry and shows how the optimal device shape is sensitive to the control algorithm employed, suggesting that both control and geometry need to be designed concurrently, rather than sequentially. A similar message is conveyed in the latter part of Chapter 14, which examines the sensitivity of an optimal WEC farm layout to the array control strategy employed. Again, a strong interaction between optimal layout and control is evidenced, especially for closely spaced devices.

13 Position and Force Constraint Analysis and Optimization

13.1 Introduction

Even at this early stage of wave energy development, it is largely accepted that most devices will have two modes of operation: the "normal" mode of operation, where power is converted from the waves into a usable form (water pressure, electricity, etc.), and a further "survivability" mode, which the device will need to enter when the wave excitation forces become too great for the motion and force constraints to be simultaneously satisfied. However, the development of supervisory control to configure a device into survival mode is not the focus here; though, indeed, there have been a number of methods proposed, many of which are particular to specific device types. Rather, the focus here is on the interplay between the wave excitation force and the physical constraints of the wave energy device, namely, the force and displacement constraints, and it will be assumed that, where system constraints cannot be satisfied, the wave energy device must enter survival mode. We note that there may also be velocity constraints related to limitations of rate of movement of various system components. While velocity constraints can also be included in the formulation framework, we focus on the main constraints of force and position, for brevity.

The suggested interplay between the force and displacement (amplitude) constraints and the wave excitation force is illustrated conceptually by Fig. 13.1. A situation that satisfies the constraints for a given wave excitation force, F_e, is indicated by an intersection of the force and position sets, and this is the region within which the optimal constrained control problem can be solved. In general, larger position and force constraints are indicated by circles of greater radii, while larger excitation forces are represented by a greater distance between circle centers, indicating the increasing difficulty of simultaneously satisfying displacement and force constraints. Ideally, we would like to identify the functions $f_1(\,)$, $f_2(\,)$, and $f_3(\,)$ so that the construct in Fig. 13.1 can be generated for a specific wave energy device. In addition, it would be beneficial if some indication of the energy landscape could be provided so that, where constraints are not immediately satisfied, some guidance is available to allow captured energy to be maximized by virtue of a constraint relaxation.

The *feasibility* problem, as identified in the preceding paragraph, has, to date, not been addressed. A number of researchers address the issue of constraints and design a controller that actuates power take-off (PTO) force in order to satisfy constraints, usually focusing on position constraints alone. Some more recent studies do examine

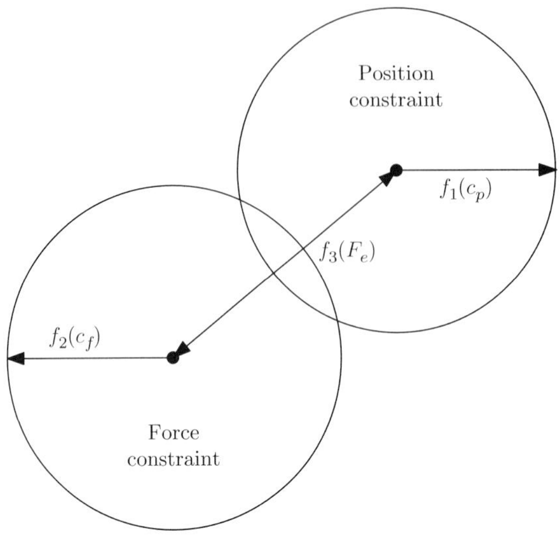

Figure 13.1 Illustration of the interplay between excitation force, F_e, position constraint, c_p, and force constraint, c_f.

both force and displacement constraints, for example, [238, 239]. However, such studies presuppose a particular control design method. For example, amplitude constraints, based on a frequency domain approach, are addressed in [53, 240, 241], while a number of authors have utilized the model predictive control (MPC) framework to maximize energy capture subject to constraints, for example, [110]. The work reported in [134] is interesting in that it uses MPC to solve for the optimal velocity profile of the wave energy device, subject to amplitude constraints, while then examining the PTO force required to achieve this velocity profile, where the PTO force is also subject to constraints.

The framework utilized here is based on the discretization of the equations of motion of the wave energy converters (WECs) by means of the approximation of the forces and the velocities with a linear combination of basis functions, as suggested in Chapter 10. A special case is considered, in which truncated Fourier series are used for the approximation (consistent with the representation used in Chapters 10 and 11). This case is particularly interesting because the results are in harmony with the frequency domain theory of wave energy conversion, as in [66]. The discretization was initially motivated by the study of an optimal controller for the maximization of the energy converted by the device, subject to constraints, as shown in Chapter 10. For brevity, only the theoretical details crucial to the development of the constraint framework are described here. The full details of the underlying control framework are given in [158].

While it is (theoretically) possible to determine the satisfaction of system constraints for a given wave climate via extensive simulation, such an approach would be both time consuming and may not capture the complete set of circumstances. In contrast, the constraint analysis tool developed in this chapter is analytical, requiring a hydrodynamic model of the wave energy device of interest together with a specification of the force and position constraints and the excitation force in order to give an immediate conclusion regarding the satisfaction of the constraints. Furthermore, the graphical nature of the answer provides some insight into the interplay between c_f, c_p, and F_e, placed against the background of absorbed energy, and therefore provides the basis for a design tool.

13.2 Self-Reacting Point Absorber Model

The general form of device considered is a two-body self-reacting point absorber restricted to heave motion only, as depicted in Fig. 13.2 and described by the frequency domain model [136]:

$$\begin{cases} \left(i\omega\, m^A + B^A + \dfrac{S^A}{i\omega}\right) V^A = F_e^A + F_r^A - F_{pto} \\ \left(i\omega\, m^B + B^B + \dfrac{S^B}{i\omega}\right) V^B = F_e^B + F_r^B + F_{pto} \end{cases} \tag{13.1}$$

where V^A and V^B are the vertical velocities of body A and body B, respectively. The radiation force F_r is

$$\begin{bmatrix} F_r^A \\ F_r^B \end{bmatrix} = -Z \begin{bmatrix} V^A \\ V^B \end{bmatrix} \text{ with } Z = \begin{bmatrix} Z^{AA} & Z^{AB} \\ Z^{BA} & Z^{BB} \end{bmatrix}$$

where $Z = Z(\omega)$ is the radiation impedance matrix. F_e^B and F_e^B denote the excitation forces, m^A and m^B are the masses, S^A and S^B the hydrodynamic stiffness coefficients, and B^A and B^B are damping coefficients used to model linear losses due to viscous effects and frictions of mechanical components; see Section 11.2.3 for the calculation of the viscous components of B^A and B^B. The damping and stiffness coefficients can also embed the terms describing a linear model for mooring loads; for example, considering

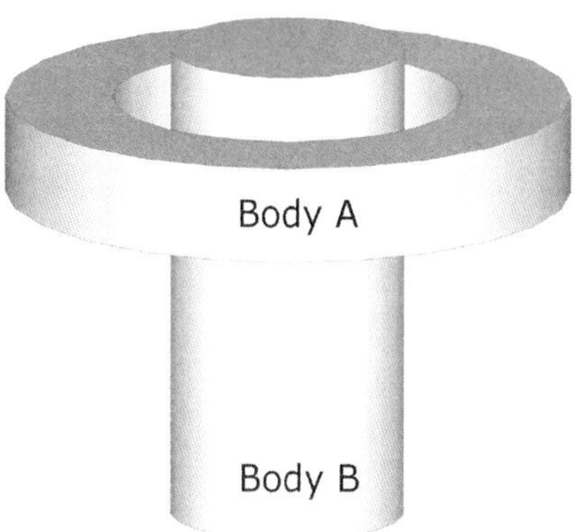

Figure 13.2 Self-reacting point absorber.

body A, this can be done by redefining both S^A and B^A as $S^A = S^A_M + S^A_H$ and $B^A = B^A_M + B^A_H$, where the subscript M denotes terms due to the mooring loads and the subscript H refers to the hydrostatic (S^A_H), viscous, and other linear mechanical dissipative effects (B^A_H). The PTO applies a force F_{pto} on both bodies with the same magnitude and opposite direction. The time domain formulation of the model in (13.1) is

$$\begin{cases} m^A \dot{v}^A(t) + B^A v^A(t) + S^A x^A(t) - f^A_r(t) = f^A_e(t) - f_{pto}(t) \\ m^B \dot{v}^B(t) + B^B v^B(t) + S^B x^B(t) - f^B_r(t) = f^B_e(t) + f_{pto}(t) \end{cases} \tag{13.2}$$

where $x^A(t)$ and $x^B(t)$ are the vertical positions of the two bodies, and $\dot{v}^A(t)$ and $\dot{v}^B(t)$ their vertical accelerations. The forces and velocities, denoted with lowercase letters in (13.2), are the inverse Fourier transform of the corresponding uppercase variables in (13.1). The radiation forces are

$$f^A_r(t) = -m^{AA}_\infty \dot{v}^A(t) - k^{AA}(t) * v^A(t) - m^{AB}_\infty \dot{v}^B(t) - k^{AB}(t) * v^B(t)$$

$$f^B_r(t) = -m^{BB}_\infty \dot{v}^B(t) - k^{BB}(t) * v^B(t) - m^{BA}_\infty \dot{v}^A(t) - k^{BA}(t) * v^A(t)$$

where the symbol $*$ denotes the convolution operator, and the parameters m^{ij}_∞ and $k^{ij}(t)$, with $i, j \in \{A, B\}$, are related to the elements of the radiation impedance matrix Z through the expressions [164]:

$$\begin{aligned} m^{ij}(\omega) &= \mathrm{Im}\{Z^{ij}(\omega)\} = m^{ij}_\infty - \frac{1}{\omega}\int_0^\infty k^{ij}(t)\, \sin(\omega t)\, dt, \\ R^{ij}(\omega) &= \mathrm{Re}\{Z^{ij}(\omega)\} = \int_0^\infty k^{ij}(t)\, \cos(\omega t)\, dt, \\ m^{ij}_\infty &= \lim_{\omega\to\infty} \mathrm{Im}\{Z^{ij}(\omega)\}. \end{aligned} \tag{13.3}$$

The energy absorbed in the interval $[0, T]$, neglecting losses, corresponds to the work performed by the PTO, that is,

$$J(T) = \int_0^T f_{pto}(t)\, (v^A(t) - v^B(t))\, dt. \tag{13.4}$$

13.2.1 Discretization

The PTO force is assumed to be such that $f_{pto}(t) \in L^2([0, T])$, where $L^2([0, T])$ is the Hilbert space of square integrable functions in the interval $[0, T]$. The velocities and the accelerations are also assumed to be square integrable, that is, $v^i(t) \in H^1([0, T])$ where $H^1([0, T])$ is the Sobolev space of square integrable functions in $[0, T]$, the derivatives of which are also square integrable. The PTO force and the velocities are then approximated, as in Chapter 10, as a linear combination of basis functions in a finite-dimensional subspace of the space $L^2([0,T])$:

$$v^i(t) \approx \hat{v}^i(t) = \sum_{n_f=1}^{N} x^i_{n_f}\, \phi_{n_f}(t), \quad i \in \{A, B\} \tag{13.5}$$

$$f_{pto}(t) \approx \hat{f}_{pto}(t) = \sum_{n_f=1}^{N^P} p_{n_f}\, \phi^P_{n_f}(t) \tag{13.6}$$

where $\{\phi_1(t), \ldots, \phi_N(t)\}$ is a basis for the finite-dimensional subspace $S^V \subset H^1([0,T])$ and $\{\phi^P_1(t), \ldots, \phi^P_{N^P}(t)\}$ is a basis for the finite-dimensional subspace $S^P \subset L^2([0,T])$. For any given set of coefficients describing the PTO force $\{p_1, \ldots, p_{N^P}\}$ and excitation forces f^A_e and f^B_e, the components of the velocities are calculated by applying the Galerkin method [242], as outlined in Section 10.2:

$$\begin{cases} \langle \hat{r}^A(t), \phi_{n_f} \rangle = 0 \\ \langle \hat{r}^B(t), \phi_{n_f} \rangle = 0 \end{cases} \quad \forall \quad n_f = 1, \ldots, N \tag{13.7}$$

where $\langle \cdot, \cdot \rangle$ denotes the inner product, defined as

$$\langle f, g \rangle = \int_0^T f(t) g(t)\, dt, \tag{13.8}$$

and $\hat{r}^A(t)$ and $\hat{r}^B(t)$ are the equations of motion (13.2) in the residual form, with the velocities and the forces approximated by using Equations (13.5) and (13.6), that is,

$$\begin{cases} \hat{r}^A(t) = m^A \dot{\hat{v}}^A(t) + B^A \hat{v}^A(t) + S^A \hat{x}^A(t) - f^A_r(t) - f^A_e(t) + f_{pto}(t) \\ \hat{r}^B(t) = m^B \dot{\hat{v}}^B(t) + B^B \hat{v}^B(t) + S^B \hat{x}^B(t) - f^B_r(t) - f^B_e(t) - f_{pto}(t) \end{cases}$$

where $\hat{x}^i(t)$ are the approximated vertical positions, which are obtained by integrating Equations (13.5). The Galerkin method is a projection method for the discretization of integral and differential equations; the solution of (13.7) is the set of the velocity components x^i_n that minimizes the difference between the velocities obtained by solving the original equations of motion (13.2) and their approximations (13.5). In other words, for any given PTO force and excitation forces, the solution of Equations (13.7) provides the best approximation of the motion of the WEC in terms of the velocity components.

Given the oscillatory nature of the problem, a truncated Fourier series, as employed in Chapters 10 and 11, is an intuitive choice as the basis for S^V and S^P. Furthermore, choosing $w_0 = 2\pi/T$, the set of functions $\{\sin(\omega_0 T), \cos(\omega_0 T), \ldots, \sin(N\omega_0 T), \cos(N\omega_0 T)\}$ form an orthogonal basis for the spaces S^V and S^P with the inner product defined in Equations (13.8). The constant term of the basis is not considered because it is assumed that all the functions have zero mean. In practice, it is assumed that the reference frames of the bodies are chosen such that the origins oscillate around their mean position with respect to the inertial reference frame.

$S_{G^{AA}}$ and $S_{G^{BB}}$ are, respectively, the Schur complements of G^{AA} and G^{BB}, and are defined as

$$S_{G^{AA}} = G^{BB} - G^{BA}{G^{AA}}^{-1}G^{AB}, \tag{13.26}$$

$$S_{G^{BB}} = G^{AA} - G^{AB}{G^{BB}}^{-1}G^{BA}. \tag{13.27}$$

It can be verified that the symmetric part of the matrix H (i.e., $(H + H^T)/2$) is positive definite; therefore, the quadratic cost function in Equation (13.22) is concave, and the global maximum of the unconstrained problem is obtained for

$$\bar{P} = (H + H')^{-1}(Q^A E^A - Q^B E^B). \tag{13.28}$$

13.3 Specification and Approximation of Constraints

The force constraint is defined as

$$\left\|\hat{f}_{pto}\right\|_{\infty} \leq F_{max}, \tag{13.29}$$

while the constraint on the relative amplitude is

$$\left\|\Delta\hat{x}\right\|_{\infty} \leq \Delta U_{max}, \tag{13.30}$$

where the infinity norm $\|\cdot\|_{\infty}$ is defined in the Appendix to this chapter (Equation (13.64)) and $\Delta\hat{x}(t) = \hat{x}^A(t) - \hat{x}^B(t)$, with

$$\hat{x}^A(t) = x_0^A + \sum_{n_f=1}^{N} \frac{b_{n_f}^A}{n_f\omega_0}(1 - \cos(n_f\omega_0 t)) + a_{n_f}^A \sin(n_f\omega_0 t), \tag{13.31}$$

$$\hat{x}^B(t) = x_0^B + \sum_{n_f=1}^{N} \frac{b_{n_f}^B}{n_f\omega_0}(1 - \cos(n_f\omega_0 t)) + a_{n_f}^B \sin(n_f\omega_0 t). \tag{13.32}$$

13.3.1 Constraint Approximation

While the infinity norm describes how the constraints are actually specified, that is, in terms of absolute limits, the infinity norm is not the easiest to manipulate mathematically and does not easily lead to geometric interpretation. The result in the chapter Appendix allows us rearrange the constraint conditions in terms of the 2-norm, which is more convenient to use.

Using the inequality in (13.66) (see the chapter Appendix), it is possible to find sufficient conditions for the *satisfaction* of the inequalities in (13.29) and (13.30) in terms of the 2-norm, Equation (13.66):

$$\left\|\hat{f}_{pto}\right\|_{\infty} \leq \sqrt{\frac{2N}{T}}\left\|\hat{f}_{pto}\right\|_2 \leq F_{max}, \tag{13.33}$$

$$\left\| \Delta \hat{x} \right\|_\infty \leq \sqrt{\frac{2N}{T}} \left\| \Delta \hat{x} \right\|_2 \leq \Delta U_{max}. \tag{13.34}$$

Inequalities (13.33) and (13.34) specify the 2-norm scaled by $\sqrt{2N/T}$ as an upper bound for the infinity norm.

Sufficient conditions for the *violation* of at least one of the constraints, also in terms of the 2-norm, can be obtained using the inequality in (13.65):

$$\left\| \hat{f}_{pto} \right\|_\infty \geq 1/\sqrt{T} \left\| \hat{f}_{pto} \right\|_2 \geq F_{max} \tag{13.35}$$

$$\left\| \Delta \hat{x} \right\|_\infty \geq 1/\sqrt{T} \left\| \Delta \hat{x} \right\|_2 \geq \Delta U_{max}. \tag{13.36}$$

Therefore, the inequalities of (13.35) and (13.36) show that the 2-norm scaled by $1/\sqrt{T}$ provides a lower bound for the infinity norm. Therefore, if the 2-norm multiplied by $1/\sqrt{T}$ violates a constraint, then the infinity norm also violates it.

A conceptual figure, showing the relative magnitudes of scaled infinity and 2-norms, for a dummy variable X_d, is shown in Fig. 13.3.

13.3.2 Constraint Calculation

The 2-norm of the force and amplitude constraints are easily calculated by applying Parseval's theorem, resulting in

$$\left\| \hat{f}_{pto} \right\|_2^2 = \frac{T}{2} \sum_{n=1}^{N} \left((a_n^p)^2 + (b_n^p)^2 \right) = \frac{T}{2} P^T P, \tag{13.37}$$

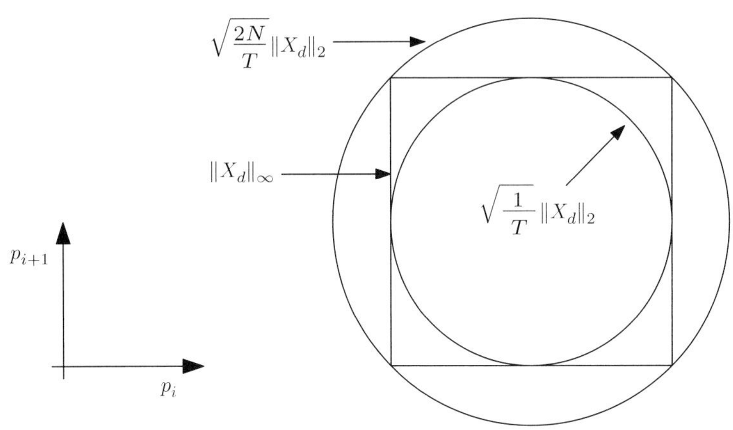

Figure 13.3 Conceptual illustration of the relationship between infinity and 2-norms.

$$\|\Delta\hat{x}\|_2^2 = \frac{T}{2}\sum_{n_f=1}^{N}\frac{1}{n_f\omega_0}\left((b_{n_f}^A - b_{n_f}^B)^2 + (a_{n_f}^A - a_{n_f}^B)^2\right)$$
$$= \frac{T}{2}(X^A - X^B)^T W^2 (X^A - X^B), \tag{13.38}$$

where the amplitude constraint is obtained by setting the initial relative position $\Delta\hat{u}(0)$ as:

$$\Delta\hat{x}(0) = \sum_{n_f=1}^{N}\frac{b_{n_f}^B - b_N^A}{n_f\omega_0}$$

to obtain a zero-mean valued relative position, and the matrix W is

$$W = \begin{bmatrix} 1/\omega_0 & 0 & \cdots & 0 & 0 \\ 0 & 1/\omega_0 & \cdots & 0 & 0 \\ \vdots & \vdots & \ddots & \vdots & \vdots \\ 0 & 0 & \cdots & 1/N\omega_0 & 0 \\ 0 & 0 & \cdots & 0 & 1/N\omega_0 \end{bmatrix}. \tag{13.39}$$

Using Equations (13.23)–(13.25), the 2-norm of the amplitude constraint in Equation (13.38) becomes a function of the vector P:

$$\|\Delta\hat{x}\|_2^2 = \frac{T}{2}\left(P^T H^T W^2 H P - 2\,Q^T\, W^2 H P + Q^T W^2 Q\right). \tag{13.40}$$

Substituting Equations (13.37) and (13.40) into inequalities (13.33) and (13.34), the sufficient conditions for satisfaction of the constraints are

$$P^T P \leq \frac{1}{N} F_{max}^2 \tag{13.41}$$

and

$$P^T H^T W^2 H P - 2\,Q^T\, W^2 H P + Q^T W^2 Q \leq \frac{1}{N}\Delta U_{max}^2, \tag{13.42}$$

while the sufficient conditions for violation of the constraints are

$$P^T P \geq 2F_{max}^2 \tag{13.43}$$

and

$$P^T H^T W^2 H P - 2\,Q^T\, W^2 H P + Q^T W^2 Q \geq 2\Delta U_{max}^2. \tag{13.44}$$

Inequalities (13.41)–(13.44) describe the constraints as a function of the PTO force only, in terms of the vector P, for given excitations E^A and E^B, contained in the vector Q.

13.3.3 Geometric Interpretation

Defining the sets

$$S_f(R_f) = \left\{P : P^T P \leq R_f\right\} \tag{13.45}$$

and

$$S_x(R_x) = \left\{P : P^T H^T W^2 H P - 2\,Q^T\,W^2 H P + Q^T W^2 Q \leq R_x\right\} \tag{13.46}$$

a sufficient condition for the satisfaction of both the force and the amplitude constraints described by (13.29) and (13.30) is

$$S_f\left(\frac{1}{N}F^2_{max}\right) \cap S_u\left(\frac{1}{N}\Delta U^2_{max}\right) \neq \emptyset; \tag{13.47}$$

that is, if the (2-norm) lower bounds on the force and amplitude constraints intersect, there is a feasible solution for force and amplitude within the constraints. Clearly, from Fig. 13.3, the condition in (13.47) is slightly conservative, since the sets corresponding to the infinity norms may intersect, even if those corresponding to the 2-norms do not.

From the upper bound on the infinity norm (Fig. 13.3) we can also specify a sufficient condition for the violation of at least one of the constraints in (13.29) or (13.30) as

$$S_f\left(2F^2_{max}\right) \cap S_x\left(2\Delta U^2_{max}\right) = \emptyset. \tag{13.48}$$

The force constraint defined by the set $S_f(R_f)$ can be interpreted as the region of the $2N$-dimensional space S^P enclosed by the *hypersphere* centered at the origin and of radius $\sqrt{R_f}$. The amplitude constraint $S_x(R_x)$ is the region of the space enclosed by the *hyperellipsoid* with axes parallel to the elements of the basis of S^P, because the matrix $H^T W^2 H$ is diagonal and with all positive elements. The center P_c of the hyperellipsoid is

$$P_c = H^{-1}Q = H^{-1}(Q^A E^A - Q^B E^B), \tag{13.49}$$

while the radii r_i are given by

$$r_i = \sqrt{\frac{R_x}{\lambda_i}} \tag{13.50}$$

where λ_i are the eigenvalues of $H^T W^2 H$, that is, the diagonal elements.

Equation (13.47) states that if the intersection between the hypersphere describing the force constraint and the hyperellipsoid describing the amplitude constraint is not the empty set, then, for the given excitation forces, the device is able to satisfy both the amplitude and the force constraints. Conversely, Equation (13.48) states that if the intersection between the two sets is the empty set, then at least one constraint will be violated.

13.4 Special Case of Single Body

A single-body heaving buoy is now considered as a special case of the self-reacting point absorber described in Section 13.2, given by the frequency domain model

$$(j\omega\, m + B + Z(\omega) + S/j\omega)\,V = F_e - F_{pto}. \tag{13.51}$$

The corresponding time domain model is

$$(m+m_\infty)\dot{v}(t)+k(t)*v(t)+B\,v(t)+S\,x(t)=f_e(t)-f_{pto}(t), \tag{13.52}$$

and the energy absorbed by the PTO, neglecting losses, is

$$J(T)=\int_0^T f_{pto}(t)\,v(t)\,dt. \tag{13.53}$$

The energy maximization problem is then discretized by following the same steps performed for the case of a self-reacting device (Section 13.2.1). That is, the excitation force, the velocity, and the PTO force are approximated using truncated Fourier series as in Equations (13.9), (13.11), and (13.12); the best approximation is then obtained by applying the Galerkin method for the minimization of the residual $\hat{r}(t)$, which amounts to solving

$$\langle \hat{r}(t),\phi_{n_f}\rangle=0 \quad \forall \quad n_f=1,\dots N, \tag{13.54}$$

and results in the linear system

$$GX=E-P. \tag{13.55}$$

The matrix G is block diagonal with 2-by-2 blocks of the form of Equation (13.20); the elements on the main diagonal are the real parts of the mechanical impedance at the N frequencies $n_f\omega_0$, as defined in Equations (13.16)–(13.19). If the matrix G is invertible, the converted energy is

$$J(P)=P^TX=-P^TG^{-1}P+P^TG^{-1}E \tag{13.56}$$

where the symmetric part of the matrix G^{-1} is positive definite, because the elements on its main diagonal (linear damping plus radiation resistance) are positive. As a consequence, the quadratic function $J(P)$ is concave and the PTO force $\bar{P}$ that maximizes $J(P)$ in Equation (13.56) is

$$\bar{P}=(G^{-T}+G^{-1})^{-1}G^{-1}E. \tag{13.57}$$

The 2-norm of the force constraint is defined as in Equation (13.37), while the 2-norm of the amplitude constraint becomes

$$\|\hat{x}\|_2^2=\frac{T}{2}\Big(P^TG^{-T}W^2G^{-1}P-2E^TG^{-T}W^2G^{-1}P+E^TG^{-T}W^2G^{-1}E\Big). \tag{13.58}$$

The set

$$S'_x(R_x)=\big\{P:P^TG^{-T}W^2G^{-1}P-2E^TG^{-T}W^2G^{-1}P \\ +E^TG^{-T}W^2G^{-1}E\le R_x\big\}, \tag{13.59}$$

describes the region of the space S^P enclosed by a hyperellipsoid centered at P_c=E; the principal axes are parallel to the elements of the basis because the matrix $G^{-T}W^2G^{-1}$ is diagonal and the radii are

$$r_i' = \sqrt{\frac{R_x}{\lambda_i'}}, \tag{13.60}$$

where λ_i' are the eigenvalues of $G^{-T}W^2G^{-1}$.

The case for N=1 (i.e., two dimensions, P_1 and P_2) is shown in Fig. 13.4; the matrix G^{-1} is of the form in Equation (13.20); thus, $G^{-T}W^2G^{-1}$ is diagonal with two coincident eigenvalues λ', and the set $S_x'(U_{max}^2)$ is a disc centered at P_c=E with radius r'=$U_{max}/\sqrt{\lambda'}$. Considering that the intersection between two disks is nonempty if the sum of the radii is larger than the distance between the centers, it follows that the sufficient condition in Equation (13.47) can be simplified to

$$F_{max} + \frac{U_{max}}{\sqrt{\lambda'}} \geq \|E\|_2. \tag{13.61}$$

Fig. 13.4 has a useful physical interpretation in that the larger bounds on amplitude and force result in larger disks for $S_x'(R_x)$ and $S_f(R_f)$, respectively, with larger (more generous) constraints increasing the likelihood of an intersection between $S_x'(R_x)$ and $S_f(R_f)$. The other important variable affecting intersection is the distance between the disk centers, which, for the one-body case, is simply the excitation force experienced

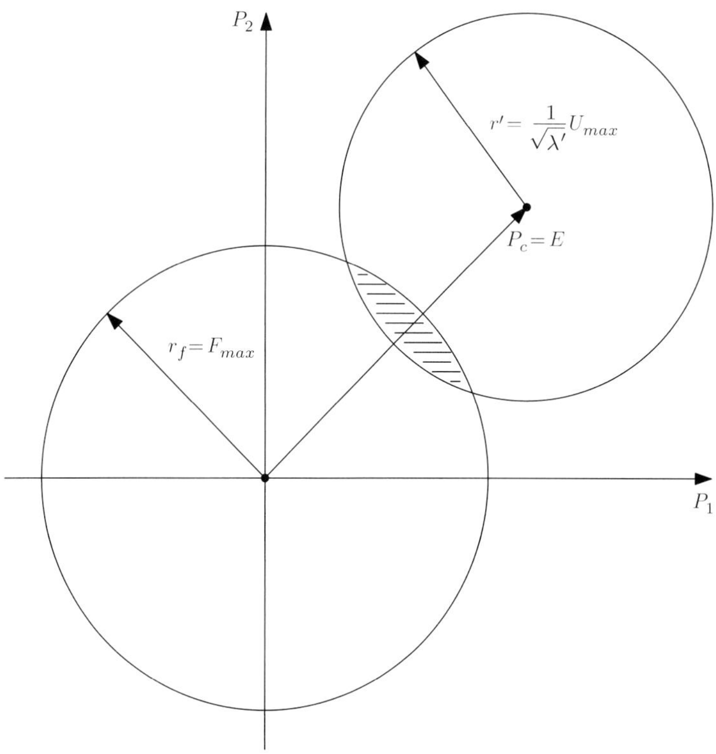

Figure 13.4 Example for one-body device, with N=1. The axes P_1 and P_2 are the components of the vector P describing the PTO force, that is, $P = [P_1\ P_2]^T$.

by the body. Therefore, for more violent sea states, there is a lower likelihood of simultaneously satisfying position and amplitude constraints, as expected.

Equation (13.48), for N=1, can also be simplified to

$$\sqrt{2}\left(F_{max} + \frac{U_{max}}{\sqrt{\lambda_I}}\right) \leq \|E\|_2. \tag{13.62}$$

The geometrical construct in Fig. 13.4 is now of the form of the originally desired specification, shown in Fig. 13.1, where we can identify the functions $f_1(\ )$, $f_2(\ )$, and $f_3(\ )$. A broad conclusion, from Fig. 13.4, is that if the circle centered at P_c contains the origin, no PTO force is required to limit the motion in order to satisfy the amplitude constraint; that is, the wave excitation force can always be contained within the amplitude range of the device.

13.5 Sample Results

Initially, in Section 13.5.1, we will take a single-body device and perform a constraint analysis for a monochromatic sea. This will allow us to develop some insight into the use of the geometrical tool and to consider it from a design perspective. While monochromatic analysis might seem restrictive, appropriate choice of the frequency, such as the resonant frequency of the device or the peak energy frequency of the incident sea, could be sufficient to determine the peak loading condition. However, a polychromatic approach for the single-body case is also considered in Section 13.5.1. Finally, we show results for a multibody device with a polychromatic sea spectrum in Section 13.5.2. In all the cases considered, the hydrodynamic parameters were calculated using the Wamit software from MIT.

13.5.1 Single Body

Here we consider a vertical cylinder with diameter 10 m, draft 25 m, and mass 4.9×10^5 kg. The incident sea is of 3 m amplitude, with a wave period of 11.4 s. PTO amplitude and force constraints are set at 3 m and 200 kN, respectively. The body has a resonant frequency of 0.63 rad/s ($\equiv$ 9.93 s).

Single-Frequency Analysis

Fig. 13.5 shows the situation, with respect to the sufficient conditions for satisfaction of the constraints in (13.61), for a frequency of 0.7 rad/s (corresponding to a period of 9 s). Clearly, there is an intersection between the circles corresponding to the amplitude and force constraints, and the optimal PTO force (for this constrained situation) is expressed by $\bar{P}_{constr}$. Fig. 13.5 also displays the optimal unconstrained PTO force, $\bar{P}$, and the contours corresponding to lines of constant energy. Therefore, it can be seen that, should we wish to relax one or the other constraint in order to increase captured power, increasing the *force* constraint would cause the $\bar{P}_{constr}$ point to move across the

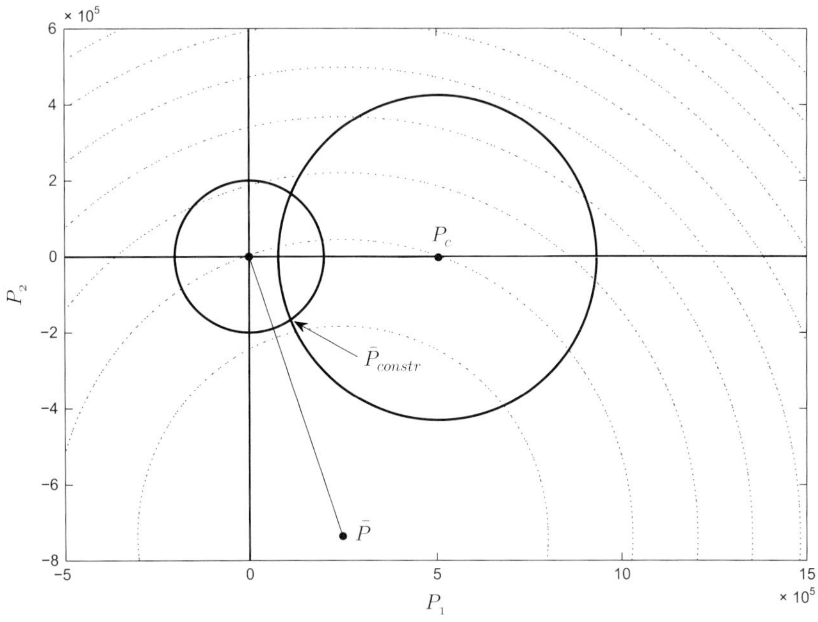

Figure 13.5 Single-body device in a monochromatic sea.

energy contours toward $\bar{P}$, while increasing the amplitude constraint yields very little extra energy.

Multiple-Frequency Analysis

While the analysis at a single frequency of Section 13.5.1 has some useful design features, real seas contain multiple frequency components, and designers need to ensure that physical constraints are satisfied for *all* active frequencies. In Fig. 13.6, we consider the same heaving cylinder in a sea state corresponding to a Bretschneider wave spectrum with parameters $H_{mo} = 6.5$ m and $T_p = 10$ s. In essence, the plane of Fig. 13.5 is extended to three dimensions, with the addition of a frequency axis.

However, this increase in dimension of the figure brings a corresponding reduction in the clarity of interpretation, but we can effectively project the figure information onto a two-dimensional plane. Fig. 13.7 shows this projection, which essentially utilizes Equations (13.61) and (13.62) for the frequency range of interest; that is, $\|E\|_2 / \left(F_{max} + \frac{U_{max}}{\sqrt{\lambda'}}\right)$ is plotted with respect to both 1 and $\sqrt{2}$. The shaded area of Fig. 13.7 is an area in which constraints are not guaranteed to be satisfied but are also not guaranteed to be violated; in essence, the true infinity norms corresponding to $\Delta\hat{u}$ and $\hat{f}_{pto}$ lie somewhere in the shaded region.

For this polychromatic case, we examine two sets of constraints:

1. $\Delta U_{max} = 2$ m, $F_{max} = 450$ kN.
2. $\Delta U_{max} = 4$ m, $F_{max} = 700$ kN.

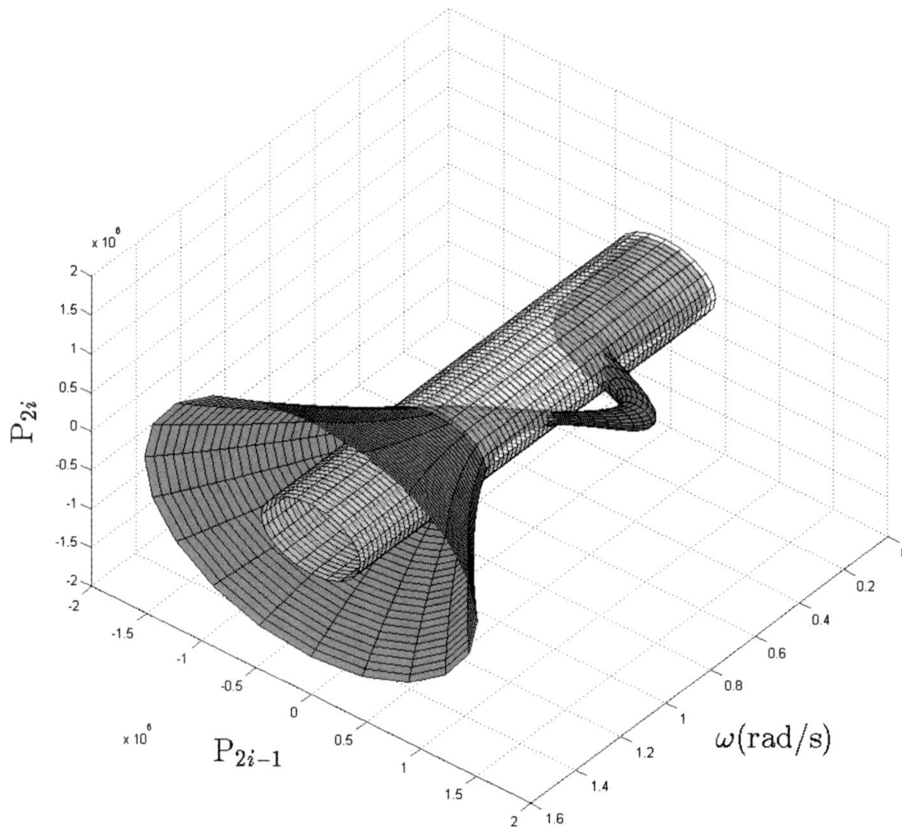

Figure 13.6 Single-body device with Bretschneider spectrum.

From Fig. 13.7, case (1) above clearly violates the constraints, while case (2) clearly (conservatively) satisfies the constraints. Further, detailed examination can now proceed by focusing on the single frequency ω^* and using a plot similar to that of Fig. 13.5, thereby improving the insight into individual constraint satisfaction and manipulation, with the energy landscape as a background available.

13.5.2 Two-Body Case

Our two-body example consists of a device as shown in Fig. 13.2, with the following dimensions:

- Torus inner diameter of 10 m
- Torus outer diameter of 20 m
- Torus draft of 8 m
- Torus mass of 1.93×10^6 kg
- Cylinder diameter of 8 m

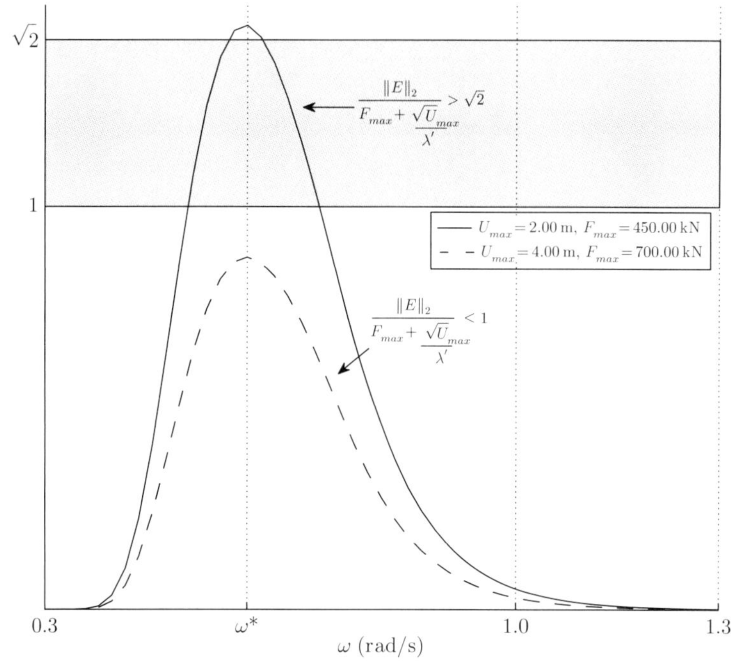

Figure 13.7 Single-body device with Bretschneider spectrum, using two-dimensional projection.

- Cylinder draft of 30 m
- Cylinder mass of 1.55×10^6 kg

The two-body case differs from the single-body case only in that the amplitude and force constraints are relative (between the two bodies) rather than absolute, and, while the general form of Fig. 13.1 is followed, $f_3(F_e)$ is not simply the distance between circles (or, in the polyspectral case, cylinders). Fig. 13.8 shows the situation for four different sets of constraints, where constraint satisfaction is guaranteed only for $\Delta U_{max} = 0.1$ m, $F_{max} = 0.45$ kN. Again, a more detailed examination can be obtained by resorting to the P_{2i}VsP_{2i-1} plane, with a focus on the peak frequency.

A characteristic difference between Figs. 13.7 and 13.8 is that the bandwidth of the two-body system is significantly narrower, reflecting the significant difference in dynamics for the single- and two-body cases, for the range of parameters considered. The discretization effect manifested in Fig. 13.8, resulting in a slightly nonsmooth spectral shape, is the result of numerical effects in Wamit, rather than the frequency points chosen for the Fourier series representation.

13.6 Conclusion

In this chapter, we have developed a procedure to examine if physical constraints will be satisfied for a particular wave energy device, under a particular set of sea conditions.

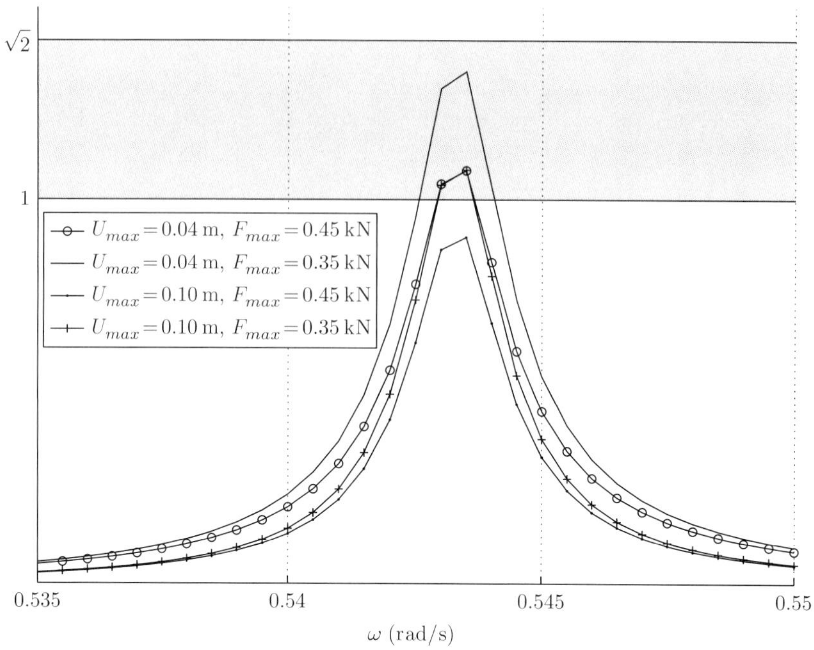

Figure 13.8 Two-body device with Bretschneider sea spectrum.

The method does not require any simulation and does not depend on the control method used to calculate the PTO force. However, the method does require hydrodynamic model parameters, which are used to evaluate the matrix G in Equation (13.14).

The method relies on a discretization of the equation of motion in frequency, using Fourier series, leading to a system of linear equations. A consequence of this discretization, for the polychromatic case, is the need to examine the constraint satisfaction at each discrete frequency. However, this examination can be done in a broad way first and then localized to the key (offending) frequency, leading to the use of a parametric plane described by the couple P_{2i}, P_{2i-1} (corresponding to a single frequency) for further investigation. If desired, the P_{2i}, P_{2i-1} plane can be employed as a design tool, showing constraint features, the energy landscape, and the location of the optimum PTO force. The superposition of these features allows the designer to make informed decisions regarding the value of relaxing (or, indeed, tightening) the constraints, in relation to energy capture. In particular, we envisage this tool being used in the techno-economical optimization of WECs, because it enables one to show the relation between characteristics of the device (maximum PTO force and oscillation amplitude), which are related to the cost, and the amount of energy produced, which is related to the revenue generated by the WEC.

The theoretical development in this chapter supports any device for which a hydrodynamic model can be obtained. The examples given show one (relative) degree of motion, that is, a single-body or a two-body WEC. While the method can be extended to multiple (relative) degrees of freedom, the intuitive feel for the geometric interpretation is likely to be diminished, though absolute confirmation of constraint satisfaction can still be established.

The transformation of the constraints from an infinity-norm representation to the more amenable 2-norm introduces some conservatism. However, given the development of sets of conditions guaranteeing both satisfaction and nonsatisfaction of the infinity norm (based on 2-norm calculations), the margin for error is relatively small, amounting to $\pm 16\%$ overall.

Finally, the development in this chapter is based on a linear hydrodynamic WEC model. While it is known that such linear approximations can suffer a lack of fidelity for large WEC motions (due to viscous and other effects), there is some agreement that linear hydrodynamic models reasonably approximate device motion over the power production region of operation [243, 244]. The interplay between force constraints, amplitude constraints, and optimal power capture in the power production region is the focus of this chapter, while constraint analysis in survival modes, where device dynamics are likely to be significantly different due to PTO locking or unloading, would need a different (and likely nonlinear) approach. Typically, survival modes are analyzed using computational fluid dynamics (CFD) models.

13.7 Appendix

The 2-norm $\|f\|_2$ of the function $f(t)$ is defined as

$$\|f\|_2 = \left(\int_0^T |f(t)|^2 dt \right)^{1/2}, \tag{13.63}$$

while the infinity-norm $\|f\|_\infty$ is defined as

$$\|f\|_\infty = \sup_{t \in [0,T]} |f(t)|. \tag{13.64}$$

A general property relating the 2-norm and the infinity-norm of a function f, for which the norm defined by Equation 13.64 exists, is [245]

$$\|f\|_2 \leq \sqrt{T} \|f\|_\infty. \tag{13.65}$$

For a zero-mean Fourier series with N frequency components, the inequality [246]

$$\|f_N\|_\infty \leq \sqrt{\frac{2N}{T}} \|f_N\|_2 \tag{13.66}$$

provides an upper bound for the infinity-norm as a function of the 2-norm.

14 Interaction between Control and Optimum System Design

14.1 Introduction

Throughout the application history of control systems, a disconnection between the tasks of physical system design and control design is often apparent. Typically, the application specialist designs the physical system, based on years of application experience, and the system is then given to the control engineer for the subsequent design of a suitable control system to optimize the operation of the device, as illustrated in Fig. 14.1(a).

However, it is unlikely that an optimal controlled system results from this sequential approach, since the geometric/structural design is fixed on entry to the control design stage. A better result can usually be obtained if the application specialist is aware of the potential and requirements of the control system, so that the control system may be taken into account at the geometric/structural design stage, as suggested in Fig. 14.1(b). Unfortunately, for a variety of reasons, often related to a lack of communication or trust between the application-specific and control domains, the sequential route is the normal one. There are a few notable exceptions, usually in relatively exotic application areas such as aerospace, where large multidisciplinary teams design such systems and there is always a quest for higher performance. For example, the General Dynamics F16 Fighting Falcon was the first aircraft to feature *relaxed static stability* in the knowledge that a feedback control system would stabilize the system, resulting in superior aerodynamic performance than an aircraft with inherent positive static stability across the full operational envelope [247].

This chapter examines two scenarios where cooperative design between the designers of wave energy systems and control engineers can result in performance in excess of what can be achieved using the sequential design philosophy of Fig. 14.1(a). In Section 14.2 the interplay between the control system and the optimal (single) device geometry is examined, while Section 14.3 considers the interaction between the control system and the optimal layout of wave energy converters (WECs) in wave farms.

14.2 Sensitivity of Optimal Wave Energy Device Geometry to Control

In general, the WEC geometry is designed for the prevailing sea conditions in a particular location, with control subsequently used to improve the energy conversion

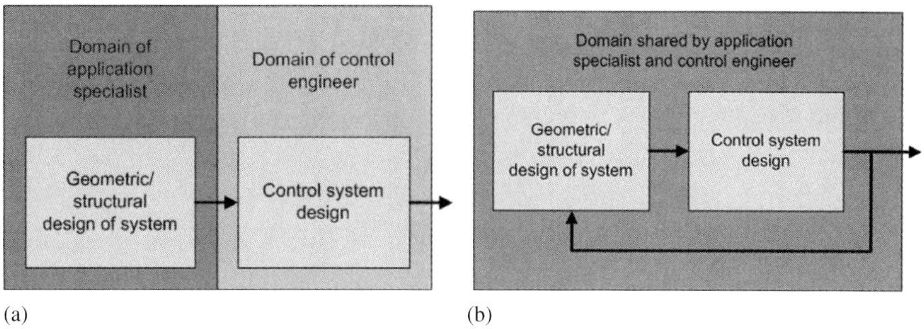

Figure 14.1 Sequential (a) and integrated design (b) philosophies.

performance of the device for sea states other than the design sea state. This usually involves matching the device frequency response, or response amplitude operator (RAO), to the predominant sea spectrum.

However, particular control strategies subsequently employed, while doing their best to broaden the RAO bandwidth, have limited capability to produce an overall optimal system. Two cases in point highlight the issue. Latching control [248, 249], which delays the motion of a WEC by locking its position at the extreme (upper and lower) points of motion, has the ability to slow down the WEC in sea periods longer than the design period. On the other hand, declutching control [250] has the opposite effect – for intervals of the motion cycle, the WEC is "unloaded," allowing the response to be speeded up in sea periods shorter than the design period. Both latching and declutching produce composite device/control ("closed loop") systems, which are non-symmetrical with respect to the predominant sea state.

Aiming to improve wave energy extraction, a number of studies have been carried out on device geometry optimization of WECs *independently* of the control system [107, 251–254]. However, a preliminary study, examining the influence of latching control on the overall geometric optimization of a heaving cylinder [255], suggests that the control system can have a significant influence on the optimal device geometry.

In this section, a focus is placed on the implications of geometry optimization in the absence of control, and the ultimate effect of the incorporation of control influence at the device geometry optimization stage, as illustrated in Fig. 14.2. A complete analysis for a WEC employing a variety of control strategies is presented. The control strategies studied are latching control, declutching control, and an optimal numerical controller [256], as documented in Chapter 10. The aim is to examine how different optimal WEC geometries evolve, depending on the different WEC control strategies employed, and the differences in overall power absorption characteristics of WEC geometries optimized with and without control considerations.

14.2.1 Mathematical Modeling

14.2.1.1 Equations of Motion

Here we consider a generic point absorber with a generic power take-off (PTO) system, as illustrated by Fig. 14.3. The point absorber is represented as a truncated vertical

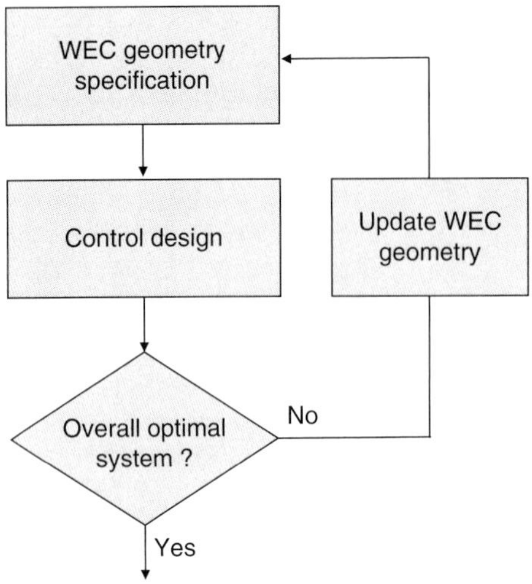

Figure 14.2 Device optimization informed by the control system employed.

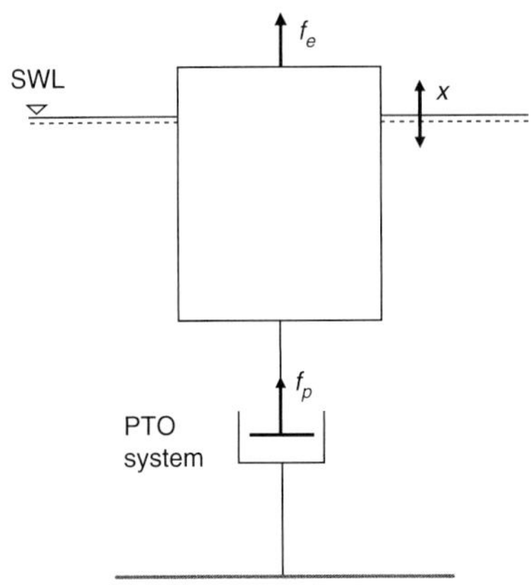

Figure 14.3 Schematic of the generic heaving floating body.

cylinder with radius r and draft d. As in previous chapters, linear hydrodynamic theory and heave oscillatory motion are assumed. In such a case, the motion of the floating body can be described, as in Equation (4.27), by the Cummins' equation [63],

$$[M + M_r(\infty)]\ddot{x}(t) + \int_0^{\infty} K(t-\tau)\dot{x}(\tau)d\tau + S_h x(t) = f(t)\,, \tag{14.1}$$

where $x(t)$ is the vertical position of the body; M is the body mass; $M_r(\infty)$ is the infinite-frequency added mass coefficient, defined with the asymptotic values of the added masses at infinite frequency; S_h is the hydrodynamic stiffness coefficient; and $f(t)$ is the total external force acting on the body. The kernel of the convolution or fluid memory term, $K(t-\tau)$, is expressed as [63]

$$K(t-\tau) = \frac{2}{\pi}\int_0^{\infty} B_r(\omega)\cos[\omega(t-\tau)]d\omega \tag{14.2}$$

where $B_r(\omega)$ is the radiation damping coefficient and ω is the wave frequency. Assuming the body is freely floating in equilibrium, the body mass is equal to the mass of the displaced water in free flotation, that is,

$$M = \rho \pi r^2 d, \tag{14.3}$$

where ρ is the water density. The external forces acting on the body,

$$f(t) = f_e(t) + f_p(t), \tag{14.4}$$

are the wave excitation force $f_e(t)$ and the PTO force $f_p(t)$. The excitation force is calculated as $f_e(t) = \mathcal{F}^{-1}\{F_e(\omega)\eta(\omega)\}$, where $\eta(\omega)$ is the Fourier transform of the wave elevation and $F_e(\omega)$ is the excitation force transfer function.

If we consider a generic PTO system with a linear damper (coefficent $B_p \in \mathbb{R}_+$), then the PTO force is given by

$$f_p(t) = -B_p\,\dot{x}(t)\,, \tag{14.5}$$

where $\dot{x}(t)$ is the vertical velocity of the body.

Absorption of Wave Power

In order to absorb energy from the waves, a point absorber must be equipped with a PTO system, as depicted in Fig. 14.3. Thus, the extracted energy and the mean extracted power by the WEC over a time range T_0 are, respectively, calculated as

$$J = -\int_0^{T_0} \dot{x}(t) f_p(t) dt, \tag{14.6}$$

$$P_a = \frac{J}{T_0}, \tag{14.7}$$

where f_p is the force applied by the PTO system.

The natural (undamped) resonant period of the device is defined as

$$T_r = 2\pi\sqrt{\frac{M + M_r(\infty)}{S_h}}\,. \tag{14.8}$$

Recall the two conditions for maximizing wave energy conversion in monochromatic waves, that is, the phase and amplitude conditions [66], as described in Equations (6.36) and (8.9). The phase condition is satisfied when the velocity of the body is in phase with the wave excitation force, which naturally occurs when the natural resonant period of the device (T_r) is equal to the wave period. The amplitude condition indicates that there is an optimal value for the damping factor B_p in Equation (14.26). Assuming the phase condition is satisfied, then the optimal damping factor is equal to the radiation damping coefficient B_r at the incident wave frequency [66].

Polychromatic waves (an approximation of real waves) are not defined by a single frequency in the time domain. Furthermore, the wave power varies on several time scales: from wave to wave (seconds), according to sea states (hours), and according to seasonal variations (months). Thus, the device geometry is usually designed for the prevailing sea conditions in a particular location, and a frequency that best characterizes the wave spectrum is selected. In this case, the natural resonant period T_r and the damping factor B_p can be tuned to the peak period (T_p) or the energy period of the spectrum (T_e). Control systems are usually subsequently employed to improve the energy conversion performance of the device for sea states other than the design sea state.

14.2.2 Optimization of the WEC Geometry

Optimization Problem

Here our aim is to examine the different optimal geometries of a heaving cylinder that are obtained when the geometry optimization process is done without any control strategy employed on the WEC, and also when different control strategies are implemented. From Equation (14.8), the parameters that define the natural period of the device are the mass M, the infinite added mass coefficient $M_r(\infty)$, and the hydrodynamic stiffness coefficient S_h. Such parameters are modified according to the radius r and draft d of the heaving cylinder.

Thus, the optimization of the device (Fig. 14.3) can be defined as: *determine the natural resonant period T_r or, equivalently, the radius r and draft d that maximize the average absorbed power P_a (14.7) by the WEC with and without control strategies.* Three different control strategies are considered in the analysis: latching control, declutching control, and an optimal (model predictive control (MPC)-like) controller.

Latching Control

As described in Chapter 9, the principle of latching control is to lock the motion of the body at the moment its velocity vanishes and to wait for a favorable situation to release the body again. As a result, the body velocity is forced to be in phase with the wave excitation force. The determination of the duration that the body is locked (latching duration, T_L) is the problem to be solved.

Latching control "slows down" the natural response of the device when the wave period is longer than the device period, in order to force the velocity of the body and the excitation force to reach their extrema (maxima or minima) at the same time.

The latching time can be determined semi-analytically for regular waves [257, 258]. However, for polychromatic waves, the concept of phase between excitation force and velocity is not well defined, in which case the optimization of the latching duration does not return a unique solution [257]. In this case, the latching duration (or, equivalently, the unlatching time) can be optimized to synchronize the peak of the velocity with the peak of the excitation force [259] or to maximize the absorbed power [249]. In addition, a number of different strategies have been proposed to determine the best unlatching time for polychromatic waves; see, for example, [260–265].

In this study, we adopt the strategy proposed by Falcão [263] and later referred to as *threshold unlatching* [264]. This strategy consists of unlatching the body at the instant when the excitation force (or other reference variable) passes a chosen threshold [264]. Here, the unlatching threshold for the wave excitation force is set to zero.

Declutching Control

Declutching control, also called *freewheeling* or *unlatching*, was considered originally in [266]. It consists of unloading the device during some parts of the power cycle, which means that the PTO force is set to zero during such instants, as described in Chapter 9.

Under declutching, the PTO is disconnected (declutched) for a period following a maximum or minimum of position (i.e., velocity going to zero). Once the device has gained momentum under this no-load condition, the PTO is again connected, loading the system. The control problem, in this case, is the determination of the declutching duration for which the PTO is disconnected. In [250], an optimal command theory based on Pontryagin's principle was proposed to determine the instants for clutching/declutching of a WEC. Two different strategies are used in [134] to determine the instants for declutching: in the first one, the body is declutched approximately $T_r/4$ before the next (predicted) peak of the excitation force, and in the second strategy, it is declutched after the excitation force passes a chosen threshold.

Declutching has also been studied in [267], which shows that declutching is an ideal strategy when the natural resonant period of the device is longer than the sea period. Effectively, loading and unloading the device allows it to "catch up" to the excitation force, which brings the device velocity into phase with the excitation force.

Following [267], we adopt declutching control for the cases when the natural resonant period of the device is longer than the sea period. Here, the instants for clutching the device are determined by the instants the excitation force reaches its extrema. In this case, the PTO system is disengaged (declutched) when the device velocity vanishes, and it is engaged (clutched) again when the excitation force achieves its maximum or minimum value.

A combination of latching and declutching control is presented in [268], and a larger increase in the energy absorption is achieved, when compared with latching or declutching implemented independently. It was also shown that declutching works better with large values of the PTO damping, whereas latching works better with small values [268].

Latching and declutching control strategies are considered as sub optimal control strategies, since they deal only with the phase condition for improving the wave energy absorption. Further improvements can be achieved if the PTO damping coefficient B_p is adjusted according to the sea state, as has been done, for example, in [267] and [269]. Both strategies implement an on/off PTO force, usually through a braking mechanism (latching) or PTO bypass (declutching). In such cases, the PTO force is defined by Equation (14.5) for the instants the PTO is on.

MPC-like Control Algorithm

The control problem for the MPC-like control algorithm is defined as follows, following the development in Chapter 10: to find the optimal profile for the PTO force that maximizes the total energy absorbed (14.6) over a time interval T_0, subjected to the equation of motion (14.1).

For this purpose, the control problem is discretized by approximating the velocity and the PTO force with a linear combination of basis functions. Such a combination results in a finite-dimension optimization problem. Following the procedure from Chapter 10, trigonometric functions are chosen as basis functions. Then, the PTO force and the velocity are, respectively, approximated with the truncated zero-mean Fourier series

$$\dot{x}(t) \approx \sum_{n_f=1}^{N/2} \nu_{n_f}^{c} \cos(n_f \omega_0 t) + \nu_{n_f}^{s} \sin(n_f \omega_0 t), \tag{14.9}$$

$$f_p(t) \approx \sum_{n_f=1}^{N/2} \phi_{n_f}^{c} \cos(n_f \omega_0 t) + \phi_{n_f}^{s} \sin(n_f \omega_0 t), \tag{14.10}$$

where ω_0 is the fundamental frequency of the Fourier series and N is the total number of components.

The best approximation of the solution for the equation of motion (14.16) is sought by applying the Galerkin method, and the result is the linear system

$$GX = P + E, \tag{14.11}$$

where X, P, and E are, respectively, the vectors of the Fourier coefficients of the velocity, PTO force, and excitation force, and G is the matrix of hydrodynamic coefficients of the system. The set of vectors are arranged as:

$$X = \left[\nu_1^c, \nu_1^s, \nu_2^c, \nu_2^s, \ldots, \nu_{\frac{N}{2}}^c, \nu_{\frac{N}{2}}^s\right]^T,$$
$$P = \left[\phi_1^c, \phi_1^s, \phi_2^c, \phi_2^s, \ldots, \phi_{\frac{N}{2}}^c, \phi_{\frac{N}{2}}^s\right]^T,$$
$$E = \left[\epsilon_1^c, \epsilon_1^s, \epsilon_2^c, \epsilon_2^s, \ldots, \epsilon_{\frac{N}{2}}^c, \epsilon_{\frac{N}{2}}^s\right]^T.$$

The matrix $G \in \mathbb{R}e^{N \times N}$ is defined as

$$G = \begin{bmatrix} \mathrm{D}_1 & \mathrm{C}_1 & 0 & \cdots & 0 & 0 \\ -\mathrm{C}_1 & \mathrm{D}_1 & 0 & \cdots & 0 & 0 \\ 0 & 0 & \ddots & & \vdots & \vdots \\ \vdots & \vdots & & \ddots & 0 & 0 \\ 0 & 0 & 0 & 0 & \mathrm{D}_{N/2} & \mathrm{C}_{N/2} \\ 0 & 0 & 0 & 0 & -\mathrm{C}_{N/2} & \mathrm{D}_{N/2} \end{bmatrix},$$

with

$$\mathrm{D}_{n_f} = B_{r,n_f}(n_f \omega_0),$$
$$\mathrm{C}_{n_f} = n_f \omega_0 \left(M + M_{r,n_f}(n_f \omega_0)\right) - \mathrm{S}_{h,n_f}/(n_f \omega_0).$$

Assuming the wave excitation force is known completely into the future and G is nonsingular, the vector of the optimal PTO force P^* is obtained by solving the following optimization problem [270]:

$$P^* = \arg\max_P(-P^T G^{-1} P - P^T G^{-1} E). \tag{14.12}$$

14.2.3 Results

Simulation Parameters

Simulation results with polychromatic waves are presented in this section. The polychromatic waves are calculated as a linear combination of monochromatic waves using a Bretschneider spectral envelope, defined by the significant wave height H_s, peak period T_p, and random phases.

The optimal geometries for each of the cases of an uncontrolled device (UD), latching controlled device (LD), declutching-controlled device (DD), and an MPC (like)-controlled device (MPCD) are determined via simulations with a set of possible values of radius r and draft d, in which 200 different cylinders are considered. As a result, a clear overview of the mapping between the absorbed power P_a and the optimization parameters can be observed. The ranges of both the radii and drafts go from 2 to 20 m, in steps of 1 m (radius) and 2 m (draft).

The coefficients $F_e(\omega)$, $B_r(\omega)$, $M_r(\infty)$, and S_h for all the cylinders are computed using the boundary element solver WAMIT [271]. The setup parameters for WAMIT are $\rho = 1025\,\text{kg/m}^3$, water depth: 60 m, gravity: 9.81 m/s^3, and frequency range: $(\pi/128 : \pi/128 : \pi)$ rad/s. For the MPCD, the simulation parameters are $N = 128$ and $\omega_0 = 0.0304$ rad/s. For all simulations, T_0 is set to 1200 s, which is the usual recording time used to obtain real statistical values of a wave field and to characterize a sea state [272].

Different criteria are used to tune the PTO damping coefficient B_p for the uncontrolled case and the latching/declutching strategies. For the uncontrolled case,

$$B_p = \sqrt{(B_r(\omega))^2 + (\omega(M + M_r(\omega)) - S_h/\omega)^2}\,, \tag{14.13}$$

which represents optimal linear damping when the body is subjected to incident regular waves [66]. Irregular waves are not defined by a single frequency in the time domain. For this reason, a frequency that best characterizes the wave spectrum should be selected. Here B_p is tuned to the peak frequency of the spectrum ω_p for each sea state.

For the latching and declutching control cases, the optimal damping for each one of the geometries and sea states is determined. In these cases,

$$B_p = D_m/100\,M, \tag{14.14}$$

and the problem to be solved is to determine the coefficient D_m, which optimizes the absorbed power for a specific geometry, control strategy, and sea state. Then, the optimal D_m coefficient determined for each of the 200 geometries is applied at the main evaluation loop (the loop that evaluates the optimal geometry). To illustrate how the coefficients D_m were calculated, Fig. 14.4 shows the evolution of the power absorbed versus D_m for some specific geometries. In these cases, $H_s = 1$ m and $T_p = 7$ s.

Simulation Results

Figs. 14.5 and 14.6 illustrate the evolution of the absorbed power P_a for different values of r and d for each of the four cases (UD, LD, DD, and MPCD). The figures also show the level curves of P_a under the surface plot, allowing easy identification of the optimal

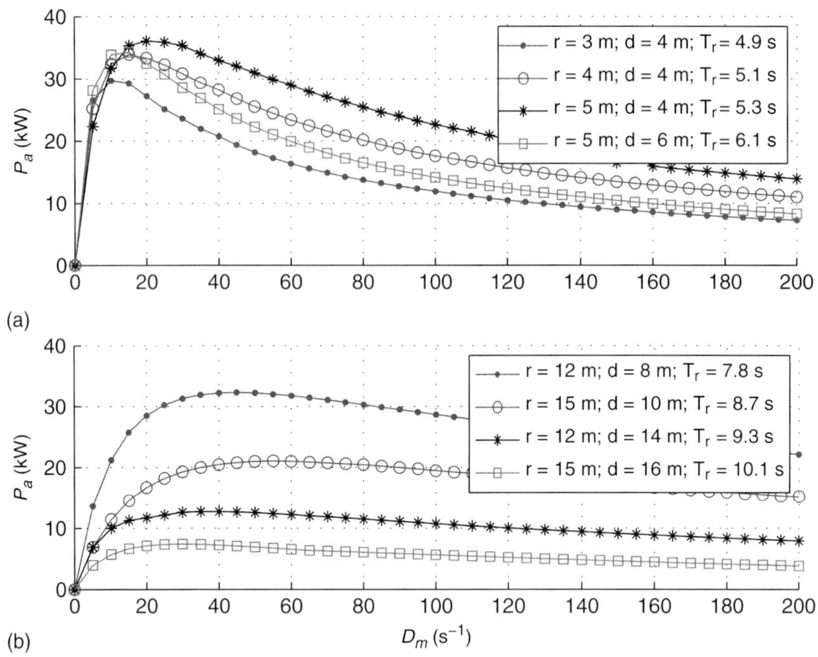

Figure 14.4 Evolution of the power absorbed versus the coefficient D_m: optimization of the optimal linear damping for the LD (a) and DD (b).

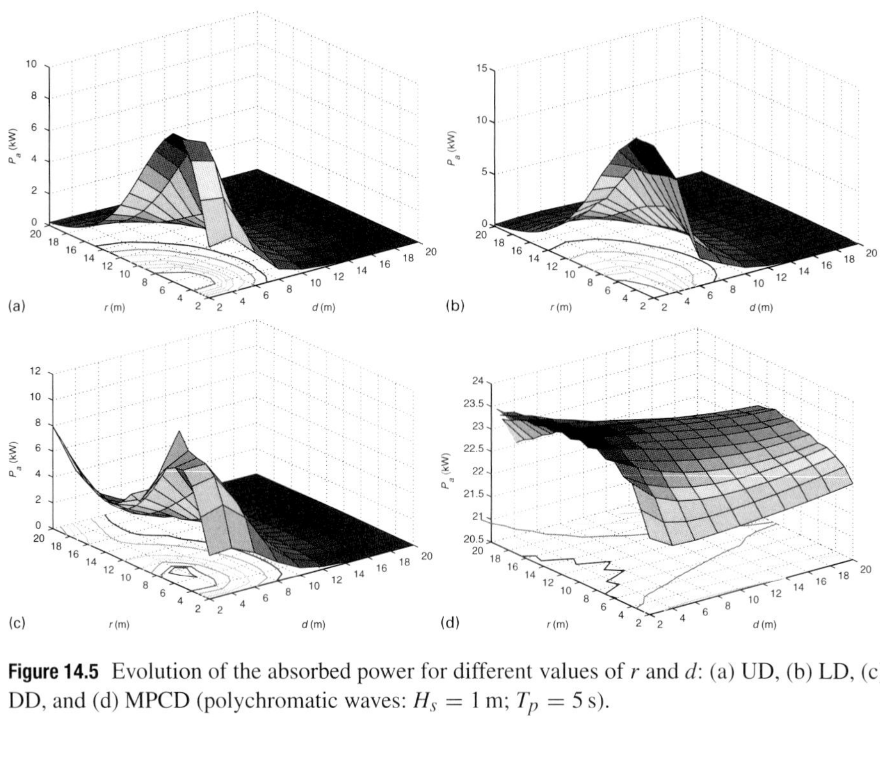

Figure 14.5 Evolution of the absorbed power for different values of r and d: (a) UD, (b) LD, (c) DD, and (d) MPCD (polychromatic waves: $H_s = 1$ m; $T_p = 5$ s).

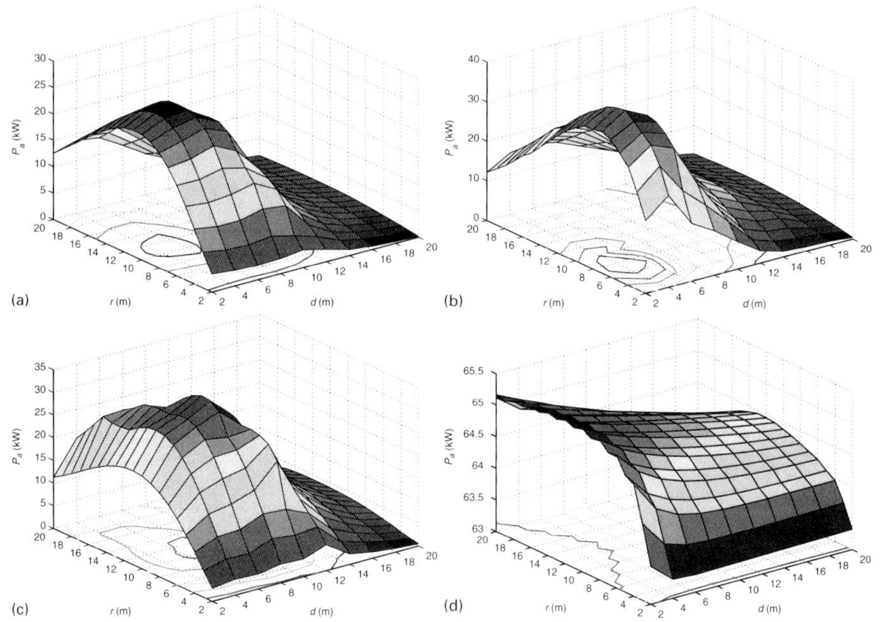

Figure 14.6 Evolution of the absorbed power for different values of r and d: (a) UD, (b) LD, (c) DD, and (d) MPCD (polychromatic waves: $H_s = 1$ m; $T_p = 7$ s).

cylinder parameters. The devices are subject to polychromatic waves with $H_s = 1$ m and peak periods of 5 s and 7 s, respectively. Table 14.1 summarizes the optimal values (r^*, d^*, T_r^*) that define the device geometry and the absorbed power P_a obtained for these cases.

The results show that the optimal geometry obtained with latching control are different from the optimal geometry obtained when no control is applied. As expected, the device natural resonant period for the uncontrolled cases is very close to the peak period T_p of the wave spectrum, whereas the optimal values of T_r are slightly smaller for the latching-controlled devices. In the latching controlled case, the optimal values for r and d are also smaller than the uncontrolled case. It is worth mentioning that larger differences between latching-controlled devices and uncontrolled devices can be found if smaller PTO damping coefficients are applied to the WEC, as shown in [255]. The optimal geometries obtained with declutching control are also slightly different from the uncontrolled case. For the MPC-controlled device, the evolution of the absorbed power is significantly different from all of the three other cases (UC, LD, and DD). The controller, in this case, optimizes the absorbed power for all geometries, regardless of the resonant period of the device and the wave period. In some respects, it can be considered that the MPC-like controller combines the best features of both latching and declutching, since it has significantly more degrees of freedom in parameterizing the control signal, compared with latching and declutching. The absorbed power, for the best geometry, is not significantly different from other geometries in the same map (Figs. 14.5(d) and 14.6(d)).

Table 14.1 Optimal Values of r, d, and T_r of the Cylinders and Absorbed Power P_a for UD, LD, DD, and MPCD (Polychromatic Waves with $H_s = 1$ m)

	$T_p = 5$ s			
	UD	LD	DD	MPCD
r^* (m)	6	2	8	10
d^* (m)	4	2	4	4
T_r^* (s)	5.55	3.68	5.91	6.24
P_a (kW)	9.38	14.67	10.82	23.58
	$T_p = 7$ s			
	UD	LD	DD	MPCD
r^* (m)	11	6	12	12
d^* (m)	6	4	8	4
T_r^* (s)	7.06	5.55	7.79	6.56
P_a (kW)	26.34	39.73	33.29	65.13

In order to illustrate the effect of each of the control strategies on the device, Fig. 14.7 shows the motion of the device controlled by latching, declutching, and the MPC-like algorithm, respectively, for a chosen geometry. In these cases, $H_s = 1$ m and $T_p = 7$ s. The MPC-like algorithm optimizes the PTO force applied to the device on a wave-to-wave time scale, resulting in more exaggerated motion of the body, and consequently more absorbed power than the latching and declutching strategies. Conceptually, LC and DC are suboptimal control methods aiming to achieve the optimum phase condition between the device velocity and the wave excitation force, but not necessarily imposing the amplitude condition.

Discussion

Table 14.2 shows the relative size of the devices optimized for irregular waves with $H_s = 1$ m and $T_p = 7$ s and the various control scenarios. For example, it is interesting that the optimal latched device has a significantly larger power/volume ratio (i.e., reduced device size) than the uncontrolled case. However, a reduced device size is not evident for all control cases, and many other considerations need to be taken into account (such as stroke length and PTO force limits) in any comparative economic assessment.

Table 14.3 shows the power absorbed for the UD, LD, DD, and MPCD for all the optimal geometries (GEO A, GEO B, GEO C, and GEO D, respectively). When we compare the power absorbed for the LD with GEO A with the power absorbed with GEO B, an improvement of 18% is obtained for GEO B. For the DD case, when we compare the power absorbed with GEO A with the power with GEO C, an improvement of 10% is obtained for GEO C. Notice that different strategies based on optimal command theory, to compute the instants for clutching/declutching, for example, [250], could be adopted to verify if greater improvements can be obtained for declutching control.

Table 14.2 Relative Size of the Various Optimized Devices and Relative Power Captured per Unit Volume

Scenario	Geometry	r (m)	d (m)	V_o (m^3)	P_a (W)	P_a/V_o
UD	GEO A	11	6	2280.80	26340	11.55
LD	GEO B	6	4	452.39	39730	87.82
DD	GEO C	12	8	3619.11	33290	9.20
MPCD	GEO D	12	4	1809.56	65130	35.99

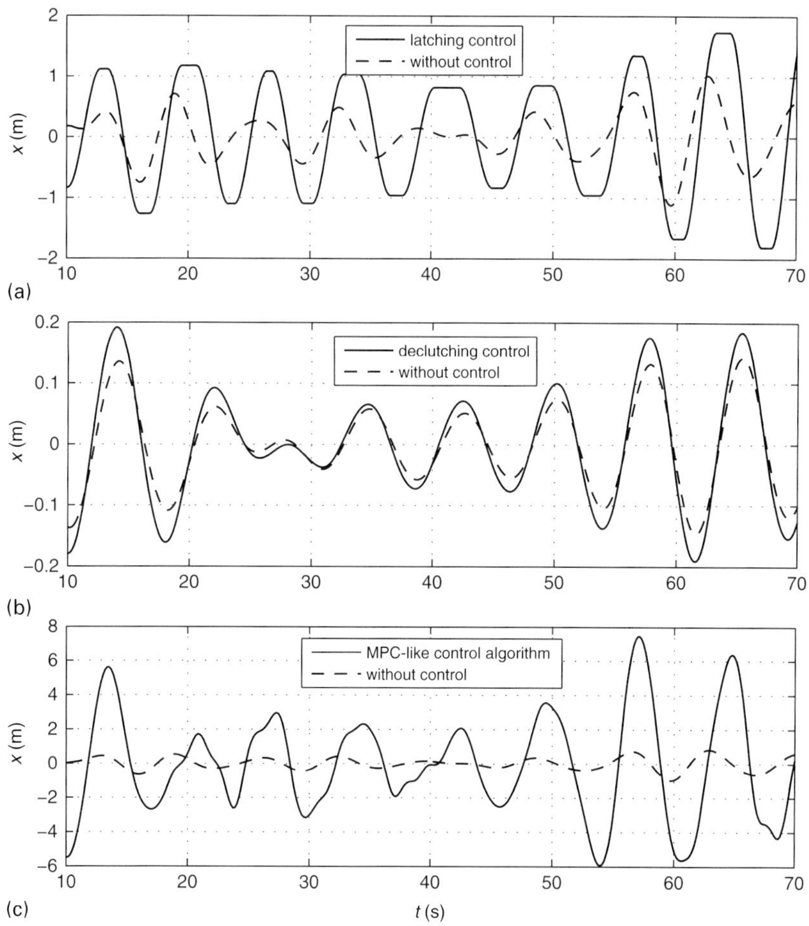

Figure 14.7 Motion of a heaving cylinder with control (solid lines) and without control (dashed lines): (a) LD ($r = 4$ m; $d = 4$ m; $T_r = 5.1$ s); (b) DD ($r = 12$ m; $d = 14$ m; $T_r = 9.3$ s); and (c) MPCD ($r = 4$ m; $d = 4$ m; $T_r = 5.1$ s).

It can also be noted that the optimal geometry obtained for the uncontrolled case (GEO A), for example, is not optimal if one wants, afterward, to apply latching control on it. The converse remark can also be stated for the optimal geometry obtained with the latching-controlled device (GEO B); the natural mechanical properties of the optimal

Table 14.3 Power Absorbed (kW) for Various Control Strategies with Different Optimal Device Geometries

	Scenario			
Geometry	UD	LD	DD	MPCD
GEO A	26.34	33.75	30.30	64.95
GEO B	17.13	39.73	19.05	64.75
GEO C	21.88	26.54	33.29	64.92
GEO D	25.55	33.10	28.02	65.12

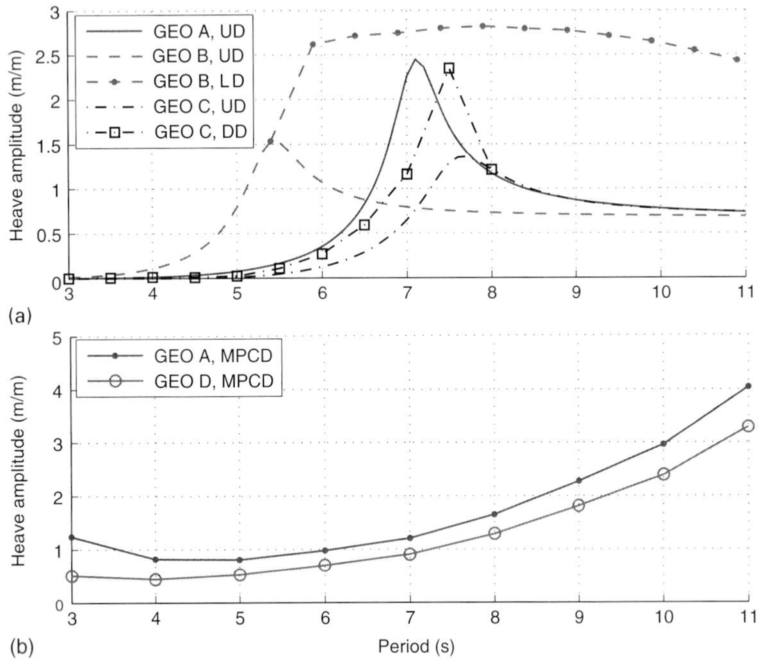

Figure 14.8 Response amplitude operators for optimal geometries (a) GEO A without control, GEO B with latching and without control, and GEO C with declutching and without control; (b) GEO A and GEO D with the MPC-like control algorithm.

shape are no longer optimal for the considered sea state, if one does not want, afterward, to apply control to the device.

Nevertheless, the MPC-like algorithm optimizes the wave energy absorption regardless of the geometry of the device. Four different geometries result in about the same amount of power. Such behavior is in accordance with one of the wave energy principles from Falnes [66], which states that a big body and a small body may produce equally large waves (and hence convert wave energy into useful energy), provided the smaller body oscillates with larger amplitude. However, it must be noted that no physical WEC constraints (force amplitude or velocity) are considered here. For practical application studies, constraints on the amplitude of the device motion or on the PTO force should be taken into account. The interested reader is referred to [279] for further details.

Fig. 14.8 compares the RAOs of the optimal geometries. Latching control extends the bandwidth of the RAO for wave periods greater than the device resonant period. Since GEO B is smaller than GEO A, the benefit of applying such a strategy is now also extended for periods between 5 and 7 s. If GEO A has been considered, the benefit of latching control would be seen only for periods longer than 7 s. The bandwidth for the declutching control is slightly broader than the UD case for wave periods shorter than the device resonant period (Fig. 14.8(a)). However, the benefit of the declutching control is seen mostly around the resonant period of the geometry, as has also been shown in [268]. On the other hand, the MPCD broadens the bandwidth of the RAO regardless of the period and geometry of the device, as illustrated by Fig. 14.8(b).

14.3 Control-Informed Optimal Array Layout for Wave Farms

As documented in Chapters 6 and 11, the effect of array interactions on the energy absorption may be either constructive or destructive, depending on whether the average energy of each unit in the array is higher or lower than the energy of isolated units. Such an effect is usually quantified by the q factor [93] (see Equation (6.121)).

For convenience, we define the q factor here again [93] as

$$q = \frac{W}{n\,W_s} \tag{14.15}$$

where W is the total energy absorbed by the array, W_s is the energy absorbed for a single isolated device, and n is the number of units in the array. If wave interactions have a constructive effect on the absorbed energy, then $q > 1$, and if the effect is destructive, $q < 1$.

A number of studies on wave energy farms have been performed over the last few decades, but the development of optimal array layouts still remains a challenging problem. From an overall economic performance perspective, the optimal layout is a function of the hydrodynamic interactions (amount of power), cost of moorings [273] and electrical interconnections [274], smoothing properties (quality of power), and maintenance costs, given a fixed capital cost for each WEC. In the following analysis, the focus is purely on the optimal hydrodynamic array layout, in terms of power maximization for a given number of WECs and the particular control systems employed.

A number of studies have investigated the influence of control strategies on wave energy array performance. Utilizing suboptimal control strategies based on linear damping [181, 266], it has been shown that the increase in the average interaction factor for irregular waves may be less than suggested by regular wave analysis, but it is still not insignificant. Different strategies based on MPC [270, 280] have shown that higher performance improvements can be achieved when the controllers of each unit run cooperatively by transmitting information to each other. However, the focus of all of these array control studies is to design the control to maximize the power capture for a *given* array layout. Such an approach follows the standard sequential design procedure, where application systems are designed by application specialists and the control system is *subsequently* added, as outlined in Fig. 14.1(a). In this study, we consider *simultaneous* design of the array layout and control system.

By virtue of conservation of energy, a WEC must interact with the waves so as to reduce the amount of wave energy that is otherwise present in the sea. The device must be able to generate waves, which interfere destructively with the sea waves [275]. Thus, in order to absorb power, a certain fraction of the excitation power arriving at the device is necessarily returned to the sea as radiated power. The radiated wave interferes with the incident wave, and the resulting wave, which propagates beyond the device, transports less power than the incident wave [275].

Control strategies for optimizing the absorbed energy modify the oscillation of each device and, consequently, the radiated wave patterns are also modified. For regular waves and optimum oscillation conditions of the body, the radiated power equals the absorbed power [66]. Therefore, it is expected that the optimal hydrodynamic array layout will be different, depending on whether the WEC is controlled or not and *how it is controlled*. In particular, some (global) control strategies can take array interactions into account [270], while other (independent) strategies do not. In the following, it will become clear that suboptimal array layouts result from optimizing the array layout for one particular controller (or no controller) and subsequently employing a different one. To illustrate the impact of such effects, the influence of three different control strategies on the optimal array layout is evaluated. The control strategies employed are passive control (PC), global control (GC), and independent control (IC). The GC and IC algorithms considered in this study were proposed in Chapter 11 [270]. Since these strategies are model-based control strategies, the formulation of the hydrodynamic model presented here follows the hydrodynamic model formulation of Chapter 11. It should be noted that a preliminary study examining the effect of control strategies on optimal wave farm layout was presented in [276], which compared global array control with passive damping for a fixed incident wave angle.

14.3.1 Equations of Motion

Linear hydrodynamic theory and heave oscillatory motion are assumed. In such a case, the motion of the floating bodies can be described, as in Chapter 11, by Cummins' equation [63] (see Equation (4.27)),

$$\mathbf{M_t}\ddot{\mathbf{x}}(t) + \mathbf{B_v}\dot{\mathbf{x}}(t) + \int_0^\infty \mathbf{K}(t-\tau)\dot{\mathbf{x}}(\tau)d\tau + \mathbf{S_h}\mathbf{x}(t) = \mathbf{f}(t)\,, \tag{14.16}$$

with $\mathbf{M_t} = [\mathbf{M} + \mathbf{M_r}(\infty)]$, $\mathbf{f}(t) = \mathbf{f_e}(t) + \mathbf{f_p}(t)$, and the kernel of the convolution term (fluid memory term) expressed as

$$\mathbf{K}(t-\tau) = \frac{2}{\pi}\int_0^\infty \mathbf{B_r}(\omega)\cos[\omega(t-\tau)]d\omega \tag{14.17}$$

where

- $\mathbf{x}(t) \in \mathbb{R}^n$ is the vector of vertical positions of the bodies;
- $\mathbf{M} \in \mathbb{R}^{n\times n}$ is the mass matrix of the system and $\mathbf{M_r}(\infty) \in \mathbb{R}^{n\times n}$ is the added mass matrix, defined with the asymptotic values of the added masses at infinite frequency;
- $\mathbf{S_h} \in \mathbb{R}^{n\times n}$ is the hydrodynamic stiffness matrix;
- $\mathbf{B_v} \in \mathbb{R}^{n\times n}$ is the viscous damping matrix, $\mathbf{B_r}(\omega) \in \mathbb{R}^{n\times n}$ is the radiation resistance matrix, and ω is the wave frequency;
- $\mathbf{f_e}(t)$ is the vector of excitation forces. The excitation force is calculated as $\mathbf{f_e}(t) = \mathcal{F}^{-1}\{\mathbf{F_e}(\omega)\eta(\omega)\}$, where $\eta(\omega)$ is the Fourier transform of the wave elevation and $\mathbf{F_e}(\omega)$ is the excitation force transfer function. The excitation power is the energy delivered by the incoming waves to each device, that is, $\mathbf{P_e}(t) = \mathbf{f_e}(t)\dot{\mathbf{x}}(t)$ [66];
- $\mathbf{f_p}(t)$ is the vector of PTO forces. $\mathbf{f_p}(t)$ is described in the next section according to the control strategy applied to the WEC.

Matrices $\mathbf{F_e}(\omega)$, $\mathbf{B_r}(\omega)$, $\mathbf{M_r}(\infty)$, and $\mathbf{S_h}$ are computed using the boundary element solver Wamit [271]. The viscous damping matrix $\mathbf{B_v}$ is given by $\mathbf{B_v} = B_v\mathbf{I_n}$, where $\mathbf{I_n}$ is the identity matrix of order n. From Lorentz linearization [187], as demonstrated in Section 11.2.3, the coefficient B_v is calculated for an isolated device as

$$B_v = \frac{\rho C_d A}{2} \frac{\int_0^T \dot{x}^2(t)|\dot{x}(t)|dt}{\int_0^T \dot{x}^2(t)dt}, \tag{14.18}$$

where ρ is the fluid density, C_d is the drag coefficient, A is the area of the body projected onto the plane orthogonal to the velocity $\dot{x}$, and T is the time interval.

In order to more easily formulate the control problem, the system equations are discretized by approximating the velocity and the PTO force with a linear combination of basis functions, as introduced in Chapter 10. Following the procedure from [270], trigonometric functions are chosen as basis functions. Then, the PTO force and the velocity are, as previously, approximated with the truncated zero-mean Fourier series

$$\dot{x}_k(t) \approx \sum_{n_f=1}^{N/2} v_k^{n_f,c} \cos(n_f\omega_0 t) + v_k^{n_f,s} \sin(n_f\omega_0 t), \tag{14.19}$$

$$f_{p,k}(t) \approx \sum_{n_f=1}^{N/2} \phi_k^{n_f,c} \cos(n_f\omega_0 t) + \phi_k^{n_f,s} \sin(n_f\omega_0 t), \tag{14.20}$$

where ω_0 is the fundamental frequency of the Fourier series; v_k and ϕ_k are, respectively, the Fourier coefficients of the velocity and the PTO force of the kth device; and N is the total number of components.

The best approximation of the solution for the equation of motion (14.16) is sought by applying the Galerkin method (see [277] for more details), and the result is the linear system

$$\mathcal{G}\mathcal{X} = \mathcal{P} + \mathcal{E} \tag{14.21}$$

Global Control

The GC strategy is aware of the full array system and has information on the motion of all the WECs in the array. In this case, the PTO forces are obtained by solving the optimal energy absorption problem, defined as: *maximize the performance function (14.25) subject to (14.21).*

It is assumed that the wave excitation force is known completely into the future, or can be forecast, and $\mathcal{G}$ is nonsingular. Then the optimal force coefficients $\mathcal{P}^*$, which maximise the array absorbed energy, are obtained by solving the optimization problem [270]:

$$\mathcal{P}^* = \arg\max_{\mathcal{P}} \left(-\mathcal{P}^T\mathcal{G}^{-1}\mathcal{P} - \mathcal{P}^T\mathcal{G}^{-1}\mathcal{E}\right) \tag{14.29}$$

where $\mathcal{P}$, $\mathcal{G}$, and $\mathcal{E}$ are defined in Equation (14.21).

Independent Control

The IC strategy assumes that each device is equipped with its own individual controller and excitation force estimator [270]. Each controller uses the model of a single isolated unit, having no communication with other devices in the array or any knowledge of them. Therefore, the optimal PTO force coefficients are obtained by solving each individual optimization problem for the kth body,

$$P_k^* = \arg\max_{P_k} \left(-P_k^T G_s^{-1} P_k - P_k^T G_s^{-1} \bar{E}_k\right), \tag{14.30}$$

with cost functions

$$W_k = -P_k{}^T X_k, \tag{14.31}$$

where $\mathrm{G}_s \in \mathbb{R}^{N\times N}$ is analogous to the matrix $\mathcal{G}$ in Equation (14.21) but now calculated using the hydrodynamic coefficients of a single isolated device, and $\bar{\mathrm{E}}_k$ is the excitation force measured by the estimator on body k. It is assumed that the excitation force estimator on each device cannot discern the excitation force due to incoming waves from the radiation generated by other bodies (for more details, see [270]) and thus estimates the total excitation experienced by the body.

14.3.4 Sensitivity of the Optimal Array Layout to Different Control Strategies

This section examines the sensitivity of the *optimal* array layout to the PC, GC, and IC strategies. The aim is to investigate how the controllers modulate the optimal distance d between the units, and the optimal array orientation β. In particular, the aim is to show that the optimal array layout is highly dependent on the control strategy employed.

Simulation with regular waves and irregular waves is performed. The regular waves are defined by wave height H and period T_w. The irregular waves are calculated as a linear combination of regular waves using a Bretschneider spectral envelope, defined by significant wave height H_s and peak period T_p. The optimal layout parameters are determined via simulations with a complete set of possible values of d/r and β, where r is the device radius. As a result, a clear overview of the mapping between the q factor and the optimization parameters (d/r and β) can be observed.

Table 14.4 Set of Cylinders Considered

WEC geometry	G1	G2	G3
Radius r (m)	4	6.25	8
Draft h (m)	10	4	2.5
Resonance T_r (s)	7.1	5.4	5

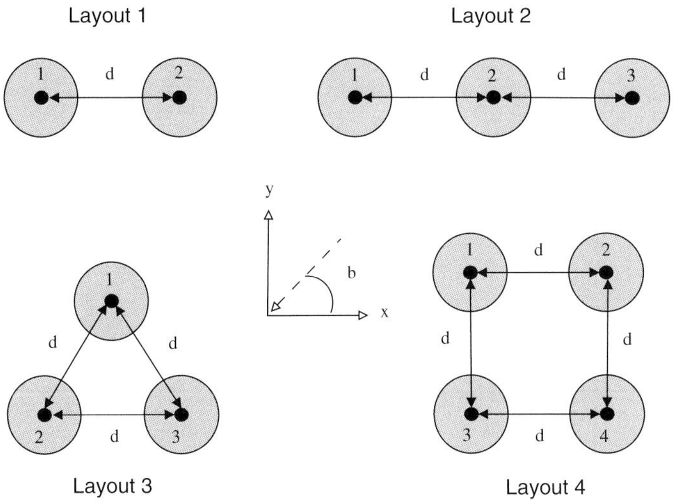

Figure 14.9 Top view of array layouts and of incident wave direction β.

Candidate Array Layouts

Fig. 14.9 illustrates the array layouts considered, where β is the incident wave direction. Layout 1 (L1) has two floating bodies; layout 2 (L2) is a linear array of three bodies; layout 3 (L3) is composed of three bodies placed at the vertices of an equilateral triangle; while layout 4 (L4) is a square array of four floating bodies. All arrays are composed of heaving cylinder devices, with a number of different device geometries considered. Each geometry has approximately the same volume ($\approx 160\pi$ m^3), but a different radius r, draft h, and resonance period T_r, as described in Table 14.4.

Influence of Inter device Spacing

Here the aim is to investigate how the optimal values of the *separating distance* change as different control strategies are considered. For each simulation, the direction of propagation of the waves β is constant. In order to avoid shadowing effects, a different value of β is considered depending on the configuration of the arrays: $\beta = 90°$ for linear configurations (L1 and L2), $\beta = 0°$ for L3, and $\beta = 45°$ for L4.

Initially, the simplest array layout L1, geometry G2, and regular waves ($H = 1$ m; $T_w = 9$ s) are considered. Fig. 14.10 shows the evolution of the q factor and the total energy absorbed W versus d/r for $\beta = 90°$. From Fig. 14.10, it is clear that different optimal d/r values are obtained for different control strategies, especially between the

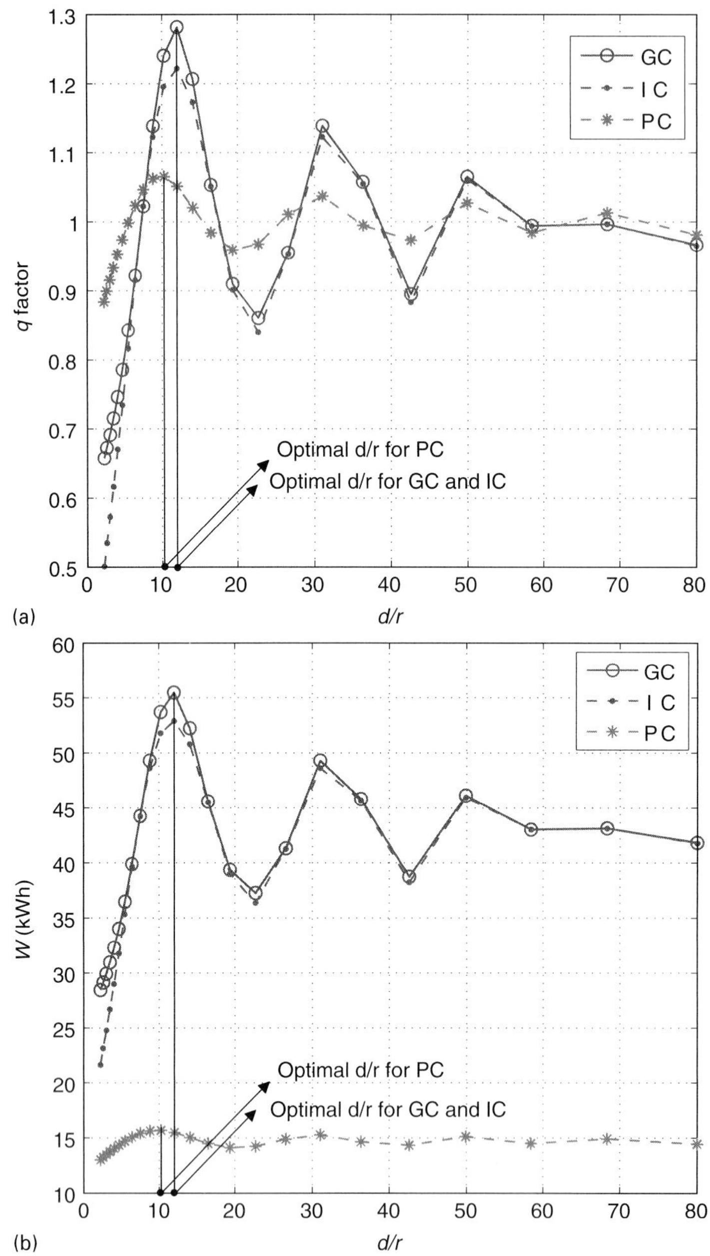

Figure 14.10 L1, G2: (a) Evolution of the q factor and (b) evolution of the total absorbed energy by the array versus d/r for constant $\beta = 90^\circ$ (regular wave: $H = 1$ m; $T_w = 9$ s).

Table 14.5 L1: Optimal Values of d/r (regular wave: $H = 1$ m and $\beta = 90°$)

Geometry	$T_w = 5$ s			$T_w = 7$ s		
	IC	GC	PC	IC	GC	PC
G1	4	4	4	10	10	10
G2	3.5	3.5	3.5	6.4	6.4	5.4
G3	3	3	3	5.4	5.4	4.6
Geometry	$T_w = 9$ s			$T_w = 11$ s		
	IC	GC	PC	IC	GC	PC
G1	17	17	17	29.3	29.3	24.5
G2	12	12	10.2	16.5	16.5	14.1
G3	9.5	9.5	8.3	12.8	12.8	11

GC and PC strategies. It is also clear that maximizing the q factor also maximizes the total absorbed energy for a certain incident wave regime and control strategy. Although GC and IC show lower q factor values than PC for some d/r ratios (Fig. 14.10(a)), the active controllers (GC and IC) produce higher captured energy levels than the passive controller (Fig. 14.10(b)).

Table 14.5 summarizes the simulation results for the simplest array layout, considering the three geometries in Table 14.4. From Table 14.5, it can be noted that the optimal values of d/r are larger for geometries with strong radiation properties (e.g., G1, which has a large d/r ratio) and higher when the periods of the incident wave are longer. Comparing the results for GC and PC, larger inter-device spacing is required with GC for optimal configuration, but when $T_w \rightarrow T_r$ the optimal distances equalize. When $T_w \rightarrow T_r$, the devices are at resonance with the waves and the control strategies exert no phase control effort.

Fig. 14.11 shows the evolution of the q factor versus d/r for $\beta = 90°$ and *irregular* waves ($H_s = 1$ m; $T_p = 9$ s). Note that the q factor is more sensitive to GC (and IC) than PC. Actually, both GC and IC actively modify the dynamics of each device, resulting in more exaggerated motion of the bodies and more absorbed energy than PC, as illustrated by Fig. 14.10(b). As a result, the active controllers cause more constructive or destructive effects than PC when different d/r ratios are considered. In addition, the evolution of the q factor for the IC strategy is not significantly different from GC (same optimal values for d/r) and is omitted from the following analysis.

Table 14.6 summarizes the optimal d/r values for the array layout L1 considering the three geometries and different periods of incident *irregular* waves. The relative difference between the optimal d/r values for GC and PC is greater for longer periods and smaller when $T_p \rightarrow T_r$, as the control strategies do not exert much effort in such a case.

Tables 14.7–14.9 summarize the optimal d/r values for GC and PC, respectively, for array layouts L2, L3, and L4, also for the irregular wave case. The results in

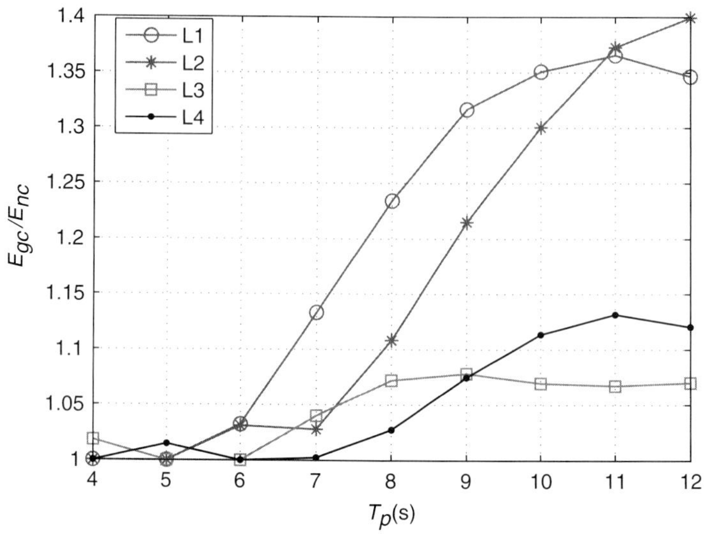

Figure 14.17 Ratio between E_{gc} and E_{nc} (irregular wave: $H_s = 1$ m).

the ratio E_{gc}/E_{nc} for the four layout configurations. Clearly, the energy capture in the control aware case is superior to the "control ignorant" case ($E_{gc}/E_{nc} \geq 1$ in all cases), *with an improvement of up to 40% in some cases*. For linear arrays, in particular, the average benefit of control aware layout design (across all T_p) is more than 19% and 16% for L1 and L2 layouts, respectively. We note that such performance improvements are significantly greater than any that can be obtained by improved control design ([146] reports an improvement of 8.2% by using model predictive control in comparison to a simple controller) or extension of IC to GC ([270] reports an average improvement of 10–20% for GC over IC).

In a wave farm deployment, the study of the control-informed optimal array layout should consider the mean wave direction and the statistical sea states occurring in a specific site. Then, the best layout can be defined after running simulations for a different set of wave parameters, for example, significant wave height and peak period.

14.4 Conclusion

This chapter considers the implications of incorporating the control system at both the WEC geometry design stage and the array layout stage. Clearly, both the optimal geometry and the optimal layout are strong functions of the prevailing wave climate, detailed in terms of predominant wave direction and wave height/period. In addition, the concept of optimality must be measured against a particular metric, and in both the geometry and layout studies, energy capture was the metric employed. In a commercial deployment, there are likely to be many other significant factors that influence both the optimal geometry and layout, mostly related to the overall economic benefit of a wave energy project. In such a more holistic context, the influence of the control strategy may play a more minor role in influencing the optimal geometry and layout, but there is certainly no disadvantage in considering the control system at both the layout and geometry design stages.

Considering energy capture as the primary metric, the following conclusions can be drawn. It is clear that different optimal geometry results are obtained depending on the type of control strategy employed on the WEC. However, the effects of constraints on the PTO force or on the displacement of the device have not been taken into account in this study. Such constraints may influence the results, especially in the MPC case, which results in more exaggerated motion of the body and requires higher PTO forces than in the latching or declutching cases. The interested reader is referred to [279].

An optimal geometry obtained with latching control is highly control dependent and shortens the original natural resonance period of the device, usually designed for the predominant wave period. The optimal geometry obtained with a declutching-controlled device is slightly different from an uncontrolled device, and its resonance period is longer than the uncontrolled case. In these cases, an overall optimal system can be obtained if the latching/declutching control design is considered at the geometry optimization stage. It may be noted that the overall device size, influencing capital costs, may be reduced for an optimal geometry using latching control. An optimal controller (like the one considered here) optimizes the wave energy absorption regardless of the geometry of the device. For this case, it has been shown that the same amount of power can be obtained, whether or not the control design is incorporated at the geometry optimization stage. The control-informed geometric optimization of a real device is specific to the WEC, PTO, sea state(s), and system constraints. For the moment, we simply make the point, through the results presented, that the nature of the control system *must* be considered when designing the WEC itself, if an overall optimal wave energy system is to be achieved.

Regarding WEC array layout, consideration of the control scheme is not the only deciding factor in optimal WEC array layout. Other factors may have high priority, including mooring and electrical infrastructure configurations, which have major economic impact. However, the particular control strategy employed clearly does have a causal effect on optimal array separating distances and needs to be considered, together with these other design factors.

It is well known that control schemes play an important role in the performance improvement of WECs. However, this chapter demonstrates that it is vital to consider the control system to be employed at the array layout stage, if the full economic potential of a wave farm is to be realized. In particular, it is demonstrated that the performance improvement in energy capture from optimal control-informed layout can be up to 40%, which is very significant, considering the struggle for wave energy to demonstrate economic viability.

Part VII

In Closing

15 Conclusion

We have often heard that writing a book is one of the best learning experiences an author will have. We have certainly found this to be true in our case. It has also become clear to us that there is still much we do not know. This entire book is based on linear theory. We assume small-amplitude waves and small-amplitude oscillations. We have therefore been able to work with linear differential equations and first-order convolutions throughout. We have also been able to use linear superposition of displacements and forces, just as we have been able to apply forward and inverse Fourier transformations when required. We have been able to invoke linear-state space approximations and matrix algebra results wherever they seemed appropriate.

While linear theory has convincingly demonstrated its success in capturing the essence of phenomena that would be difficult to comprehend otherwise, as the field moves forward, there will come a time when our questions become more sophisticated and require answers that are more specific and more intricate. We are witnessing such progress already. Indeed, just the last 2–3 years have seen a rapidly growing interest in nonlinear modeling and nonlinear control of wave energy devices.

Most nonlinear modeling approaches that are somewhat general in scope tend to build on the linear solutions. Many such approaches use a perturbation type technique (e.g., [60]), that "expands" the actual solution with respect to a perturbation parameter (typically small). Alternatively, one could use multiple-scale expansion approaches [8] to provide similar results. Solutions are typically obtained to increasing orders of accuracy, each subsequent-order solution building on the previous-order solutions. Thus, solutions that are accurate to second order contain functions of the first-order solution, while solutions accurate to third order will contain both second- and first-order solutions. First-order solutions are just our linear solutions, however. Thus, the linear theory-based approaches discussed in this book can be utilized as the basic building blocks for the higher-order solutions required for nonlinear problems. Therefore, a good grasp of the linear theory-based approaches of this book may be essential for anyone hoping to sift through and quantify the different effects of multiple nonlinearities.

In this book, we reviewed some historically and otherwise important wave energy devices and the principles driving their action. We then discussed how control could enhance the conversion rates and thereby bring costs down, and argued for a formal analysis structure, so that complex design decisions are still possible when it is not possible to proceed on the basis of intuition alone. We discussed the challenges that wave energy devices pose for control, particularly, control that would enable the best performance in changing wave conditions. Such control is precluded by the very action the device must perform in order to absorb energy, unless we can find a way to provide predictions of the device oscillations and exciting force some time into the future. We illustrated this difficulty through mathematical arguments. Subject to the small-amplitude assumption, and linear superposition, we studied ways to quantify the wave input, beginning with the regular wave. We also considered statistical representations such as the spectral density, which is very commonly used in design work, and wave group statistics, which can be used to predict periods of low and high wave activity. Conceivably, one could also use these to design a supervisory control system that allows switching between energy conversion and survival mode maneuvers.

We learned that the shape of the immersed volume of the device is very important. It determines the body's wave-making ability. The greater the rate of energy radiated to the far field, the greater the power that that body can absorb. For two-dimensional devices (crest-spanning terminators) in a single mode of oscillation, the best geometry is one that maximizes wave radiation in the up-wave direction. A similar effect can also be achieved by combining two oscillation modes, one symmetric and one antisymmetric. We considered how three-dimensional devices can also potentially benefit from radiation patterns such as those generated by surge, pitch, and sway oscillations. When three-dimensional devices operate at their best, they can absorb more energy than is incident on their width (diameters). We also studied arrays of heaving three-dimensional devices and discussed how, through favorable mutual hydrodynamic interactions, the units could absorb more energy than they would if they were operating in isolation.

We examined reactive control next (in Chapter 7), particularly how it may be implemented in the time domain for significant gains in energy conversion. A suboptimal strategy we used obtained up-wave measurements at a single point for use in deriving the necessary control forces. We also found, in simulations, however, that obtaining wave predictions using a more complete propagation model led to greater gains, since the wave elevations so predicted were significantly more accurate. Alternative wave prediction models were presented in Chapter 12, which are based on historical measurement only. Discrete-time linear autoregressive models were found to give good prediction of future wave elevation and, for the practical examples studied, tend to perform at least as well as harmonic models or nonlinear forecasting models based on artificial neural networks. However, where an instantaneous narrow-banded wave process can be assumed, Chapter 8 shows how instantaneous wave period and height can be estimated using an extended Kalman filter. The use of discrete-time prediction models, in Chapter 12, allows a straightforward comparison of forecasting models that use only historical information and models that additionally use up-wave sensors. For the cases considered, little advantage in using additional up-wave sensor information was observed.

One control strategy is, of course, to synthesize a controller that does not rely on future information, and in Chapter 8 such a controller is considered, which uses an instantaneous measure of wave period and height. The derived controller assumes a very simple model of the process, which has the disadvantage of limiting the control capability compared with, for example model predictive control (MPC), though the practical value of such a controller is evidenced in being more robust with regard to modeling errors. A further advantage of the simple causal philosophy considered is that amplitude constraints can be naturally considered as part of the control calculation.

Switching control was considered next, and we outlined the mathematical framework allowing latching and unlatching of the device velocity so as to maximize the absorbed power. Here, we found that our goal of maximizing the power absorbed over a length of time extending into the future meant that our optimum control formalism would require information on the exciting force from the future, even though no reactive control was used.

Since wave energy devices have relatively slow dynamics, it is possible to consider numerical optimization to synthesize control strategies. One advantage of numerical optimization approaches is that device displacement, velocity, and power take-off (PTO) force constraints can be easily considered *within* the optimization process, which is significantly different from calculating the optimal control and subsequently trying to impose limits on displacement, velocity, and/or PTO force. The optimal numerical control philosophy presented in Chapter 10 utilizes a pseudospectral approach, where the device variables (displacement and velocity) and the control force are parameterized using harmonic basis functions. This leads to both a well-conditioned numerical optimization problem and a computational saving, by virtue of a parsimonious representation of the system variables. One further advantage of this representation is that a significant portion of the radiation damping convolution calculation can be precomputed, providing a further computational saving. The basic strategy of Chapter 10, which can be used for multibody devices, is extended to wave energy converter (WEC) arrays in Chapter 11, which shows that significant benefit can be obtained by deliberate manipulation of the motion of each device, so that constructive radiation patters are generated. Such coordinated array control can result in energy gains over the case where devices are controlled independently, especially when the devices are closely spaced in the array.

The final section of the book (Part VI) deals with the broader aspects of WEC systems but considers the interaction with the control system. While certain control systems, including those reported in Chapters 8 and 10, have the theoretical capability to maintain WEC operation within physical (displacement and/or force) constraints, this can happen only if there exists a control solution that can simultaneously satisfy force and displacement constraints. When it is not possible to simultaneously satisfy force and displacement constraints, the WEC must be placed in a survival mode, which mitigates the effect of excessive wave excitation on the system. Chapter 13 provides analytical assistance in recognizing situations in which physical constraints are likely to be exceeded (without the need for extensive simulation) and also provides a design tool that shows the relative benefit of relaxing either force or displacement constraints in improving power capture. The geometric analysis of Chapter 13 shows the intuitive relationship between the force and displacement constraint sets, and the excitation force, where the degree of overlap (or separation) between the constraint sets is a direct function of the magnitude of the wave excitation force. Chapter 14 moves up a further level in the WEC system, in considering the interaction between the control system and the optimal WEC shape (in Section 14.2) and optimal WEC array layout (in Section 14.3), in an effort to obtain an overall optimal wave energy system, from a power capture perspective. Both studies suggest a significant interrelationship between the control problem and broader physical system optimization and strongly motivate the need to consider the control and physical optimization problems simultaneously or, at least, in an iterative fashion. In particular, it is demonstrated that following a (traditional) two-step sequential process of the physical system design, followed by the control design, *will not* result in an overall optimal system.

This book has a focus on the technical problem of power capture maximization from waves, from a hydrodynamic perspective. However, having a sole focus on power capture may not result in a system that mimimizes the cost of delivered wave energy. Various other aspects need to be considered, not least the economic cost of various control decisions and other judgments relating to the physcial system design, including force limits, WEC geometry, and WEC array layout (for the interested reader, the concept of total economic optimization is articulated in [278]). In addition, the book does not consider any details of the PTO system, which brings its own control problems (using the PTO to implement the calculated optimal WEC velocity or force profile) and has its own specific characteristics that impact the wider WEC control problem, including efficiency characteristics of components such as hydraulic motors and electric generators. The need to maintain an operating point in favorable regions of various device efficiency curves may limit the hydrodynamic control freedom available, and any constraints on delivered power may further limit hydrodynamic control freedom. Such output constraints are present in both electrical energy delivery, via grid constraints, and in the case of potable water production via reverse osmosis, where pressure variation in the hydraulic system is highly undesirable.

Finally, the impact of various hydrodynamic control strategies on the lifetime of various WEC components has not been considered. In various traditional regulatory control strategies, such as optimal control and MPC, it is normal to explicitly weight control energy, or control excursions, in the cost function. However, even if such terms are incorporated in WEC energy maximizing performance function, finding the appropriate weights, which can economically trade off the energy captured against the component wear implied by various control actions, is a nontrivial problem.

In spite of the aforementioned limitations of this book, we hope that it provides a solid foundation of principles and ideas to build on, as we enter the next era of WEC design, WEC array design, and WEC control design. For sure, there are many challenging problems yet to be solved, not least recognizing the true nonlinear nature of WEC hydrodynamics and PTO systems and targeting economic performance instead of a purely technical goal, so many volumes are yet to be written. However, we hope that this volume provides some assistance to researchers and practitioners alike, in the quest for economic delivery of wave energy.

References

[1] Lamb H. *Hydrodynamics*. Cambridge University Press, Cambridge, UK; 1932. 6th ed. Reprinted by Dover, 1945.

[2] Stoker JJ. *Water Waves: Mathematical Theory with Applications*. John Wiley & Sons, New York; 1958. Wiley Classics Library Edition, 1992.

[3] Kinsman B. *Wind Waves: Their Generation and Propagation on the Ocean Surface*. Prentice-Hall, Englewood-Cliffs, NJ,; 1964. Reprinted by Dover, NY, 1984.

[4] Whitham GB. *Linear and Nonlinear Waves*. John Wiley & Sons, New York; 1973.

[5] Wehausen JV. Motion of floating bodies. *Annual Review of Fluid Mechanics*. 1971;3: 237–268.

[6] Newman JN. *Marine Hydrodynamics*. MIT Press, Cambridge, MA; 1978.

[7] Lewis EV. *Principles of Naval Architecture*, vol. 2. Society of Naval Architects and Marine Engineers, Jersey City, NJ; 1988, chapter 4.

[8] Mei CC. *The Applied Dynamics of Ocean Surface Waves*. World Scientific, Singapore; 1992, chapter 7.

[9] Salter SH. Absorbing wave makers and wide tanks. In: Wiegel RL, editor. *Proc. Conference on Directional Wave Spectra Applications*, Berkeley, CA, 1981.

[10] Thorpe T. An overview of wave energy technologies: Status, performance and costs. 1998. Available at www.citeseerx.ist.psu.edu/viewdoc/download?doi=10.1.1.466.3188&rep=rep1&type=pdf.

[11] EIA. Annual Energy Outlook. U.S. Energy Information Administration, 2014. 0383(2014). Available at: www.eia.gov.

[12] IEA. World Energy Projected Outlook, 2012. Available at: www.eia.gov.

[13] Van Vuuren DP, Nakicenovic N, Riahi K, Kammen D, Modi V, Nilsson M, et al. An energy vision: The transformation towards sustainability – interconnected challenges and solutions. *Current Opinion on Environmental Sustainability*. 2012;4:18–34.

[14] Electric Power Research Institute. Mapping and Assessment of the United States Ocean Wave Energy Resource. EPRI; 2011. 1024637. Prepared by G. Hagerman and G. Scott.

[15] Ross D. *Power from Sea Waves*. Oxford University Press, Oxford, UK; 1995. First published in 1981.

[16] McCormick ME. *Ocean Wave Energy Conversion*. John Wiley & Sons, New York; 1981. Reprinted with revisions by Dover, 2007.

[17] Salter SH. Wave power. *Nature*. 1974 June;720–724.

[18] Masuda Y. Wave activated generator. In: Proc. *International Colloquium on the Exposition of Oceans*, Bordeaux, France, 1971.

[19] Bessho M. Feasibility study of a floating type wave absorber. In: *Proc. 34th Japan Towing Tank Conference*, 1973.

[20] Evans DV. A theory for wave power absorption by oscillating bodies. *Journal of Fluid Mechanics*. 1976;77(1):1–25.

[21] Newman JN. The interaction of stationary vessels with regular waves. In: *Proc. 11th Symposium on Naval Hydrodynamics*, London, 1976, pp. 491–501.

[22] Mei CC. Power extraction from water waves. *Journal of Ship Research*. 1976;20:63–66.

[23] Mehlum E, Stammes JJ. Power production based on focusing ocean swells. In: *Proc. 1st Symposium on Wave Energy Utilization*, Gothenburg, Sweden, 1979. pp. 29–35.

[24] McCormick ME. Analysis of a wave energy conversion buoy. *Journal of Hydronautics*. 1974;8(3):77–82.

[25] Whittaker TJT, McPeake FA, Barr AG. The development and testing of a wave-activated navigational buoy with a Wells turbine. *Journal of Energy Resources Technology*. 1985;107(2):268–273.

[26] Maeda H, Tanaka H, Kinoshita T. Theoretical and experimental study on wave power absorption. In: Inui T, editor. *Proc. 13th Symposium on Naval Hydrodynamics*. The Shipbuilding Association of Japan, 1980, pp. 857–876.

[27] Dean WR. On the reflection of surface waves by a submerged circular cylinder. *Proceedings of the Cambridge Philosophical Society*. 1948;44:483–491.

[28] Ursell F. Surface waves on deep water in the presence of a submerged circular cylinder I. *Proceedings of the Cambridge Philosophical Society*. 1950;46:141–152.

[29] Ogilvie TF. First- and second-order force on a cylinder submerged under a free surface. *Journal of Fluid Mechanics*. 1963;16:451–472.

[30] Evans DV, Jeffrey DC, Salter SH, Taylor JRM. Submerged cylinder wave energy device: Theory and experiment. *Applied Ocean Research*. 1979;1(1):3–12.

[31] French MJ. *Invention and Evolution: Design in Nature and Engineering*. Cambridge University Press, Cambridge, UK; 1988.

[32] Budal K, Falnes J. Wave power conversion by point absorbers. *Norwegian Maritime Research*. 1978;6(4):2–11.

[33] Budal K, Falnes J. Interacting point absorbers with controlled motion. In: Count BM, editor. *Power from Sea Waves*. Academic Press, London, 1980, pp. 381–399.

[34] Falnes J, Lillebekken PM. Budal's latching-controlled-buoy type wave power plant. In: Lewis AW, Thomas GP, editors. *Proc. Fifth European Wave Energy Conference*, Cork, Ireland, 2003. Proc. (2005).

[35] Masuda Y. Experimental full-scale results of wave power machine Kaimei in 1978. In: *Proc. 1st International Symposium on Wave Energy Utilization*, Gothenburg, Sweden, 1979. pp. 349–360.

[36] Miyazaki T. Japanese wave energy devices. In: *Proc. First European Wave Energy Symposium*, Edinburgh, UK, 1993, pp. 15–20.

[37] Whittaker TJT, McIllwaine SJ, Raghunathan S. A review of the Islay shoreline wave power station. In: *Proc. First European Wave Energy Symposium*, Edinburgh, UK, 1993, pp. 283–286.

[38] Ravindran M, Koola P. Energy from sea waves: The Indian wave energy program. *Current Science*. 1991;60:676–680.

[39] Falcao AFO. Wave energy utilization: A review of the technologies. *Renewable and Sustainable Energy Reviews*. 2010;14:899–918.

[40] Eidsmoen H. On the theory and simulation of heaving-buoy wave-energy converters with control. PhD thesis, Norwegian University of Science and Technology, 1995.

[41] Hotta H. R & D on wave power in Japan. In: *Proc. Second European Wave Energy Symposium*, 1995. Lisbon, Portugal.

[42] Washio Y, Osawa H, Imai M, Korde UA, Ito H. The offshore floating wave energy device Mighty Whale. In: *Proc. 1st Coastal Environment Science and Technology Panel Meeting–US Japan Conf. on Utilization of Natural Resources*, Hayama, Japan, 1998.

[43] Pelamis Wave Power. Pelamis: P-750 Wave Energy Converter, 2009. Available at: www.pelamiswave.com.

[44] Friis-Madsen E, Sorensen HC, Parmeggiani S. The development of a Wave Dragon 1.5 MW demonstrator. In: *Proc. 4th International Conference on Ocean Energy*, Dublin, October 2012.

[45] Hansen RH, Kramer MM. Modeling and control of the Wave Star prototype. In: *Proc. 9th European Wave and Tidal Energy Conference*, Southampton, UK, 2011, paper 163.

[46] Nolte J, Ertekin RC, Davis EP. In-Ocean experiments of a wave energy conversion device moored to an anchor and to a drogue. Journal of Ocean Technology. 2013 March;8(1):72–85.

[47] Korde UA. Development of a reactive control apparatus for a fixed two-dimensional oscillating water column wave energy device. *Ocean Engineering*. 1991;18(5):465–483.

[48] Nise N. *Control Systems Engineering*, 6th ed. John Wiley & Sons, New York; 2011.

[49] Ogata K. *Modern Control Engineering*, 6th ed. Prentice Hall/Pearson, Upper Saddle River, NJ; 2010.

[50] Dorf RC, Bishop RH. *Modern Control Systems*, 12th ed. Prentice Hall/Pearson, Upper Saddle River, NJ; 2011.

[51] Schetzen M. *The Volterra and Wiener Theories of Nonlinear Systems*. John Wiley & Sons, New York; 1980.

[52] Friedland B. *Control System Design; An Introduction to State-Space Methods*. McGraw-Hill, New York; 1986. Reissued by Dover, 2005.

[53] Evans DV. Power from water waves. *Annual Review of Fluid Mechanics*. 1981;13: 157–187.

[54] Packard A, Doyle J. The complex structured singular value. *Automatica*. 1993;29(1): 71–109.

[55] Petersen I, Ugrinovskii V, Savkin A. *Robust Control Design Using* H_∞ Methods. Springer, Berlin; 2000.

[56] Press WH, Teukolsky SA, Vetterling WT, Flannery BP. *Numerical Recipes*. Cambridge University Press, Cambridge, UK; 1988.

[57] Soeterboek R. *Predictive Control: A Unified Approach*. Prentice-Hall, Englewood Cliffs, NJ; 1992.

[58] Rao S. *Mechanical Vibrations*, 5th ed. Prentice Hall/Pearson, Upper Saddle River, NJ; 2010.

[59] Meirovitch L. *Analytical Methods in Vibrations*. Macmillan; 1967.

[60] Greenberg MD. Advanced Engineering Mathematics, 2nd ed. Prentice Hall/Pearson, Upper Saddle River, NJ; 1998.

[61] Brown J, Churchill R. *Fourier Series and Boundary Value Problems*, 8th ed. McGraw-Hill, New York; 2011.

[62] Sarpkaya T, Issacson M. *Mechanics of Wave Forces on Offshore Structures*. Van Nostrand Reinhold, New York; 1981.

[63] Cummins W. The impulse response function and ship motions. *Schiffetechnik*. 1962;47:101–109.

[64] Wehausen JV. Causality and the radiation condition. *Journal of Engineering Mathematics*. 1992;26:153–158.

[65] Wehausen JV, Laitone EV. *Surface Waves*. vol. 9 of *Handbuch der Physik*. Flugge S, editor. Springer-Verlag, Berlin; 1960.

[66] Falnes J. *Ocean Waves and Oscillating Systems: Linear Interactions Including Wave Energy Extraction*. Cambridge University Press, Cambridge, UK; 2002.

[67] Gran S. Lectures in Ocean Engineering: Waves and Wave Forces. University of Oslo and A/S Veritas, Norway; 1987.

[68] Longuet-Higgins MS. The statistical analysis of a random moving surface. *Philosophical Transactions of the Royal Society of London, Series A*. 1957;249:321–387.

[69] Longuet-Higgins MS. Statistical properties of wave groups in a random sea state. *Philosophical Transactions of the Royal Society of London, Series A*. 1984;312(1521):219–250.

[70] Goda Y. *Random Seas and Design of Maritime Structures*, 2nd ed. World Scientific, Singapore; 2000.

[71] Falcao AFO. Phase control through load control of oscillating body wave energy converters with hydraulic PTO system. *Ocean Engineering*. 2008;35:358–366.

[72] Salter SH. Recent progress on ducks. In: *Proc. 1st Symposium on Wave Energy Utilization*, Gothenburg, Sweden, 1979.

[73] Belmont MR. A lower bound estimate of the gains stemming from quiescent period predictive control using conventional sea state statistics. *Journal of Renewable and Sustainable Energy*. 2009;1:063104.

[74] Taylor JRM. High pressure hydraulics. In: Cruz J, editor. *Ocean Wave Energy: Current Status and Future Perspectives*. Springer, Berlin; 2008, pp. 189–285.

[75] Yeung RW. A singularity distribution method for free surface flow problems. Dept. of Naval Architecture, University of California, Berkeley; 1973. NA–73–6.

[76] WAMIT. WAMIT User Manual. Wamit Inc.; 2013. Version 7.

[77] HYDRAN. A Computer Program for the Hydroelastic Response Analysis of Ocean Structures. Offcoast Inc., Kailua, HI; 2012. Version 5.1.7.

[78] Ertekin RC, Riggs HR, Che XL, Du SX. Efficient methods for hydroelastic analysis of very large floating structures. *Journal of Ship Research*. 1993;37(1):58–76.

[79] Havelock T. Waves due to a floating sphere making periodic heaving oscillations. *Proceedings of the Royal Society of London A*. 1955;231(1184):1–7.

[80] Bessho M. On time-reversed velocity potentials in the theory of water waves. *Journal of Society of Naval Architects of Kansai*. 1975;159.

[81] Salter SH, Lin CP. Wide tank efficiency measurements on a model of the sloped IPS buoy. In: *Proc. 3rd European Wave Energy Conference*, Patras, Greece, 1998.

[82] Falnes J, Kuniawan A. Fundamental formulae for wave energy conversion. *Royal Society Open Science*. 2015;2(14305):1–34.

[83] Falnes J, Hals J. Heaving buoys, point absorbers and arrays. *Philosophical Transactions of the Royal Society of London, Series A*. 2012;370:246–277.

[84] Evans DV. Some analytic results for two and three dimensional wave-energy absorbers. In: Count BM, editor. *Power from Sea Waves*. Academic Press, UK; 1980, pp. 212–249.

[85] Bessho M. On boundary value problems of an oscillating body floating on water. *Memoirs of the Defense Academy of Japan*. 1968;8(1):183–200.

[86] Evans DV. Wave power absorption by systems of oscillating surface pressure distributions. *Journal of Fluid Mechanics*. 1982 1;114:481–499.

[87] Stansell P, Pizer DJ. Maximum wave power absorption by attenuating line absorbers under volume constraints. *Applied Ocean Research*. 2013;40:83–93.

[88] Falnes J, McIver P. Surface wave interactions with systems of oscillating bodies and pressure distributions. *Applied Ocean Research*. 1985;7(4):225–234.

[89] Thomas GP, Evans DV. Arrays of three-dimensional wave-energy absorbers. *Journal of Fluid Mechanics*. 1981;108:67–88.

[90] Mei CC. Hydrodynamic principles of wave power extraction. *Philosophical Transactions of the Royal Society of London, Series A*. 2012;370:208–234.

[91] Garnaud X, Mei CC. Bragg Scattering and wave-power extraction by an array of small buoys. *Philosophical Transactions of the Royal Society of London, Series A*. 2010;466:79–106.

[92] Salter SH. Development of the duck concept. In: *Proc. Wave Energy Conference*, 1978. Heathrow, UK.

[93] Budal K, Falnes J. Optimum operation of improved wave power converter. *Marine Science Communication*. 1977;3:133–150.

[94] Skyner D. Solo duck linear analysis. Technical Report, University of Edinburgh, 1987.

[95] Nebel P. Maximizing the efficiency of wave-energy plants using complex conjugate control. *Proc IMechE Part I – Journal of Systems and Control Engineering*. 1992;206(4): 225–236.

[96] Falnes J. On non-causal impulse response functions related to propagating water waves. *Applied Ocean Research*. 1995;17(6):379–389.

[97] Korde UA. On a near-optimal control approach for a wave energy converter in irregular waves. *Applied Ocean Research*. 2014;46:79–93.

[98] Belmont MR, Horwood JMK, Thurley RWF, Baker J. Filters for linear sea-wave prediction. *Ocean Engineering*. 2006;33(17–18):2332–2351.

[99] Korde UA. Near-optimal control of a wave energy device using deterministic-model driven incident wave prediction. *Applied Ocean Research*. 2015;53:31–45.

[100] Gradshteyn IS, Ryzhik IM. *Table of Integrals, Series, and Products*, 5th ed. Jeffrey A, editor. Academic Press, San Diego, CA; 1994.

[101] Naito S, Nakamura S. Wave energy absorption in irregular waves by feedforward control system. In: Evans DV, Falcão AF de O, editors. *Proc. IUTAM Symp. Hydrodynamics of Wave Energy Utilization*. Springer Verlag, Berlin; 1985, pp. 269–280.

[102] Dannenberg J, Naaijen P, Hessner K, den Boom H, Reichert K. The On board Wave and Motion Estimator OWME. In: *Proc. International Society of Offshore and Polar Engineers Conference, Beijing*, China, 2010.

[103] Wu G. Direct simulation and deterministic prediction of large-scale nonlinear wave field. PhD thesis, Massachusetts Institute of Technology, 2004.

[104] Korde UA. Performance of 2 two-body heaving axisymmetric wave energy converters under control in irregular waves. In: *Proc. 11th European Wave and Tidal Energy Conference*, Nantes, France, 2015.

[105] Falnes J. Wave-energy conversion through relative motion between two single-mode oscillating bodies. In: *Proc. ASME Offshore Mechanics and Arctic Engineering Conference*, Lisbon, Portugal, 1998.

[106] Fusco F, Ringwood J. A Study of the prediction requirements in real-time control of wave energy converters. *IEEE Trans on Sustainable Energy*. 2012;3(1):176–184.

[107] Babarit A, Clement AH. Optimal latching control of a wave energy device in regular and irregular waves. *Applied Ocean Research*. 2006;28:77–91.

[148] Weinstock R. *Calculus of Variations*. McGraw-Hill, New York; 1952. Reprinted by Dover, 1974.

[149] Stengel RF. *Optimal Control and Estimation*. John Wiley & Sons, New York; 1986. Reprinted by Dover, 1994.

[150] Korde UA, Schoen MP, Lin F. Strategies for time-domain control of wave energy devices in irregular waves. In: *Proc. 11th International Society of Offshore and Polar Engineers (ISOPE) Conference*, Stavanger, Norway, 2002.

[151] Widder DV. *Advanced Calculus*. Prentice-Hall, Englewood Cliffs, NJ; 1947. Reprinted by Dover, 1989.

[152] Gelfand IM, Fomin SV. *Calculus of Variations*. Translated by Silverman RA. Prentice-Hall, Englewood Cliffs, NJ; 1963. Reprinted by Dover, 1991.

[153] Korde UA. Latching control of deep water wave energy devices using an active reference. *Ocean Engineering*. 2002;29:1343–1355.

[154] Eidsmoen H. Optimum control of a floating wave-energy converter with restricted amplitude. *Journal of Offshore Mechanics and Arctic Engineering*. 1996;118(2):96–102.

[155] Hals J, Falnes J, Moan T. Constrained optimal control of a heaving buoy wave energy converter. *Journal of Offshore Mechanics and Arctic Engineering*. 2011;133(1):1–15.

[156] Li G, Belmont MR. Model predictive control of sea wave energy converters. Part I: a convex approach for the case of single device. *Renewable Energy*. 2014;69:453–463.

[157] Finlayson BA, Scriven LE. The method of weighted residuals – a review. *Applied Mechanics Reviews*. 1966 Sep;19(9):735–748.

[158] Bacelli G, Ringwood JV, Gilloteaux JC. A control system for a self-reacting point absorber wave energy converter subject to constraints. In: *Proc. IFAC World Congress*, Milan, 2011, vol. 18, pp. 11387–11392.

[159] Herber DR, Allison JT. Wave energy extraction maximization in irregular ocean waves using pseudospectral methods. In: *Proc. 2013 ASME Design Engineering Technical Conference*, No. DETC2013-12600, ASME, Portland, OR, 2013.

[160] Elnagar G, Kazemi MA, Razzaghi M. The pseudospectral Legendre method for discretizing optimal control problems. *IEEE Transactions on Automatic Control*. 1995;40: 1793–1796.

[161] Bacelli G. Optimal control of wave energy converters. PhD thesis, National University of Ireland Maynooth, 2014.

[162] Bacelli G, Ringwood JV. Nonlinear optimal wave energy converter control with application to a flap-type device. In: *IFAC World Congress*, Cape Town, South Africa, 2014.

[163] Budal K, Falnes J. A resonant point absorber of ocean-wave power. *Nature*. 1975 Aug;256(5517):478–479. (With corrigendum in Vol.257, p.626).

[164] Ogilvie T. Recent progress toward the understanding and prediction of ship motions. In: *Proc. Fifth Symposium on Naval Hydrodynamics*, Bergen, Norway, 1964.

[165] Damaren CJ. Time-domain floating body dynamics by rational approximation of the radiation impedance and diffraction mapping. *Ocean Engineering*. 2000 Jun;27(6):687–705.

[166] Bacelli G, Balitsky P, Ringwood JV. Coordinated control of arrays of wave energy devices – benefits over independent control. *IEEE Transactions on Sustainable Energy*. 2013;4(4):1091–1099.

[167] Carcas M. The pelamis wave energy converter. Ocean Power Delivery Ltd., 2010:10.

[168] Whittaker T, Folley M. Nearshore oscillating wave surge converters and the development of Oyster. *Philosophical Transactions of the Royal Society of London Series A: Mathematical, Physical and Engineering Sciences*. 2012;370(1959):345–364.

[169] Draper M. More than just a ripple: Ocean Power Technologies sets its sights high. *Refocus*. 2006;7(1):54–56.

[170] Sandvik CM. Wave-to-Wire Model of the wave energy converter Bolt2: Control and power extraction with an all-electric power take-off system. PhD thesis, Norwegian University of Science and Technology, 2012.

[171] Garnaud X, Mei CC. Wave-power extraction by a compact array of buoys. *Journal of Fluid Mechanics*. 2009;635:389–413.

[172] Antonutti R, Hearn GE. Optimisation of Point-Absorber Arrays. In: *Proc. 9th European Wave and Tidal Energy Conference (EWTEC)*, Southampton, UK, 2011.

[173] Garnaud X, Mei CC. Comparison of wave power extraction by a compact array of small buoys and by a large buoy. *IET Renewable Power Generation*. 2010 Nov;4(6):519–530.

[174] Falnes J. Radiation impedance matrix and optimum power absorption for interacting oscillators in surface waves. *Applied Ocean Research*. 1980 Apr;2(2):75–80.

[175] Justino PAP, Clément AH. Hydrodynamic performance for small arrays of submerged spheres. In: *Proc. 5th European Wave and Tidal Energy Conference (EWTEC)*, Cork, Ireland, 2003, pp. 266–273.

[176] Child BMF, Venugopal V. Non-optimal tuning of wave energy device arrays. In: *Proc. 2nd International Conference on Ocean Energy (ICOE)*, Brest, France, 2008.

[177] Bellew S, Stallard T, Stansby PK. Optimisation of a heterogeneous array of heaving bodies. In: *8th European Wave and Tidal Energy Conference (EWTEC)*, Uppsala, Sweden, 2009.

[178] De Backer G, Vantorre M, Beels C, De Rouck J, Frigaard P. Power absorption by closely spaced point absorbers in constrained conditions. *IET Renewable Power Generation*. 2010 Nov;4(6):579–591.

[179] Ricci P, Saulnier JB, Falcão AF de O. Point-absorber arrays: A configuration study off the Portuguese West-Coast. In: *7th European Wave and Tidal Energy Conference (EWTEC)*, Porto, Portugal, 2007.

[180] Folley M, Whittaker TJT. The effect of sub-optimal control and the spectral wave climate on the performance of wave energy converter arrays. *Applied Ocean Research*. 2009;31(4):260–266.

[181] Cruz J, Sykes R, Siddorn P, Taylor RE. Estimating the loads and energy yield of arrays of wave energy converters under realistic seas. *IET Renewable Power Generation*. 2010 Nov;4(6):488–497.

[182] Westphalen J, Bacelli G, Balitsky P, Ringwood JV. Control strategies for arrays of wave energy devices. In: *9th European Wave and Tidal Energy Conference (EWTEC)*, Southampton, UK, 2011.

[183] Bacelli G, Ringwood J. Constrained control of arrays of wave energy devices. *International Journal of Marine Energy*. 2013;3–4(0):e53–e69.

[184] Douglas JF, Gasiorek JM, Swaffield JA. *Fluid Mechanics*, 5th ed. Pearson/Prentice Hall, Upper Saddle River, NJ; 2005.

[185] Bellew S, Stallard T. Linear modelling of wave device arrays and comparison to experimental and second order models. In: *International Workshop for Water Waves and Floating Bodies*, Harbin, China, 2010, pp. 2–5.

[186] Bhinder MA, Babarit A, Gentaz L, Ferrant P. Assessment of Viscous Damping via 3D-CFD modelling of a floating wave energy device. In: *9th European Wave and Tidal Energy Conference (EWTEC)*, Southampton, UK, 2011.

[187] Terra GM, van de Berg WJ, Maas LRM. Experimental verification of Lorentz' linearization procedure for quadratic friction. *Fluid Dynamics Research*. 1995;36:175–188.

[188] Folley M, Whittaker TWT, van't Hoff J. The design of small seabed-mounted bottom-hinged wave energy converters. In: *7th European Wave and Tidal Energy Conference (EWTEC)*, Porto, Portugal, 2007.

[189] Fusco F, Ringwood JV. A Study of the prediction requirements in real-time control of wave energy converters. *IEEE Transactions on Sustainable Energy*. 2012 Jan;3(1):176–184.

[190] Kwakernaak H, Sivan R. *Linear Optimal Control Systems*, vol. 172. Wiley-Interscience, New York; 1972.

[191] Folley M, Whittaker T. The adequacy of phase-averaged models for modelling wave farms. *ASME Conference Proceedings*. 2011;2011(44373):663–671.

[192] Babarit A. Impact of long separating distances on the energy production of two interacting wave energy converters. *Ocean Engineering*. 2010;37(8–9):718–729.

[193] Babarit A, Borgarino B, Ferrant P, Clément A. Assessment of the influence of the distance between two wave energy converters on energy production. *IET Renewable Power Generation*. 2010 Nov;4(6):592–601.

[194] Korde UA. Control system applications in wave energy conversion. In: *Proc. OCEANS 2000 MTS/IEEE Conference and Exhibition*, Providence, RI; 2000;3:1817–1824.

[195] Falnes J. A review of wave-energy extraction. *Marine Structures*. 2007;20:285–201.

[196] Tedd J, Knapp W, Frigaard P, Kofoed JP. Turbine control strategy including wave prediction for overtopping wave energy converters. CA-OE Workshop, Uppsala, 2005.

[197] Tedd J, Frigaard P. Short term wave forecasting, using digital filters, for improved control of wave energy converters. In: *Proc. International Offshore and Polar Engineering Conference*, Lisbon, Portugal, 2007, pp. 388–394.

[198] Frigaard P, Brorsen M. A time-domain method for separating incident and reflected irregular waves. *Coastal Engineering*. 1995;24:205–215.

[199] Fusco F. Forecasting requirements in the optimal control of wave energy converters. EE/2009/2/JVR. Dept. of Electronic Engineering, National University of Ireland Maynooth, 2009.

[200] Korde UA. Efficient primary energy conversion in irregular waves. *Ocean Engineering*. 1999;26:625–651.

[201] Schoen MP, Hals J, Moan T. Wave prediction and fuzzy logic control of wave energy converters in irregular waves. In: *Proc. 16th Mediterranean Conference on Control and Automation*, Ajaccio, Corsica, France, 2008, pp. 767–772.

[202] Van Den Boom H. OWME project makes technological breakthrough. *Marin Report*. 2009.

[203] Fusco F, Gilloteaux JC, Ringwood J. A Study on prediction requirements in time-domain control of wave energy converters. In: *Proc. Control Applications in Marine Systems (CAMS)*, Rostock, Germany, 2010, pp. 405–410.

[204] Cruz J. *Ocean Wave Energy: Current Status and Future Perspectives*. Springer-Verlag; 2008,

[205] Fusco F, Ringwood JV. Linear Models for short term wave forecasting. In: *Proc. World Renewable Energy Congress (WRECX)*, Glasgow, UK, 2008, pp. 1030–1035.

[206] Fusco F, Ringwood JV. A Study on short-term sea profile prediction for wave energy applications. In: *Proc. 8th European Wave and Tidal Energy Conference (EWTEC)*, Southampton, UK, 2009, pp. 756–765.

[207] Nolan G, Ringwood JV, Holmes B. Short term wave energy variability off the west coast of Ireland. In: *Proc. 7th European Wave and Tidal Energy Conference (EWTEC)*, Porto, Portugal, 2007.

[208] Fusco F. Short-Term wave forecasting as a univariate time series problem. EE/2009/3/JVR. Dept. of Electronic Engineering, National University of Ireland Maynooth, 2009.

[209] Harvey AC. *Forecasting, Structural Time Series Models and the Kalman Filter*. Cambridge University Press, Cambridge, UK; 1989.

[210] Young PC, Pedregal DJ, Tych W. Dynamic harmonic regression. *Journal of Forecasting*. 1999;18:369–394.

[211] Box GEP, Jenkins GM, Reinsel GC. *Time Series Analysis: Forecasting and Control*. Prentice-Hall, Englewood Cliffs, NJ; 1994.

[212] Shook DS, Mohtadi C, Shah SL. Identification for long-range predictive control. *IEE Proceedings D (Control Theory and Applications)*. 1991;138(1):75–84.

[213] Norgaard M, Ravn O, Poulsen NK, Hansen LK. *Neural Networks for Modelling and Control of Dynamic Systems*. Springer, Heidelberg, Germany; 2000.

[214] Wilamowski BM, Iplikci S, Kaynak O, Efe MO. An algorithm for fast convergence in training neural networks. *IEEE Transactions on Neural Networks*. 2001:1778–1782.

[215] Mackay DJ. Introduction to Gaussian Processes. In: *Proc. Neural Networks and Machine Learning Conference*, 1997, pp. 133–165.

[216] Robert CP, Casella G. *Monte Carlo Statistical Methods*. Springer, Heidelberg, Germany; 2004.

[217] Isermann R. *Fault-Diagnosis Systems: An Introduction from Fault Detection to Fault Tolerance*. Springer, Heidelberg, Germany; 2006.

[218] Esteves R, Valente A, Costa M, Reis F, Azevedo EB. Dados direccionais de agitação marítima nos Açores (Projecto CLIMAAT-CLIMA e Meteorologia dos Arquipélagos Atlânticos). 6s Jornadas Portuguesas de Engenharia Costeira e Portuária, PIANC, Funchal; 2009.

[219] Azevedo EB, Rodrigues MC. Projectos CLIMAAT & CLIMARCOST Clima e Meteorologia dos Arquipélagos Atlânticos, Clima Marítimo e Costeiro. Workshop Intenacional sobre Clima e Recursos Naturais nos Países de Língua Portuguesa, WSCRA08. 2008;.

[220] Barrera C, Azevedo EB, Rueda MJ, Gelado MD, O L. Real-time monitoring network in the Macaronesian region as a contribution to the Coastal Ocean Observations Panel (COOP). *Journal of Operational Oceanography*. 2008;1:59–69.

[221] Ochi MK. *Applied Probability and Stochastic Processes*. Wiley Inter-Science, New York; 1990.

[222] Bernhard HP. A tight upper bound on the gain of linear and nonlinear predictors for stationary stochastic processes. *IEEE Transactions on Signal Processing*. 1998 Nov;46(11):2909–2917.

[223] Shannon CE. The mathematical theory of communication. *Bell System Technical Journal*. 1948;(27):379–423.

[224] Hong X, Billings SA. Time Series Multistep-ahead predictability estimation and ranking. *Journal of Forecasting*. 1999;18:139–149.

[225] Tedd J, Frigaard P. Short term wave forecasting, using digital filters, for improved control of wave energy converters. In: *Proc. International Offshore and Polar Engineering Conference*, Lisbon, Portugal, 2007, pp. 388–394.

[226] Belmont MR, Horwood JMK, Thurley RWF, Baker J. Shallow angle wave profiling LIDAR. *Atmospheric and Oceanic Technology*. 2007;24(6):1150–1156.

[227] Halliday JR, Dorrell DG, Wood AR. An application of the fast fourier transform to the short-term prediction of sea wave behaviour. *IEEE Transactions on Renewable Energy*. 2011 June;36(6):1685–1692.

[228] Belmont M, Horwood J, Thurley R, Baker J. Filters for linear sea-wave prediction. *Ocean Engineering*. 2006 Dec;33(17–18):2332–2351.

[229] Voronovich V, Holmes B, Thomas GP. A preliminary numerical and experimental study of wave prediction. In: *Proc. 6th European Wave and Tidal Energy Conference (EWTEC)*, Glasgow, UK, 2005, pp. 535–542.

[230] Monk K, Conley D, Lopes M, Zou Q. Pneumatic power regulation by wave forecasting and real-time relief valve control for an OWC. In: *Proc. European Wave and Tidal Energy Conference Series*, Aalborg, Denmark, 2013.

[231] Paparella F, Monk K, Winands V, Lopes M, Conley D, Ringwood JV. Up-wave and autoregressive methods for short-term wave forecasting for an oscillating water column. *IEEE Transactions on Sustainable Energy*. 2015;6(1):171–178.

[232] Fusco F, Ringwood JV. Short-term wave forecasting for real-time control of wave energy converters. *IEEE Transactions on Sustainable Energy*. 2010 July;1(2):99–106.

[233] Korde UA. Up-wave surface elevation for smooth hydrodynamic control of wave energy conversion in irregular waves. In: *Proc. OCEANS'13 MTS/IEEE*, San Diego, 2013.

[234] Shook DS, Mohtadi C, Shah SL. Identification for long-range predictive control. *IEE Proceedings D – Control Theory and Applications*.1991;138:75–84.

[235] Fusco F, Ringwood JV. A study on the prediction requirements in real-time control of wave energy converters. *IEEE Transactions on Sustainable Energy*. 2012;3(1):176–184.

[236] Paparella F, Monk K, Winands V, Lopes M, Conley D, Ringwood JV. Benefits of up-wave measurements in linear short-term wave forecasting for wave energy applications. In: *Proc. IEEE Multi-Conference on Systems and Control*, Antibes, France, 2014.

[237] Fusco F, Ringwood JV. A model for the sensitivity of non-causal control of wave energy converters to wave excitation force prediction errors. In: *Proc. 9th European Wave and Tidal Energy Conference (EWTEC)*, Southampton, UK, 2011.

[238] Kovaltchouk T, Rongere F, Primot M, Aubry J, Ahmed HB, Multon B. Model predictive control of a direct wave energy converter constrained by the electrical chain using an energetic approach. In: *European Wave and Tidal Energy Conference*, Nantes, 2015, pp. 07D3-1-1-07D3-1-10.

[239] Tona P, Hoai-Nam N, Sabiron G, Creff Y. An efficiency-aware model predictive control strategy for a heaving buoy wave energy converter. In: *European Wave and Tidal Energy Conference*, Nantes; 2015, pp. 07D3–2–1–07D3–2–10.

[240] Pizer D. Maximum wave-power absorption of point absorbers under motion constraints. *Applied Ocean Research*. 1993;15(4):227–234.

[241] Falnes J. Maximum wave-energy absorption by oscillating systems consisting of bodies and water columns with restricted or unrestricted amplitudes. In: *Proc. International Offshore and Polar Engineering Conference*, Seattle, WA, 2000, pp. 420–425.

[242] Boyd JP. *Chebyshev and Fourier Spectral Methods*. Dover, Mineola, NY; 2001.

[243] Bjarte-Larsson T, Falnes J. Laboratory experiment on heaving body with hydraulic power take-off and latching control. *Ocean Engineering*. 2006 May;33(7):847–877.

[244] Weller SD, Stallard TJ, Stansby PK. Experimental measurements of the complex motion of a suspended axisymmetric floating body in regular and near-focused waves. *Applied Ocean Research*. 2013;39(0):137–145.

[245] Atkinson K, Han W. *Theoretical Numerical Analysis: A Functional Analysis Framework*. Springer, Heidelberg, Germany; 2005.

[246] Timan AF. *Theory of Approximation of Functions of a Real Variable*. Courier Dover Publications, New York; 1994.

[247] Abzug MJ, Larrabee EE. *Airplane Stability and Control: A History of the Technologies That Made Aviation Possible*, 2nd ed. Cambridge University Press, Cambridge, UK; 2002.

[248] Budal K, Falnes J. A resonant point absorver of ocean waves. *Nature*. 1975;256:478–479.

[249] Babarit A, Clément AH. Optimal latching control of a wave energy device in regular and irregular waves. *Applied Ocean Research*. 2006 April;28(2):77–91.

[250] Babarit A, Clément AH. Application of the optimal command method to the control of SEAREV wave energy converter: A study on the influence of time constants on the efficiency of latching control. In: *European Control Conference*, Budapest, Hungary, 2009.

[251] Alves M, Traylor H, Sarmento A. Hydrodynamic optimization of a wave energy converter using a heave motion buoy. In: *Proc. 6th International Conference on Wave and Tidal Energy*, Porto, Portugal, 2007.

[252] McCabe AP, Aggidis GA, Widden MB. Optimizing the shape of a surge-and-pitch wave energy collector using a genetic algorithm. *Renewable Energy*. 2010;35(12):2767–2775.

[253] McCabe AP. Constrained optimization of the shape of a wave energy collector by genetic algorithm. *Renewable Energy*. 2013;51:274–284.

[254] Kurniawan A, Moan T. Optimal geometries for wave absorbers oscillating about a fixed axis. *Journal of Oceanic Engineering*. 2013;38(1):117–130.

[255] Gilloteaux JC, Ringwood JV. Control-informed geometric optimisation of wave energy converters. In: *Proc. IFAC Conference on Control Applications in Marine Systems (CAMS)*, Rostock, Germany, 2010, pp. 399–404.

[256] Bacelli G, Ringwood JV. Numerical optimal control of wave energy converters. *IEEE Transactions on Sustainable Energy*. 2015 April;6(2):294–302.

[257] Babarit A, Duclos G, Clément AH. Comparison of latching strategies for a heaving wave energy device in random sea. *Applied Ocean Research*. 2004 July;26(5):227–238.

[258] Nolan G, Ringwood JV, Leithead WE, Butler S. Optimal damping profiles for a heaving buoy wave-energy converter. In: *Proc. 15th International Offshore and Polar Engineering Conference*, Seoul, Korea, 2005.

[259] Hals J, Bjarte-Larsson T, Falnes J. Optimum reactive control and control by latching of a wave-absorbing semisubmerged heaving sphere. *ASME Conference Proceedings*. 2002 Jan;2002(36142):415–423.

[260] Budal K, Falnes J, Iversen LC, Lillebeken PM, Oltedal G, Hals T, et al. The Norwegian wave-power buoy project. In: *Proc. 2nd International Symposium on Wave Energy Utilization*, Trondheim, Norway, 1982, pp. 323–344.

[261] Eidsmoen H. Tight-moored amplitude-limited heaving buoy wave energy converter with phase control. *Applied Ocean Research*. 1998 June;20(3):157–161.

[262] Korde UA. Phase control of floating bodies from an on-board reference. *Applied Ocean Research*. 2001 Oct;23(5):251–262.

[263] Falcão AFO. Phase control through load control of oscillating-body wave energy converters with hydraulic PTO system. *Ocean Engineering*. 2008;35(3):358–366.

[264] Lopes MFP, Hals J, Gomes RPF, Moan T, Gato LMC, Falcão AFO. Experimental and numerical investigation of non-predictive phase-control strategies for a point-absorbing wave energy converter. *Ocean Engineering*. 2009;36(5):386–402.

[265] Feng Z, Kerrigan EC. Latching control of wave energy converters using derivative-free optimization. In: *Proc. IEEE 52nd Annual Conference on Decision and Control*, Florence, Italy, 2013, pp. 7474–7479.

[266] Folley M, Whittaker T. The control of wave energy converters using active bipolar damping. *Proceedings of the Institution of Mechanical Engineers, Part M: Journal of Engineering for the Maritime Environment*. 2009 Nov;223(4):479–487.

[267] Teillant B, Gilloteaux JC, V RJ. Optimal damping profile for a heaving buoy wave energy converter. In: *IFAC Conference on Control Applications in Marine Systems (CAMS)*, Rostock, Germany, 2010.

[268] Clément AH, Babarit A. Discrete control of resonant wave energy devices. *Philosophical Transactions of the Royal Society A: Mathematical,Physical and Engineering Sciences*. 2012;370(1959):288–314.

[269] Feng Z, Kerrigan EC. Declutching control of wave energy converters using derivative-free optimization. In: *Proc. 19th IFAC World Congress*, Cape Town, South Africa, 2013, pp. 7647–7652.

[270] Bacelli G, Balitsky P, Ringwood JV. Coordinated control of arrays of wave energy devices – benefits over independent control. *IEEE Transactions on Sustainable Energy*. 2013;4(4):1091–1099.

[271] Wamit, Inc . Wamit User Manual Versions 6.4, 6.4PC and 6.3S, 6.3S-PC. Wamit Inc. and Massachusetts Institute of Technology, 1998–2006.

[272] Labeyrie J. Stationary and transient states of random seas. *Marine Strucutures*. 1990;3:43–58.

[273] Gao Z, Moan T. Mooring system analysis of multiple wave energy converters in a farm configuration. In: *Proc. 8th European Wave and Tidal Energy Conference*, Uppsala, Sweden, 2009, pp. 509–518.

[274] Sharkey F, Bannon E, Conlon M, Gaughan K. Maximising value of electrical networks for wave energy converter arrays. *International Journal of Marine Energy*. 2013;1:55–69.

[275] Falcão AFO. Modelling and control of oscillating-body wave energy converters with hydraulic power take-off and gas accumulator. *Ocean Engineering*. 2007;34:2021–2032.

[276] Balitsky P, Bacelli G, Ringwood JV. Control-influenced layout optimization of arrays of wave energy converters. In: *ASME 2014 33rd International Conference on Ocean, Offshore and Arctic Engineering*, San Francisco, CA. American Society of Mechanical Engineers; 2014, pp. V09BT09A022.

[277] Bacelli G, Ringwood JV, Gilloteaux JC. A control systyem for a self-reacting point absorber wave energy converter subject to constraints. In: *Proc. IFAC World Congress*, Milan, Italy, 2011, pp. 11387–11392.

[278] Costello R, Teillant B, Weber J, Ringwood JV. Techno-economic optimisation for wave energy converters. In: *Proc. International Conference on Ocean Energy*, Dublin, Ireland, 2012.

[279] Garcia-Rosa PB, Bacelli G, Ringwood JV. Control-Informed Geometric Optimization of Wave Energy Converters: The Impact of Device Motion and Force Constraints. *Energies*. 2015;8(12):13672–13687.

[280] Li G, Belmont MR. Model predictive control of sea wave energy converters–Part II: The case of an array of devices. *Renewable Energy*. 2014;68:540–549.

Index